Sea Surveying

(Illustrations)

Sea Surveying

(Illustrations)

Edited by

A. E. INGHAM

Department of Land Surveying, North East London Polytechnic

A Wiley–Interscience Publication

JOHN WILEY & SONS

London · New York · Sydney · Toronto

Library of Congress Cataloging in Publication Data:

Ingham, Alan E.
Sea surveying.

"A Wiley–Interscience publication."
Bibliography: p.
1. Hydrographic surveying. I. Title.

VK591.I48 526.9′9 74–3066

ISBN 0 471 42729 2

Printed in Great Britain by
William Clowes & Sons, Limited
London, Beccles and Colchester

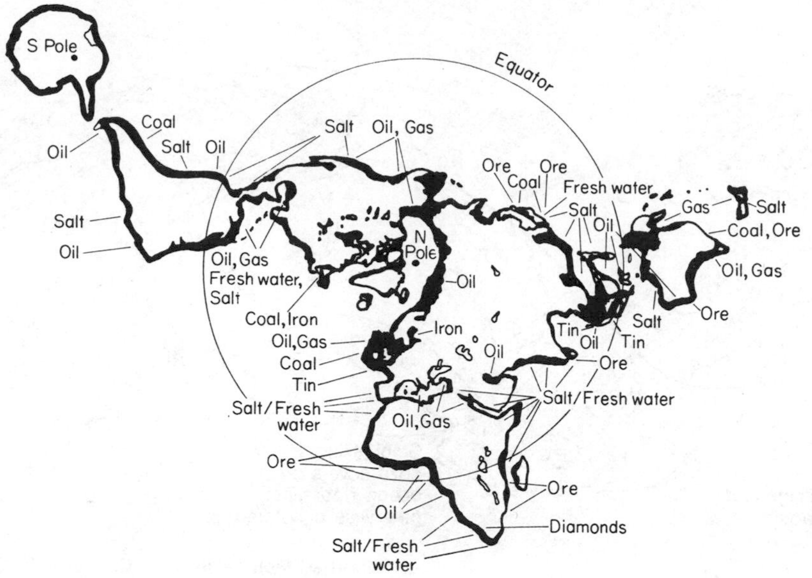

Figure 1. The continental shelf areas of the world and areas of exploitation of resources by man.

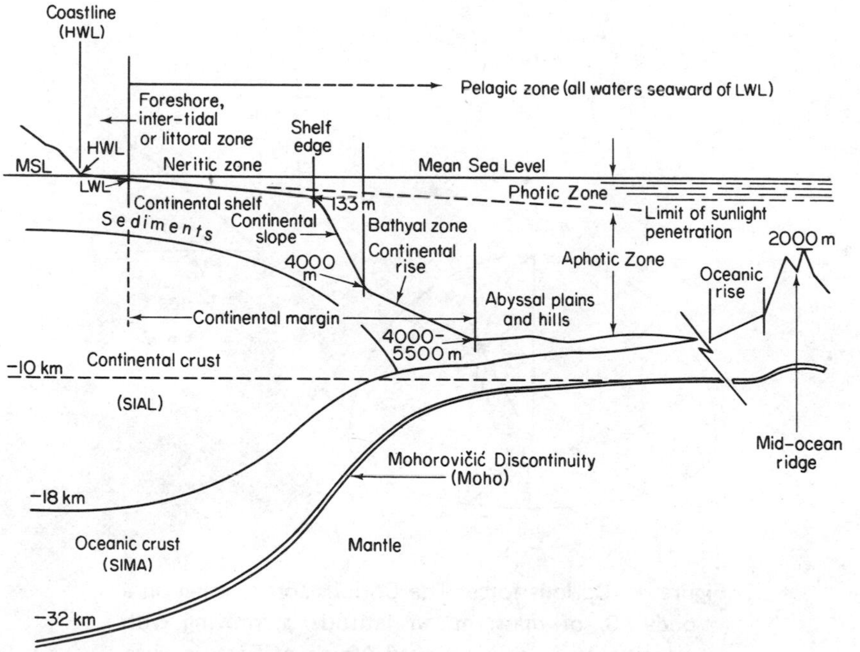

Figure 2. A cross-section of the earth's surface between coast and mid-ocean showing average depths of major features (not to scale). A feature common in the W. Pacific ocean is omitted — the *trench*, which may be over 10,000 m deep, is normally associated with steep, mountainous coasts.

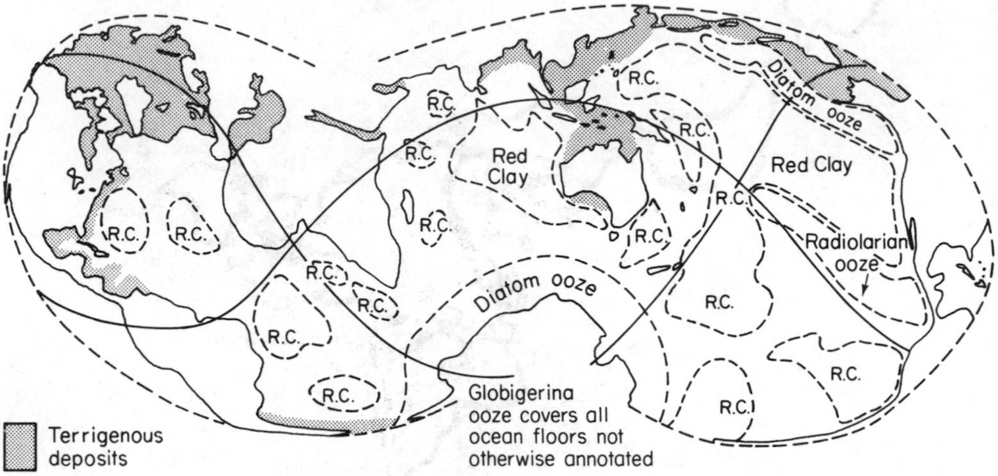

(Modified Mollweide projection, after F. C. Fuglister)

Figure 3. The principal sedimentary deposits in the deep ocean.

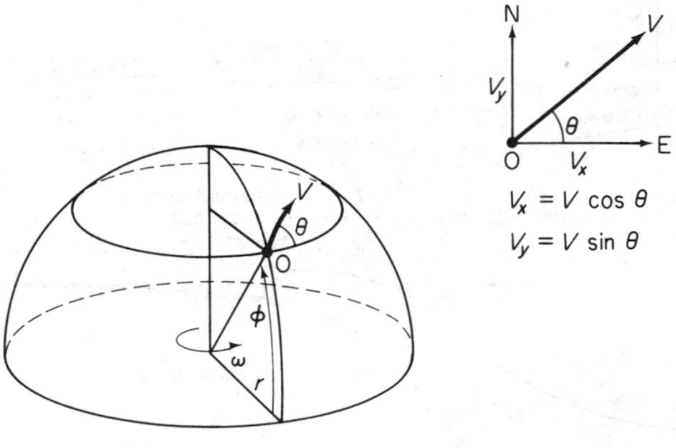

Figure 4. Coriolis force. The Coriolis force acting on a body, 0, of mass, m, in latitude ϕ, moving with velocity, V, in a direction θ North of East, is given by the expression: force = $2\omega V \sin \phi \, m$ (ω is the angular rate of rotation of the earth).

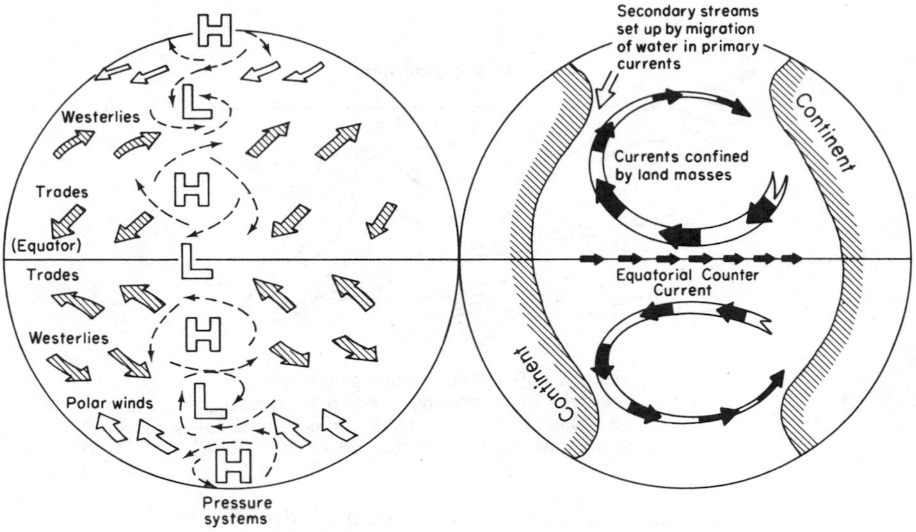

Figure 5. The wind, pressure and current systems of the world, shown schematically.

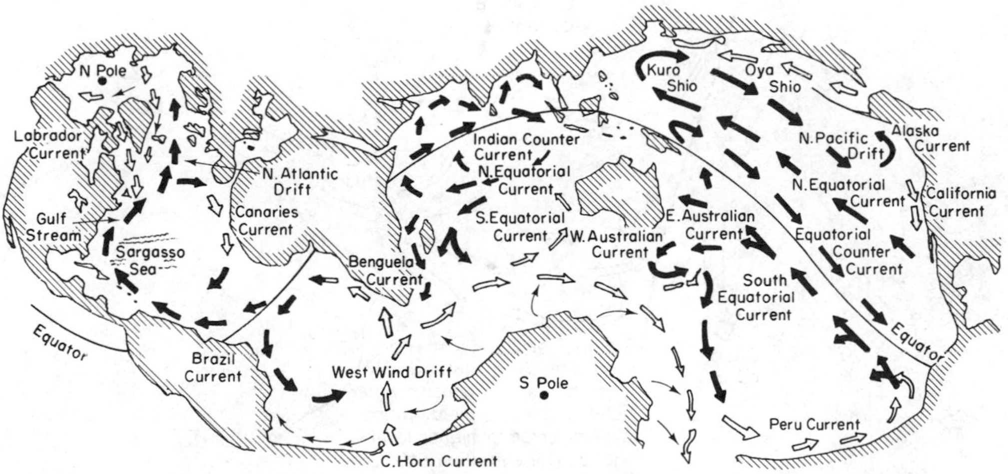

Figure 6. Ocean currents.

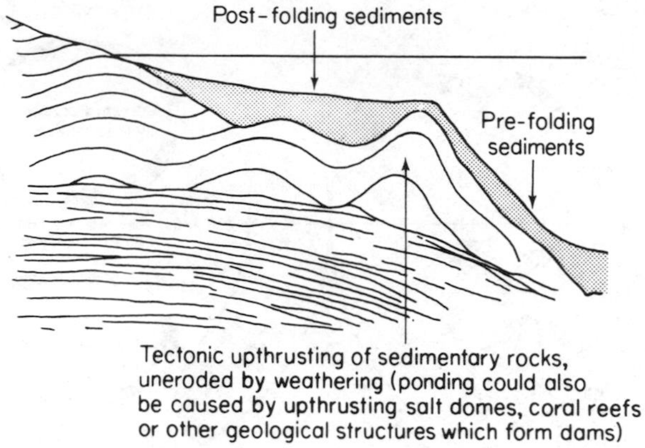

Figure 7. The ponding of sedimentary deposits to form shelves.

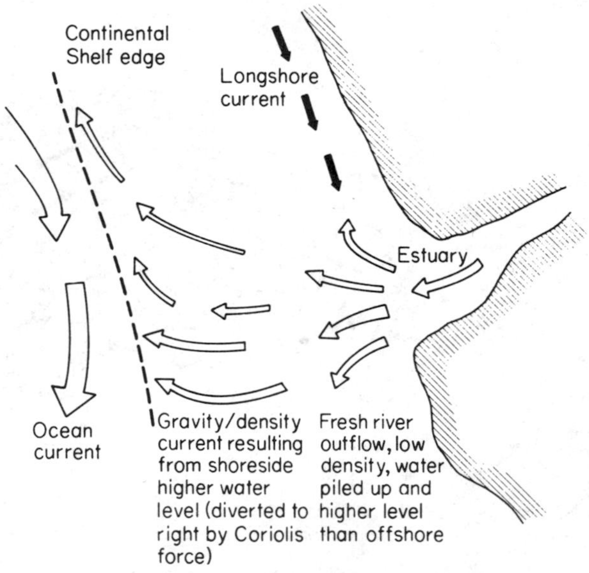

Figure 8. An example of current patterns which may be found off a continental land-mass.

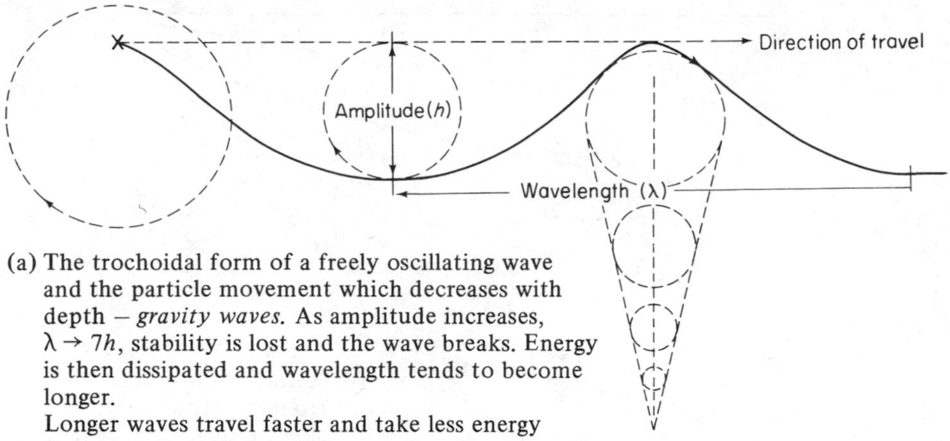

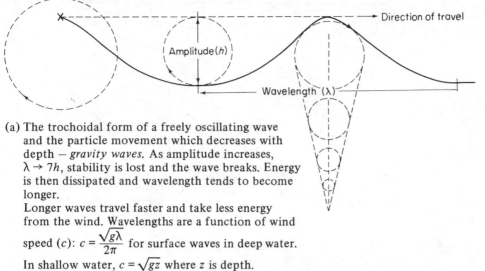

(a) The trochoidal form of a freely oscillating wave and the particle movement which decreases with depth – *gravity waves.* As amplitude increases, $\lambda \to 7h$, stability is lost and the wave breaks. Energy is then dissipated and wavelength tends to become longer.
Longer waves travel faster and take less energy from the wind. Wavelengths are a function of wind speed (c): $c = \frac{\sqrt{g\lambda}}{2\pi}$ for surface waves in deep water.
In shallow water, $c = \sqrt{gz}$ where z is depth.

(b) Water movement then extends to seabed, where friction of seabed slows wave-troughs. Crests overtake troughs until breaking occurs, roughly when $\lambda = z$.

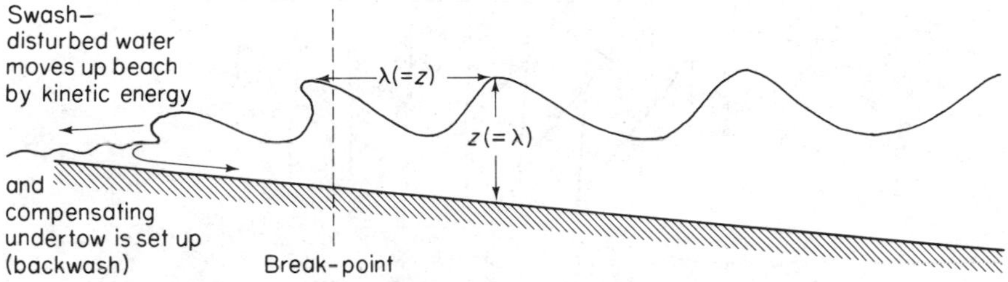

Particle velocity decreases with depth; wavelength decreases; energy remains constant; amplitude increases; steepness increases, particle movement becomes more ovoid; mass transport increases.

(c) Undertow may be accompanied, or replaced, by a rip current.

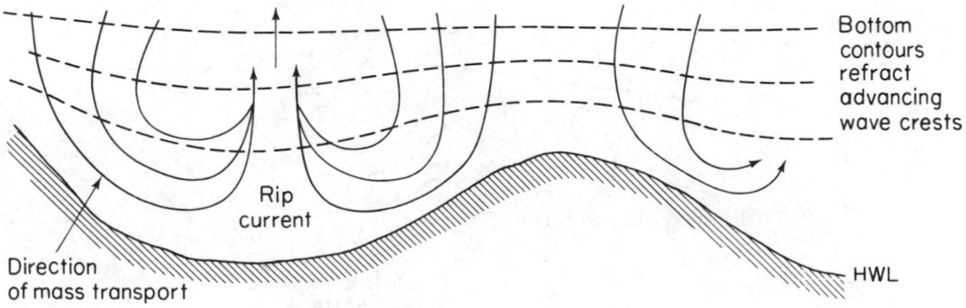

Figure 9. Wave translation on reaching inshore waters.

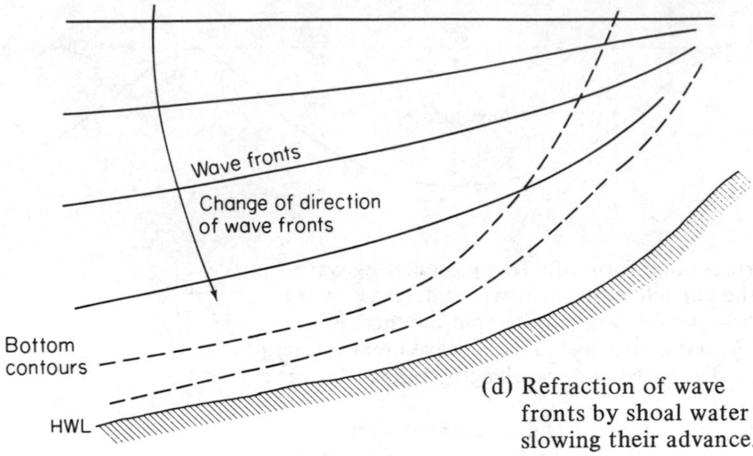

(d) Refraction of wave fronts by shoal water slowing their advance.

(e) Superposition is the meeting of two wave trains which proceed without modification.

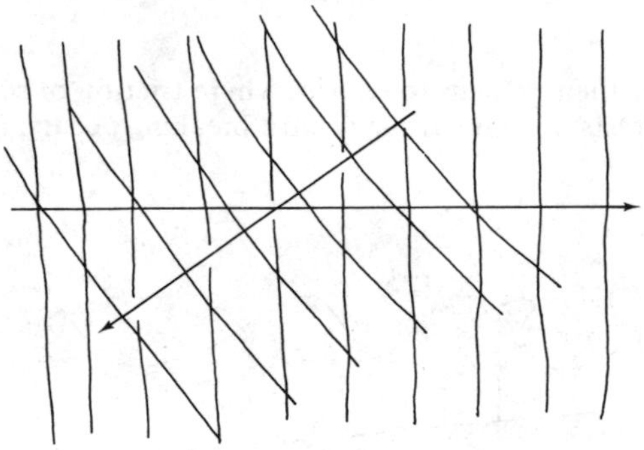

(f) Diffraction is caused when a wave train is obstructed by a feature of a size comparable with that of the wave front.

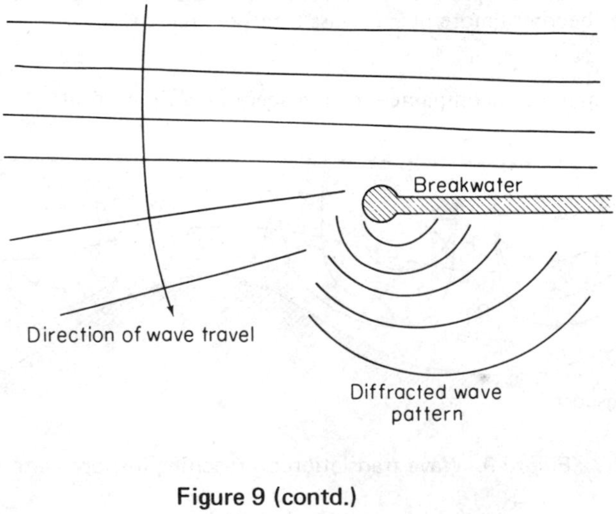

Figure 9 (contd.)

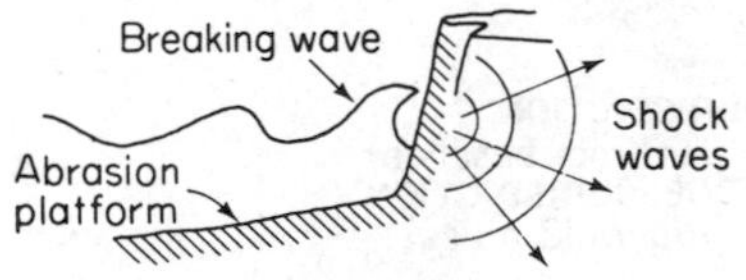

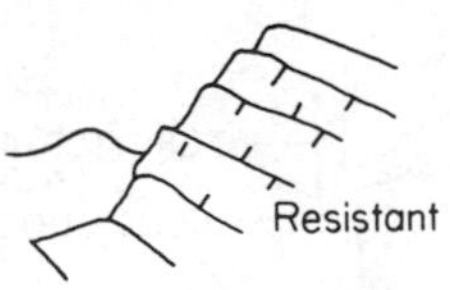

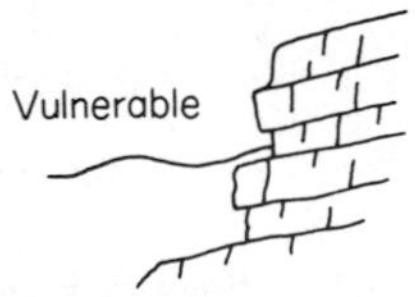

(a) Rock/cliff shoreline. Compressed air trapped in breaking wave 'explodes' and weakens cliff face and causes undermining and land slips. Cliff material so removed is abraded by waves until particles are small enough to be transported away.

The direction of the bedding planes of the cliff affect erosion suffered by the cliff.

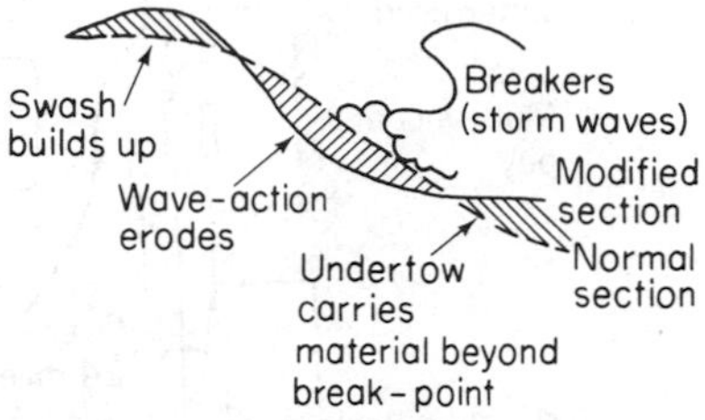

(b) Shingle beach.

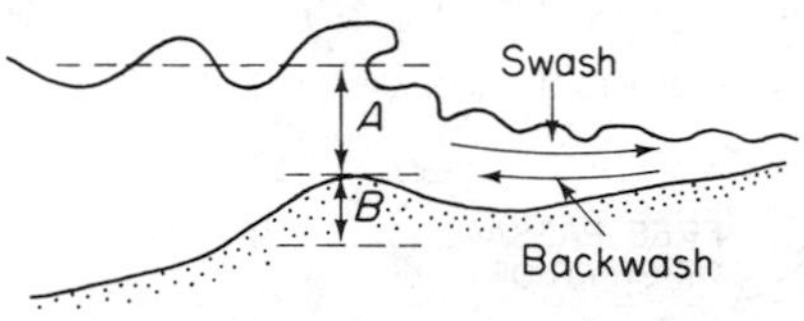

(c) Sandy shoreline: When swash and backwash are self-cancelling, a break-point bar is set up and $A \triangleq 2B$.

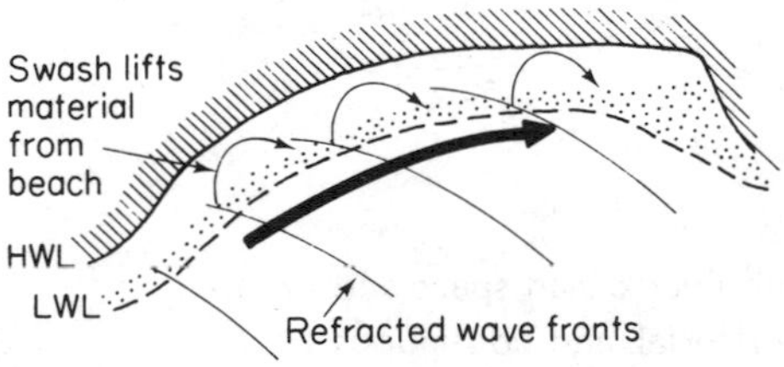

Longshore current is set up, transporting material until an obstruction is encountered.

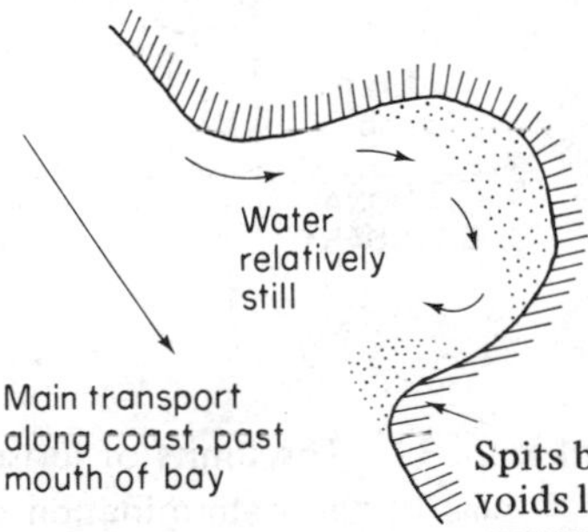

Material deposited at undisturbed parts of coast.

Spits build up in voids left by currents. Storms curve the spit and deposit further material in their lee.

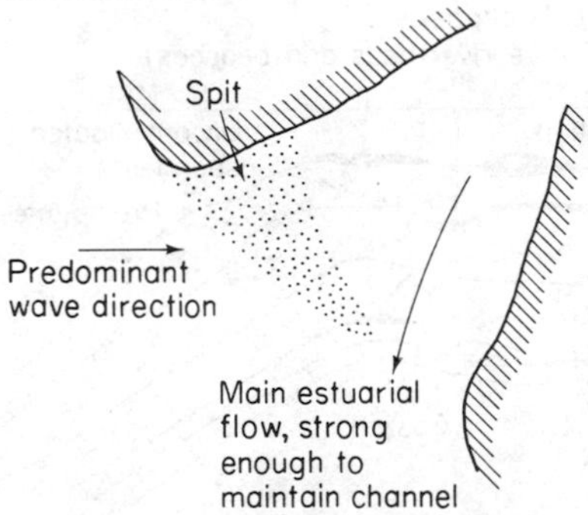

Where estuarial flow is weak, a 'baymouth bar' is constructed, the bay is 'embayed' and the river seeks an outlet at some other point on the coast.

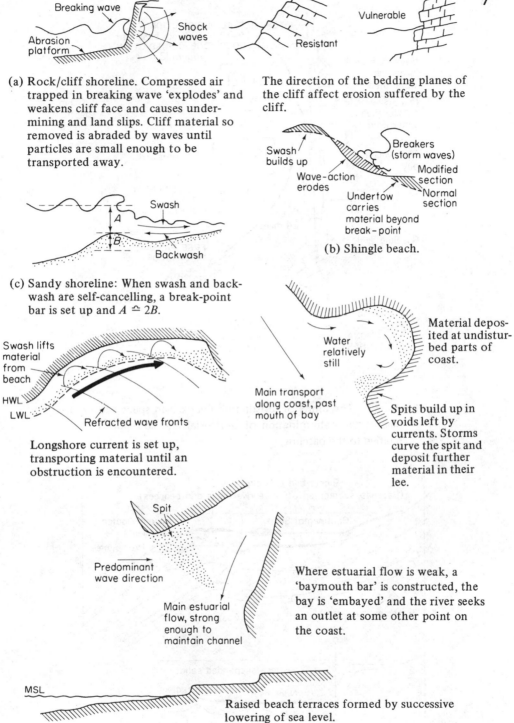

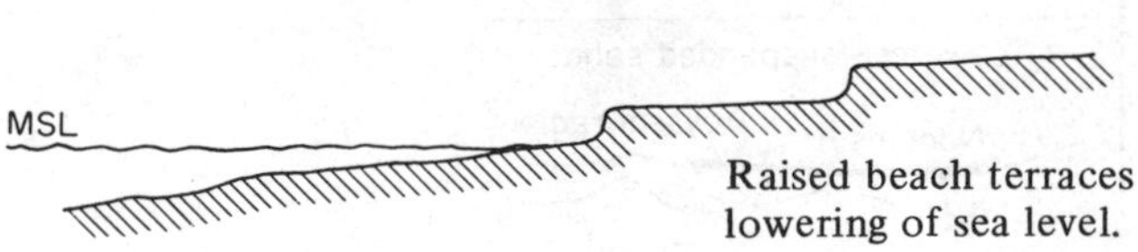

Raised beach terraces formed by successive lowering of sea level.

Figure 10. Wave effect on coastal features.

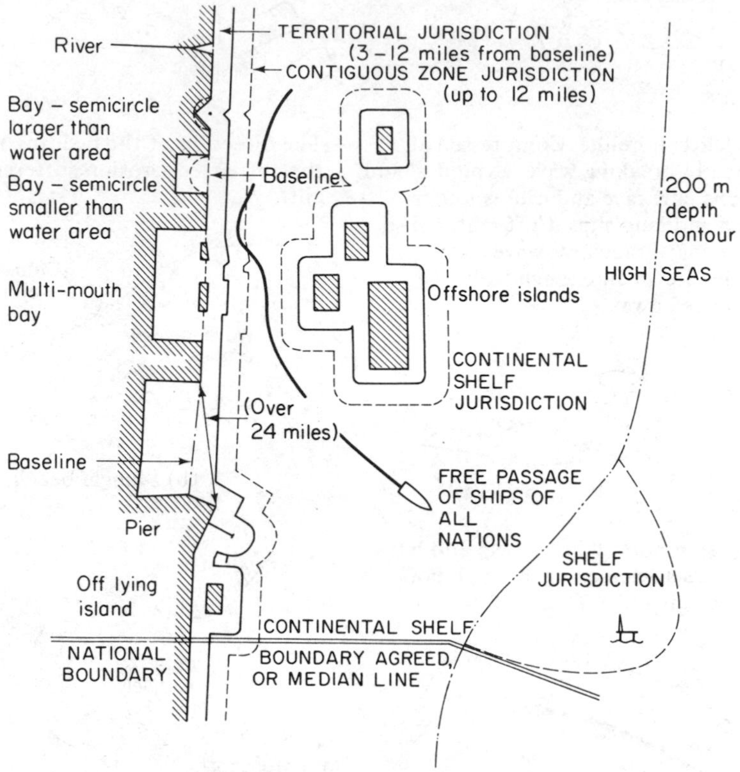

Figure 11. The limits of jurisdiction over ocean space and examples of the determination of territorial and contiguous zones relative to the baseline.

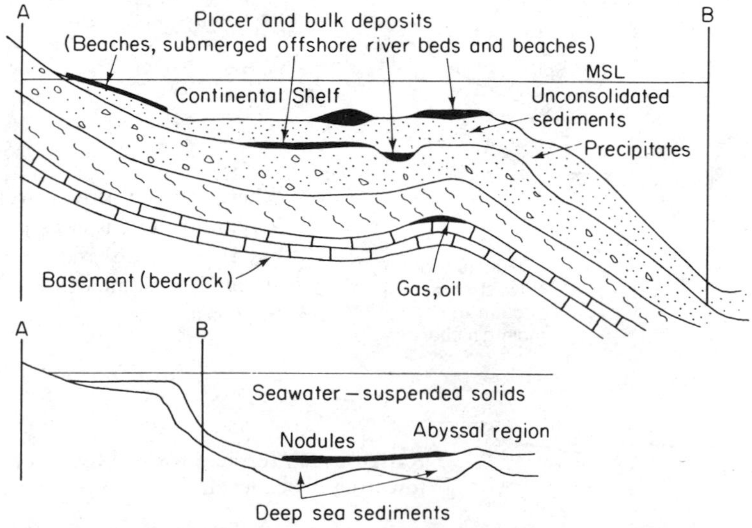

Figure 12. The location of mineral resources offshore.

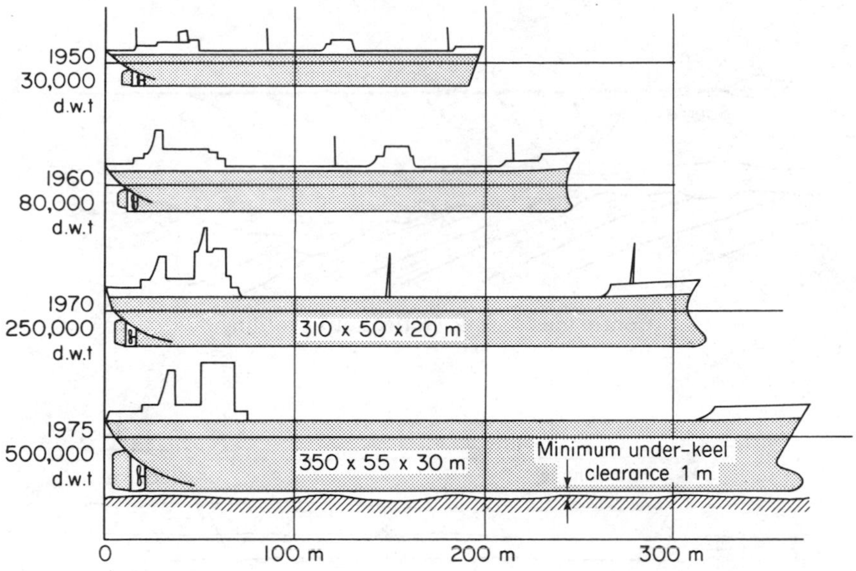

Figure 13. Tanker sizes in the past twenty-five years.

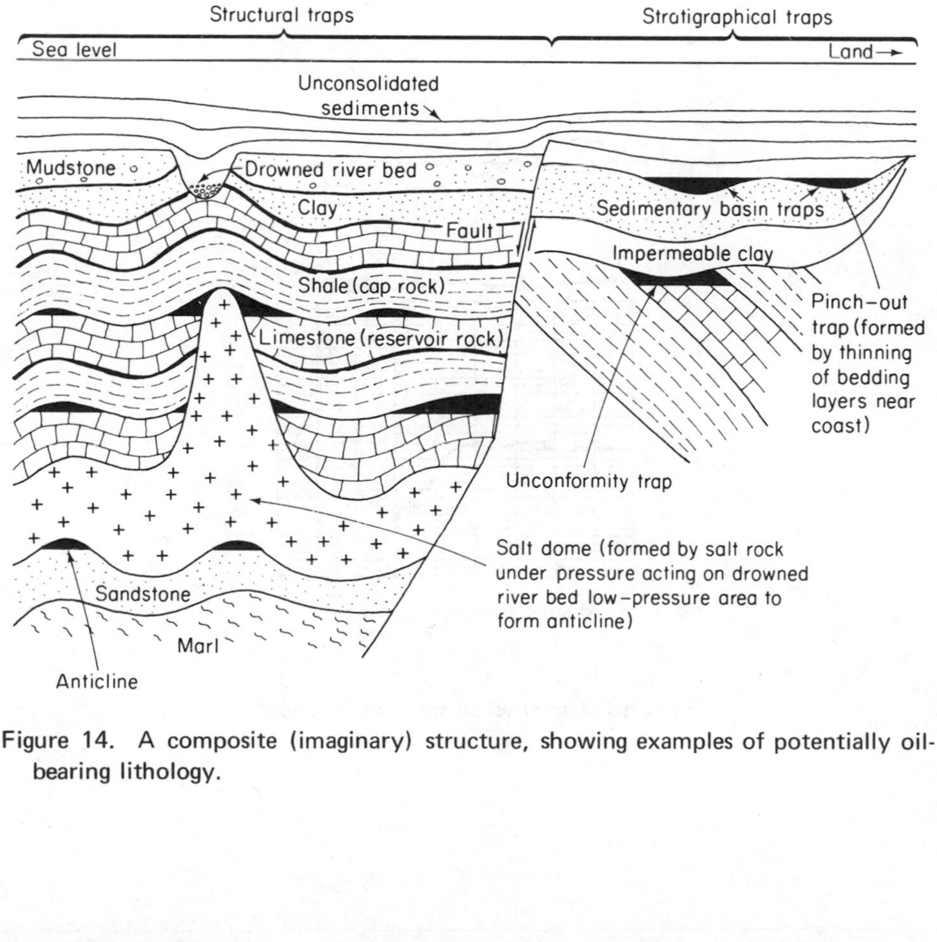

Figure 14. A composite (imaginary) structure, showing examples of potentially oil-bearing lithology.

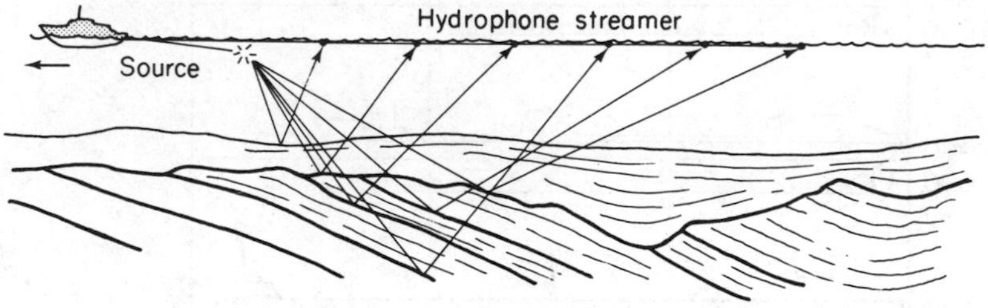

Continuous seismic (reflection) profiling

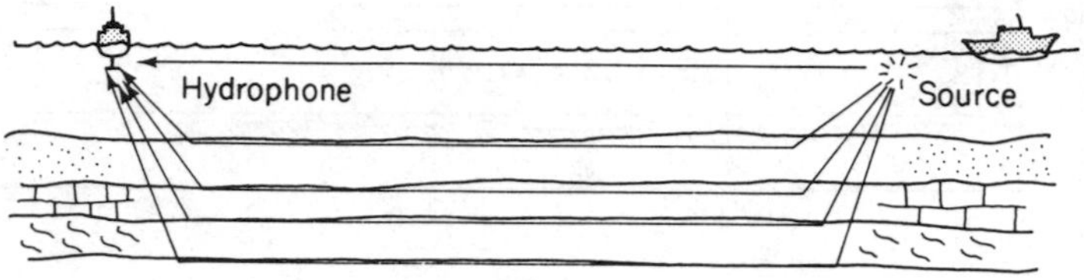

Refraction shooting

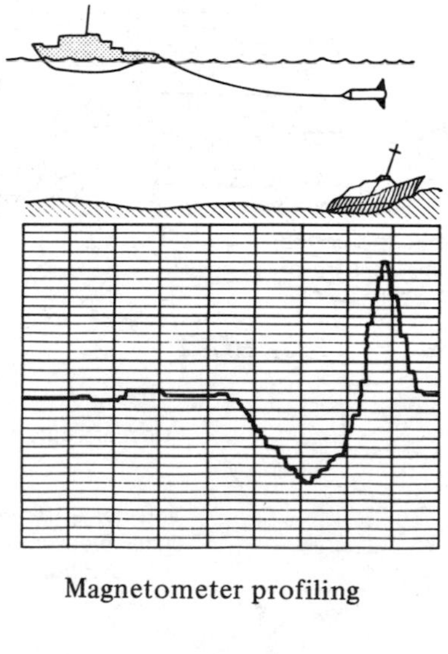

Magnetometer profiling

Figure 15. Geophysical survey techniques.

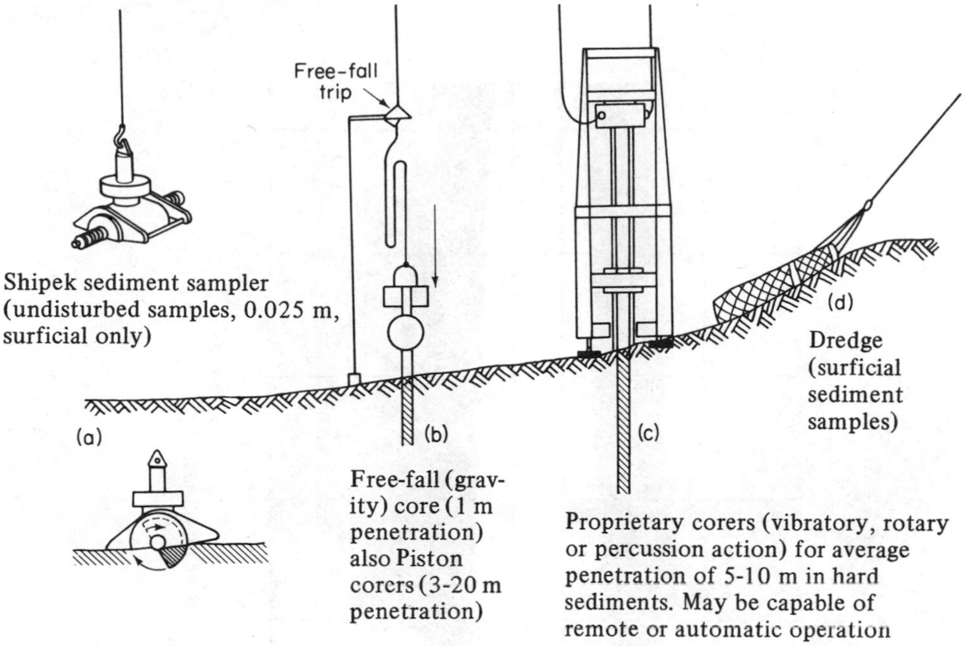

Figure 16. Geological survey techniques.

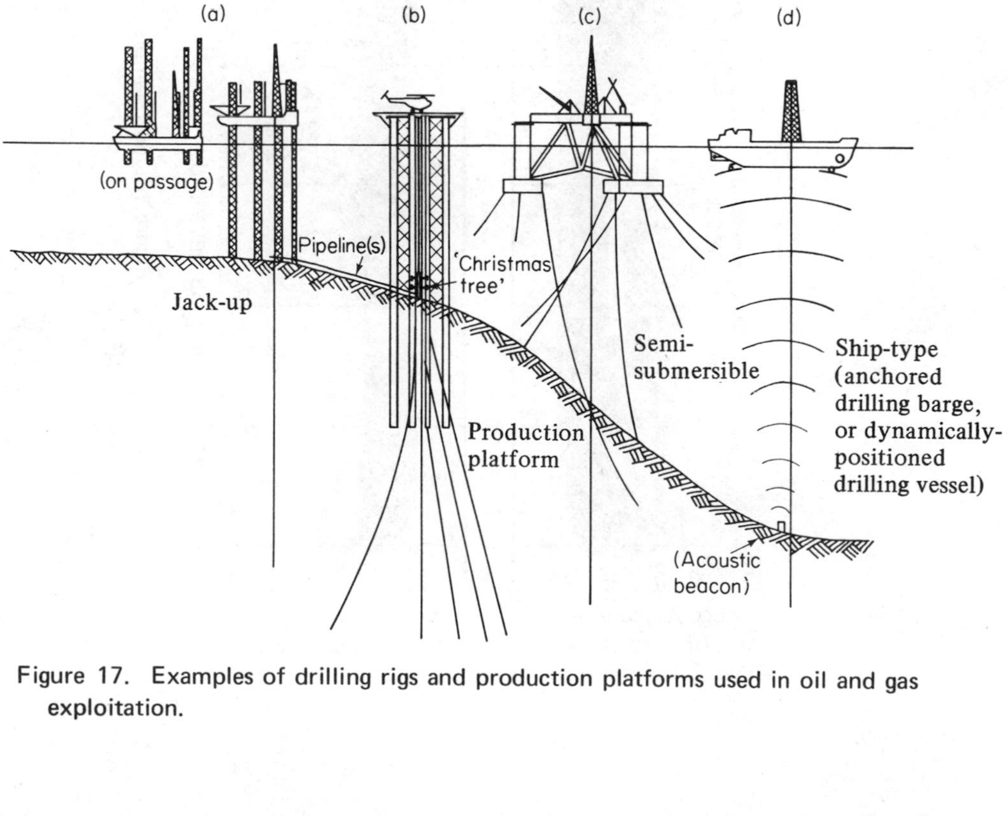

Figure 17. Examples of drilling rigs and production platforms used in oil and gas exploitation.

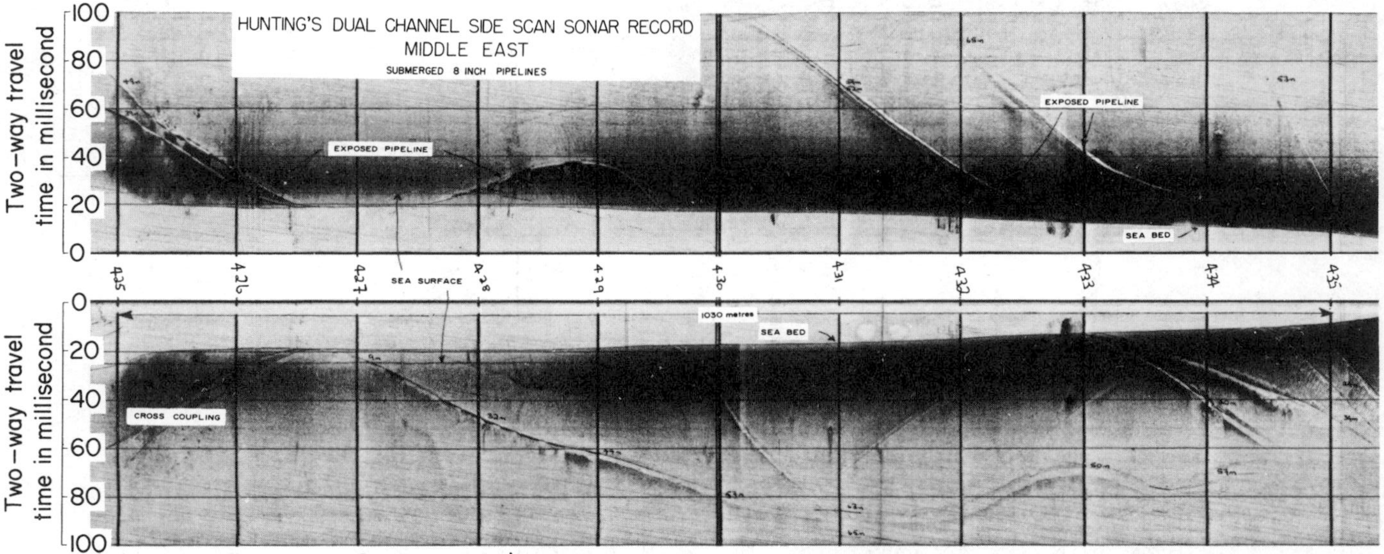

(Courtesy Hunting Surveys Ltd.)

Figure 18. (a) Example of side-scan record over pipelines.

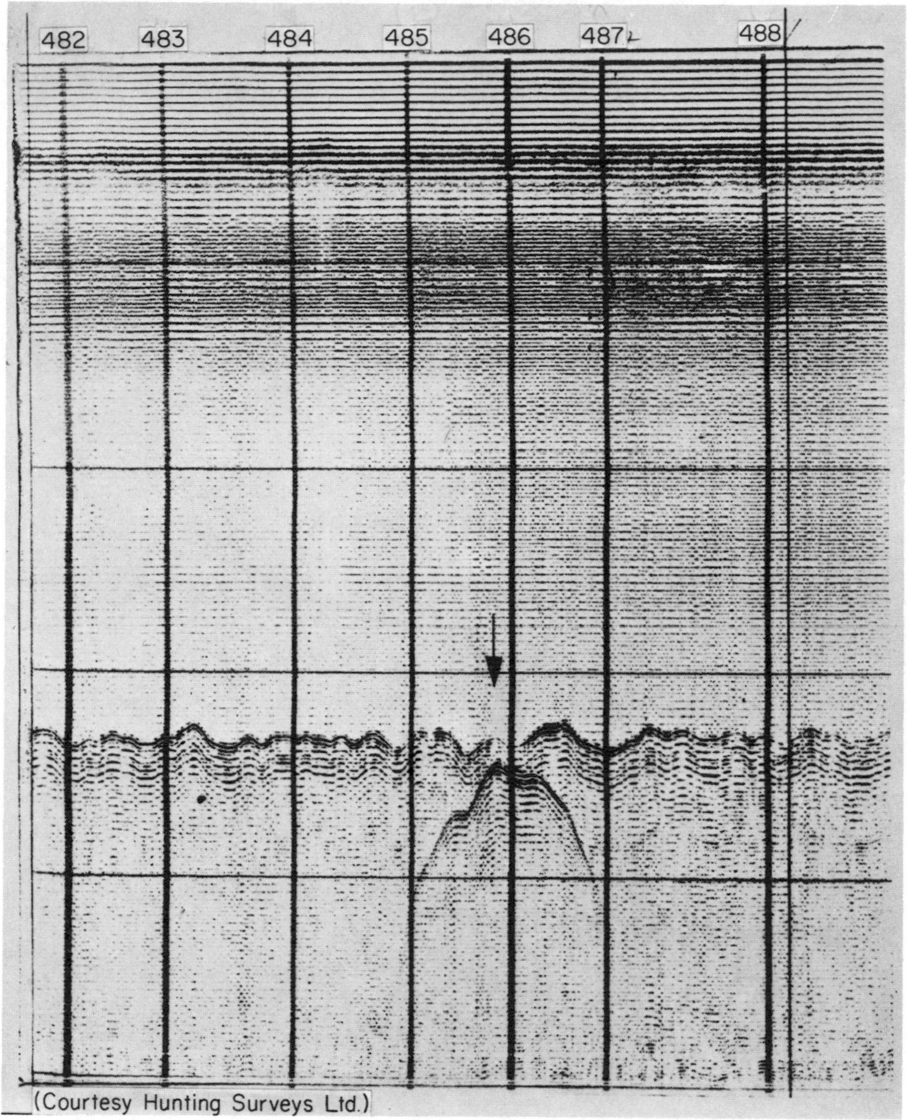

(b) Example of profiling record over a pipeline.

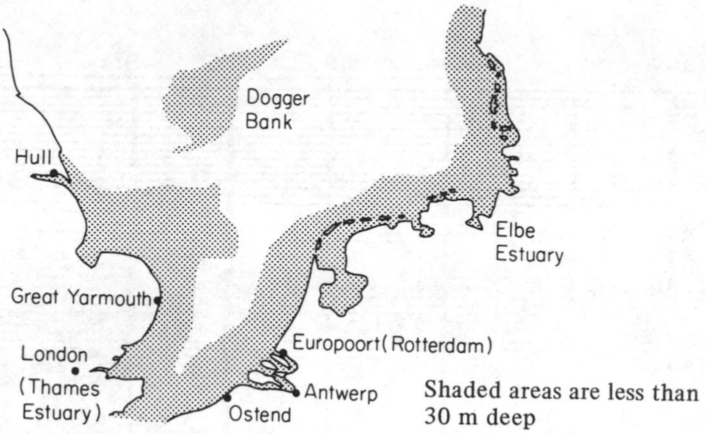

Figure 19. The areas of the Southern North Sea hazardous to deep-draught tanker operation.

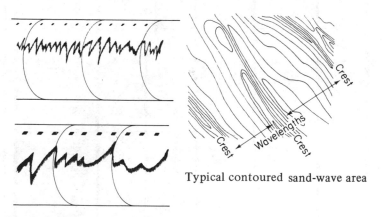

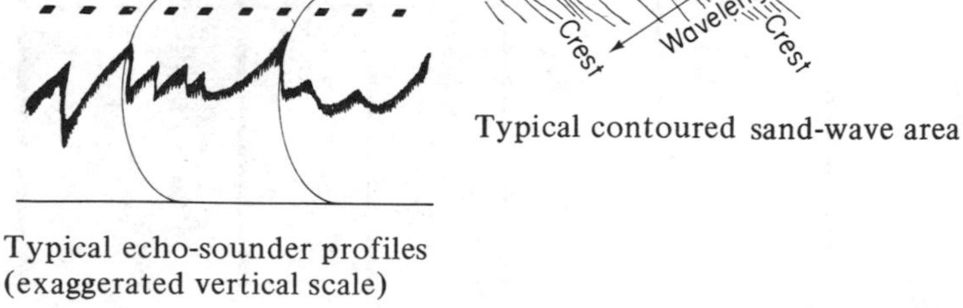

Figure 20. The sand-wave problem.

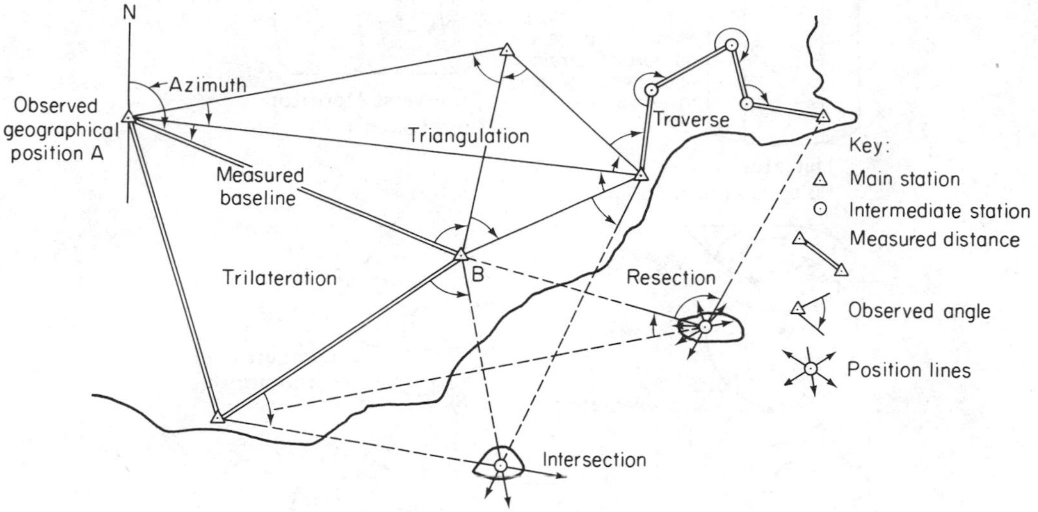

Figure 21. Shore control techniques.

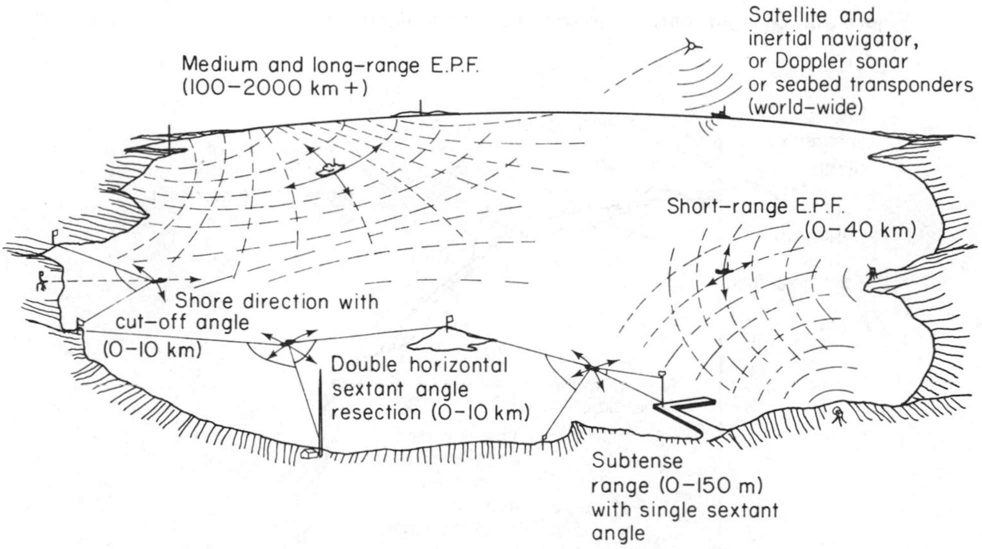

Figure 22. Position fixing techniques offshore.

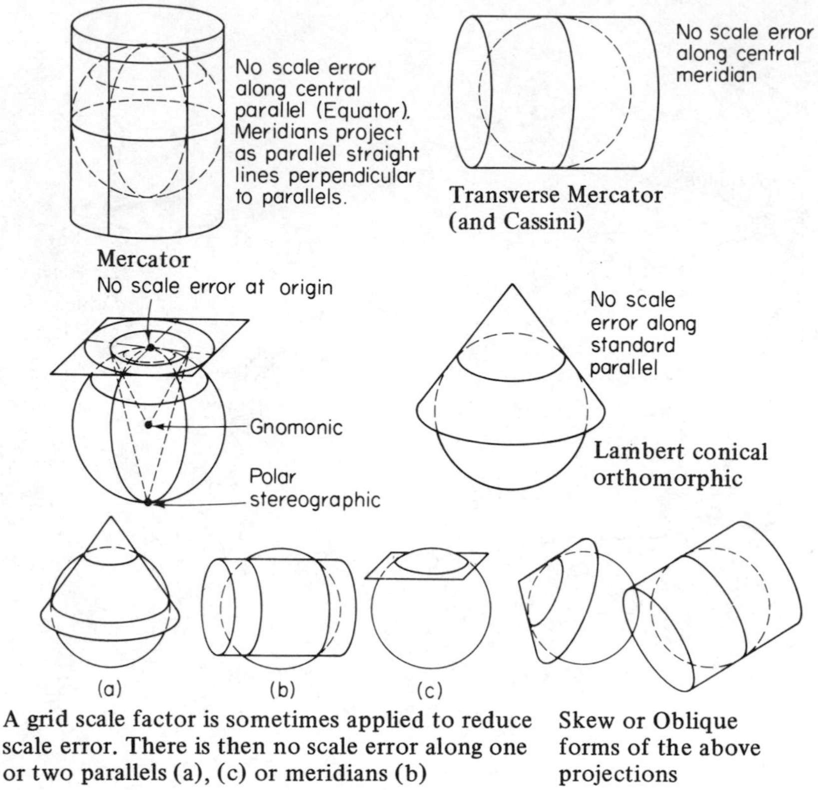

Figure 23. Diagrammatic representation of projections useful to the surveyor.

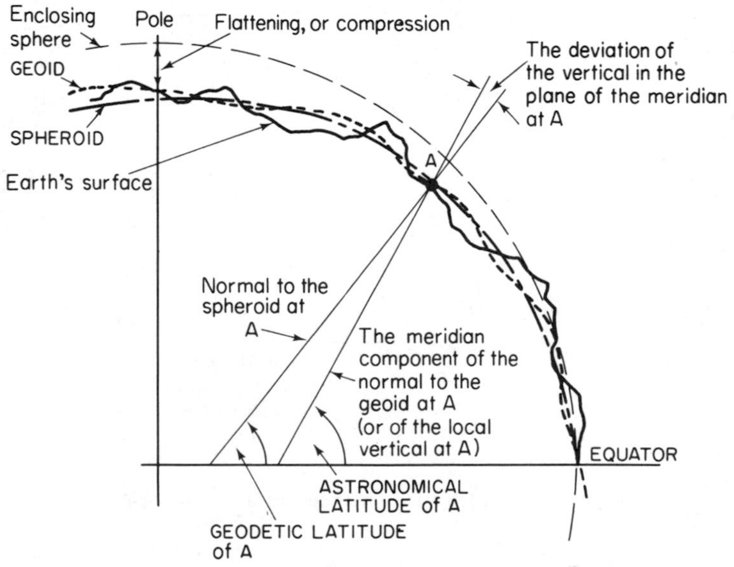

Figure 24. The surfaces of geodesy.

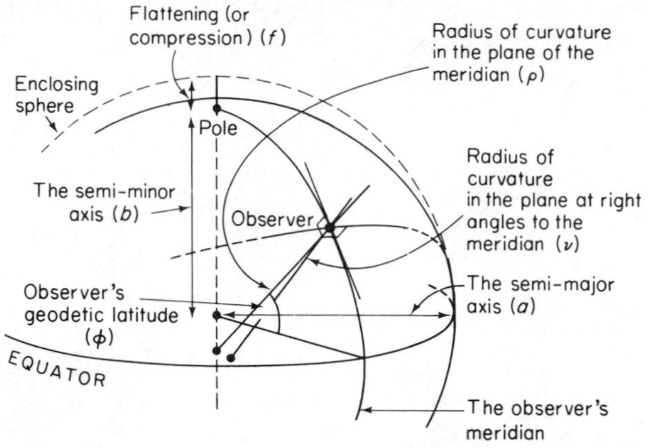

Figure 25. The constants and variables of the spheroid.

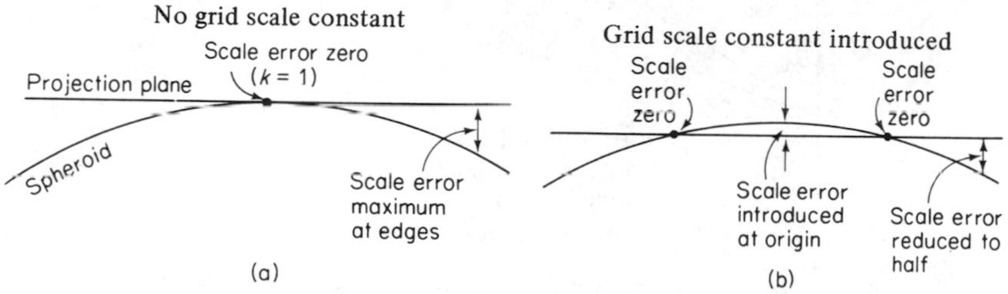

Figure 26. The effect of introducing a grid scale constant.

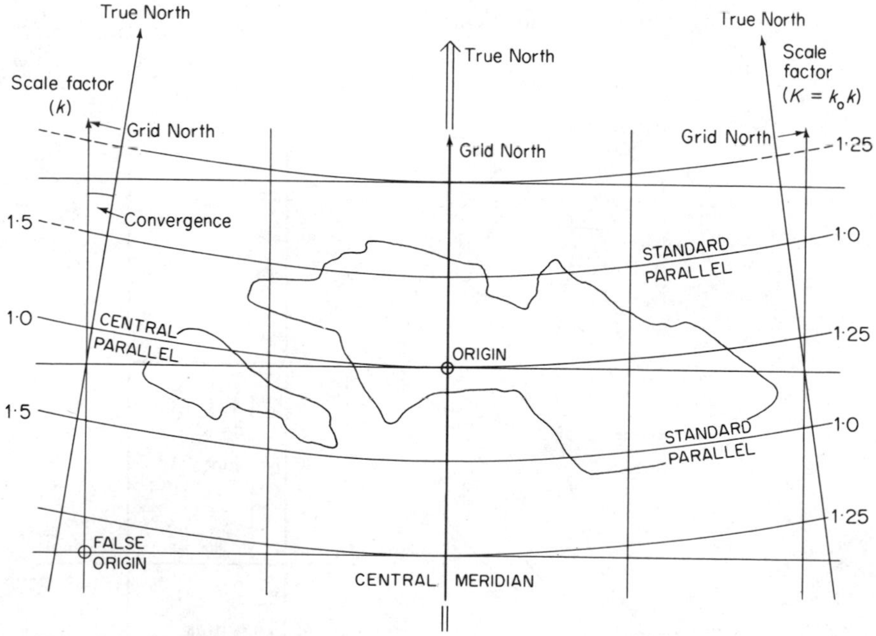

Figure 27. The relationship between spheroid, projection and grid. The example shows meridians as they would plot on an orthomorphic conical projection using one central parallel (left) or two standard parallels (right).

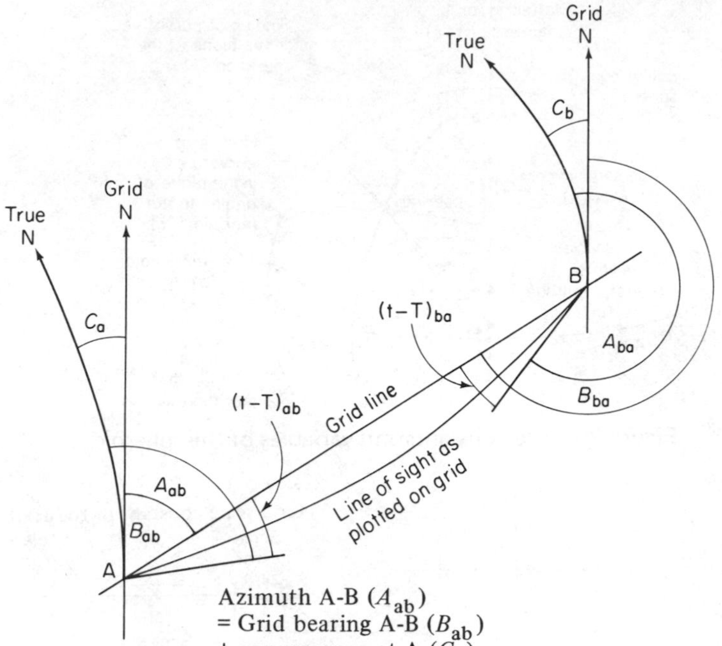

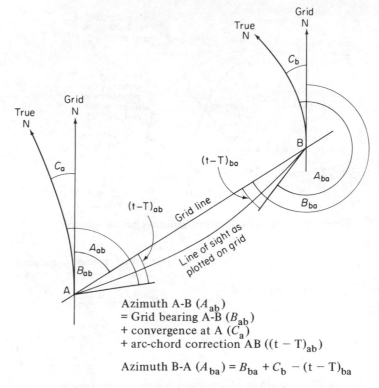

Azimuth A-B (A_{ab})
= Grid bearing A-B (B_{ab})
+ convergence at A (C_a)
+ arc-chord correction AB ($(t - T)_{ab}$)

Azimuth B-A $(A_{ba}) = B_{ba} + C_b - (t - T)_{ba}$

Figure 28. The relationship between azimuth, grid bearing, convergence and arc to chord correction.

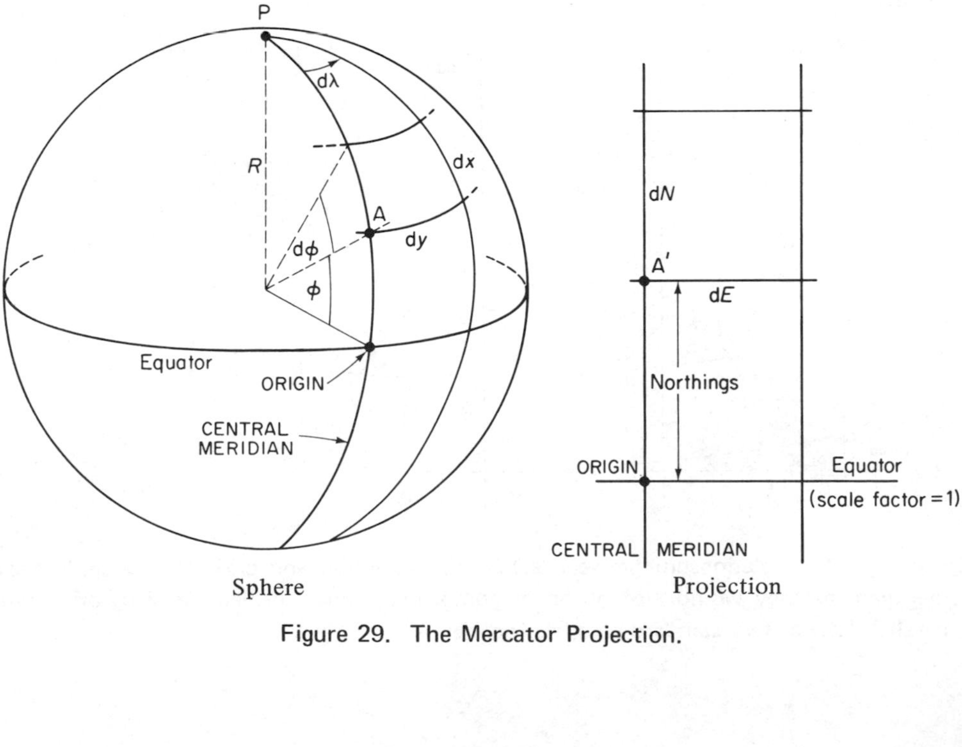

Figure 29. The Mercator Projection.

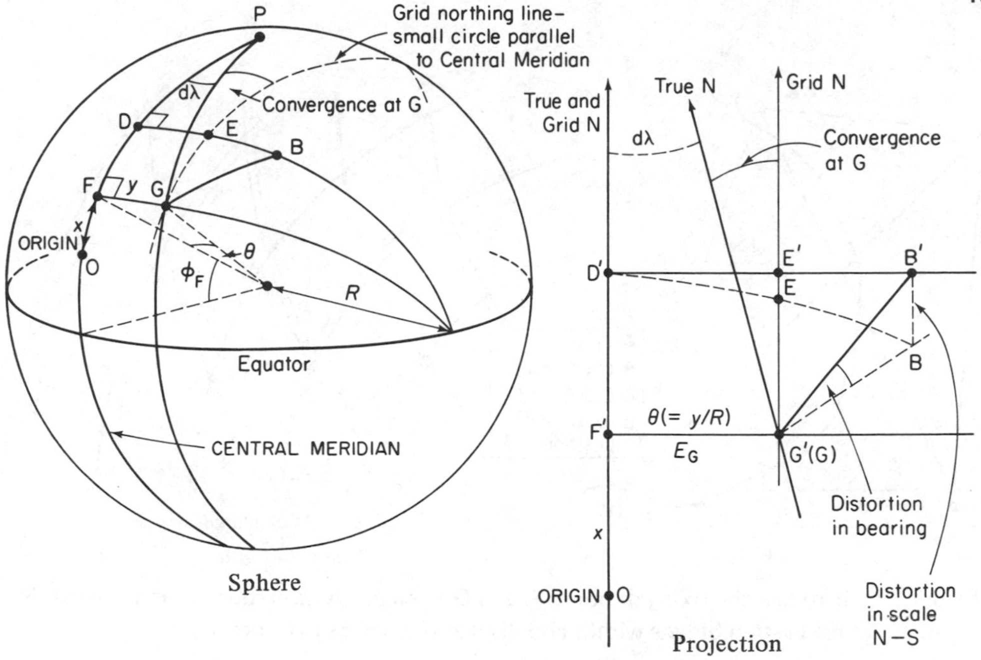

Figure 30. The Cassini-Soldner Projection.

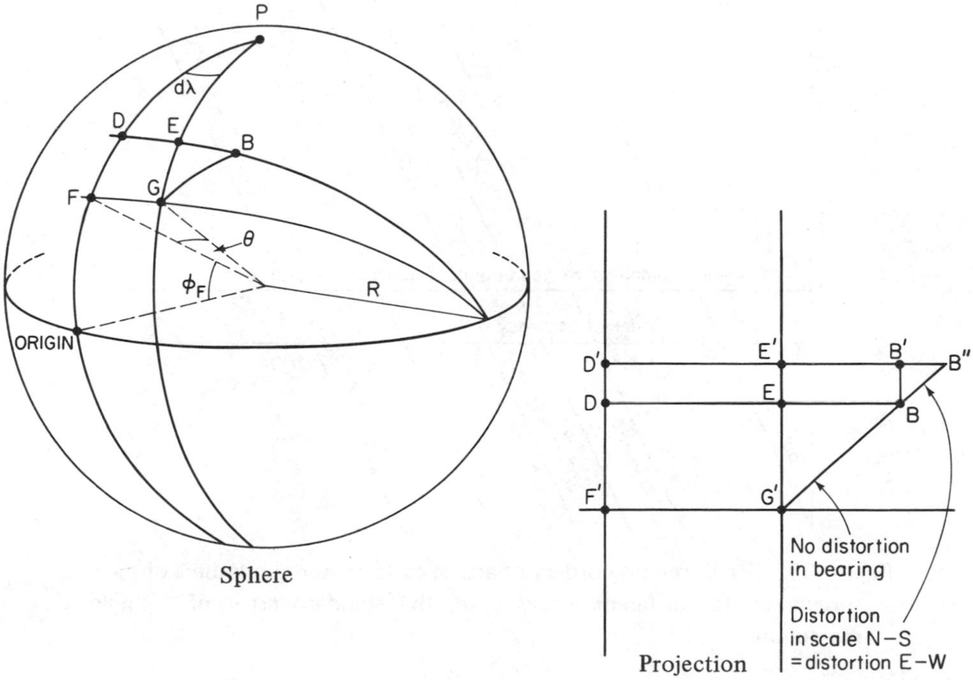

Figure 31. The Transverse Mercator Projection.

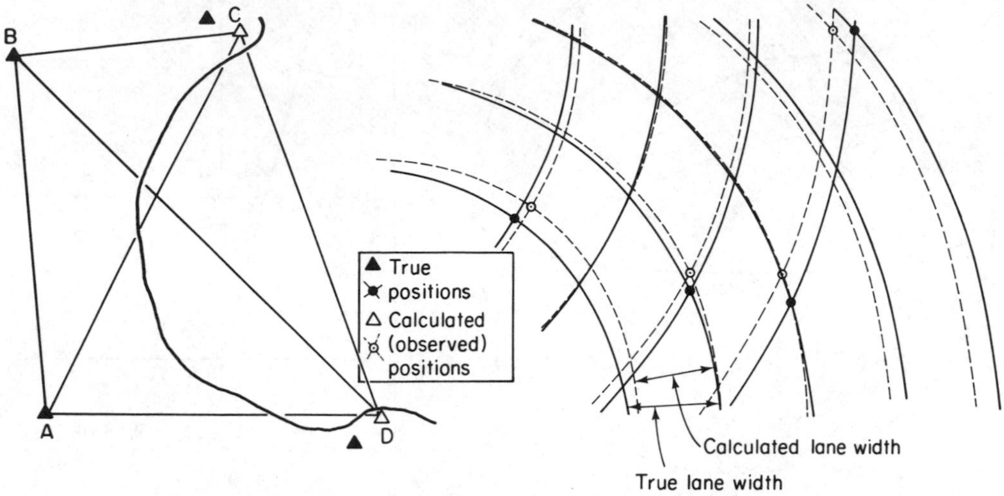

Figure 32. Errors in the fixing of points C and D from geodetic stations A and B result in incorrect calibration of lane widths and displaced patterns offshore.

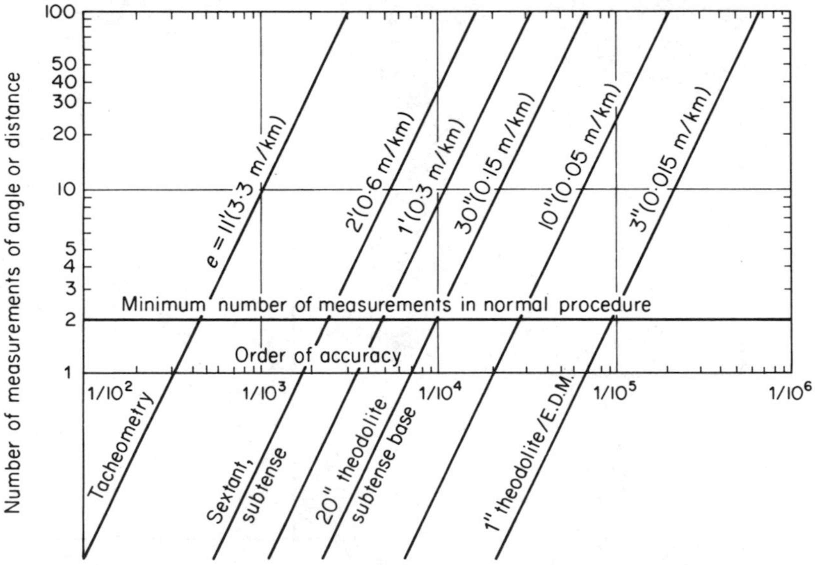

Figure 33. Graph relating orders of accuracy to required numbers of measurements, for different values of *e*, the standard error of a single measurement.

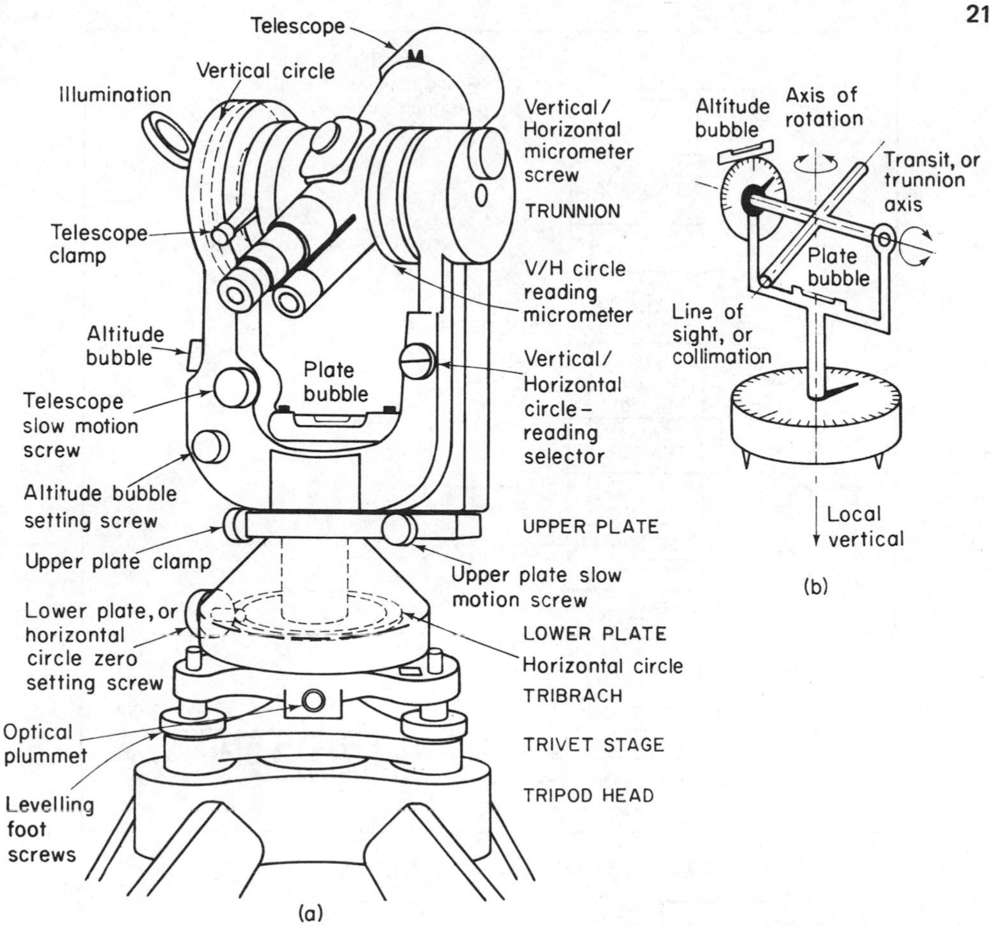

Figure 34. (a) Features of the modern theodolite.
(b) The axes of the theodolite and basic features.

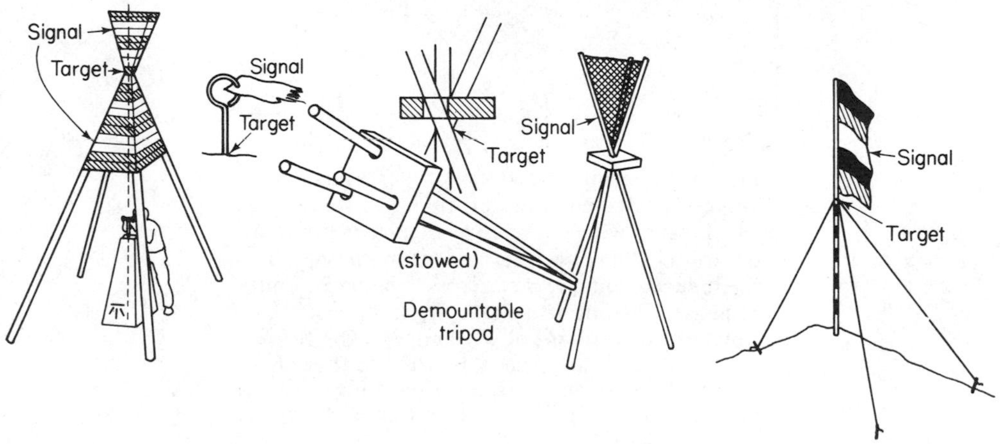

Figure 35. Some features of opaque targets.

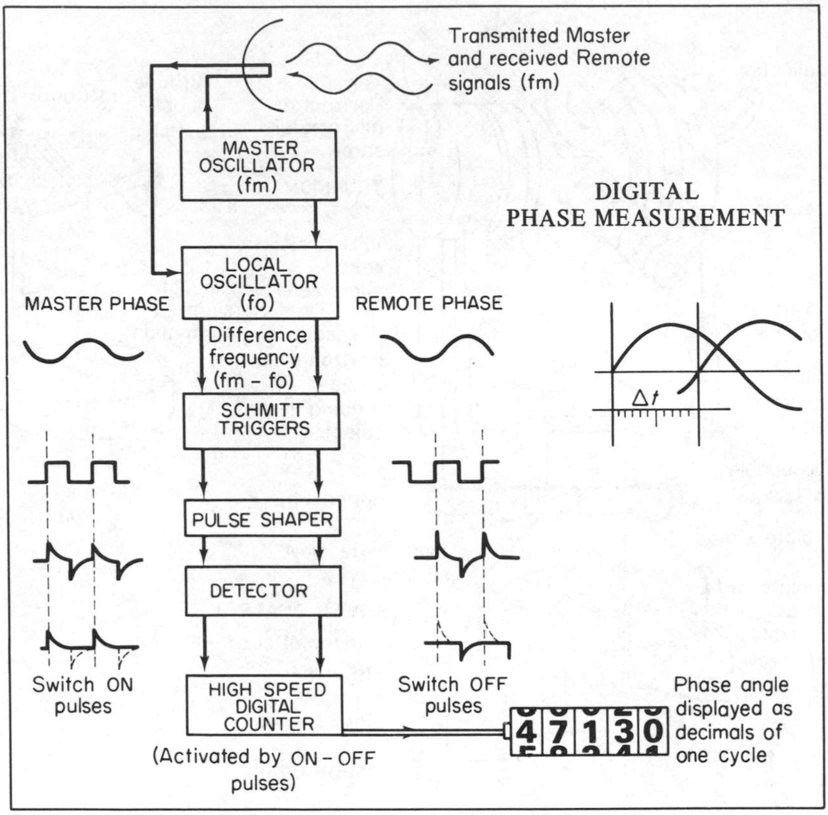

ELECTROMECHANICAL RESOLVER

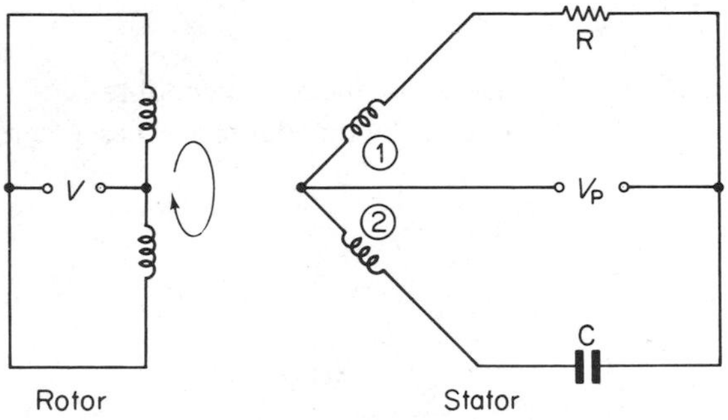

The rotor, having data voltage V across its coils, induces a voltage in the stator coils of value which varies with the relative angular position of the rotor. The voltage across R is in phase with the induced voltage, that across C being 90° out of phase. The output voltage, V_p has phase which is the resultant of V_R and V_C. The rotor is rotated until the phase V is equal to that of V_P, and its angular position is then a measure of the phase difference as a fraction of one cycle.

Figure 36. Basic methods of phase measurement.

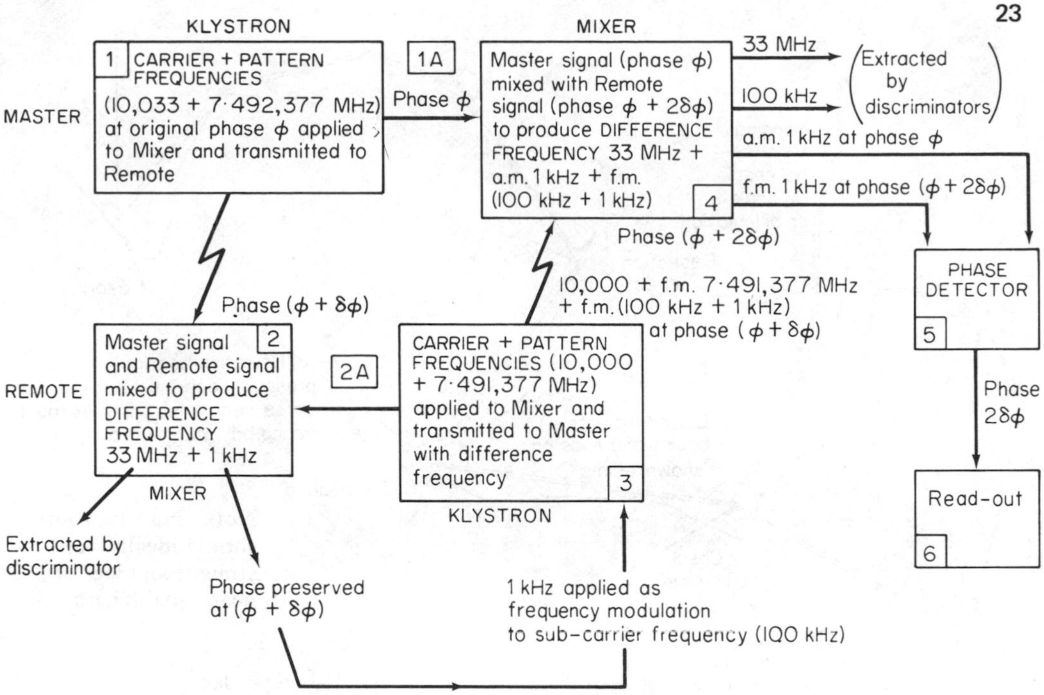

Note: a.m. = amplitude modulated; f.m. = frequency modulated.

Figure 37. Block diagram of the significant features in the operation of the Tellurometer system.

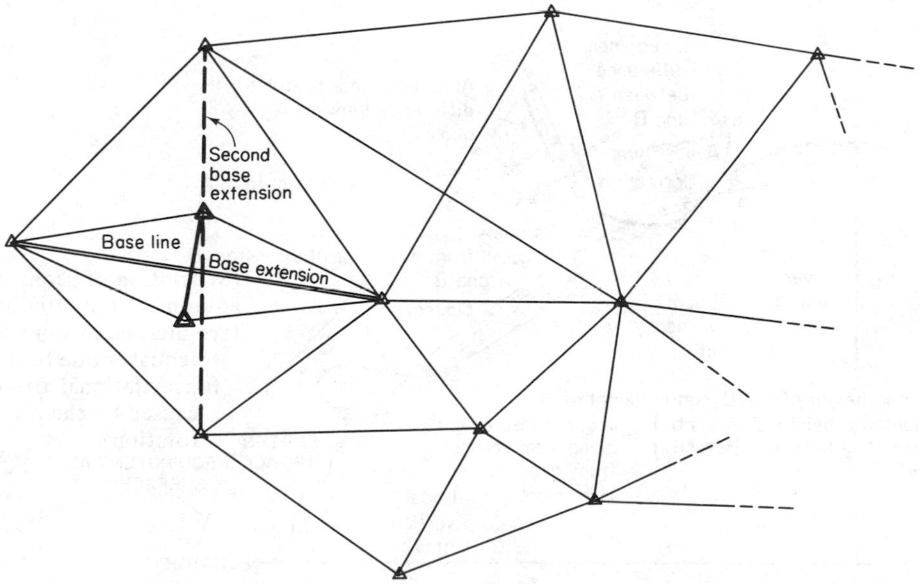

Figure 38. An example of base extension triangulation.

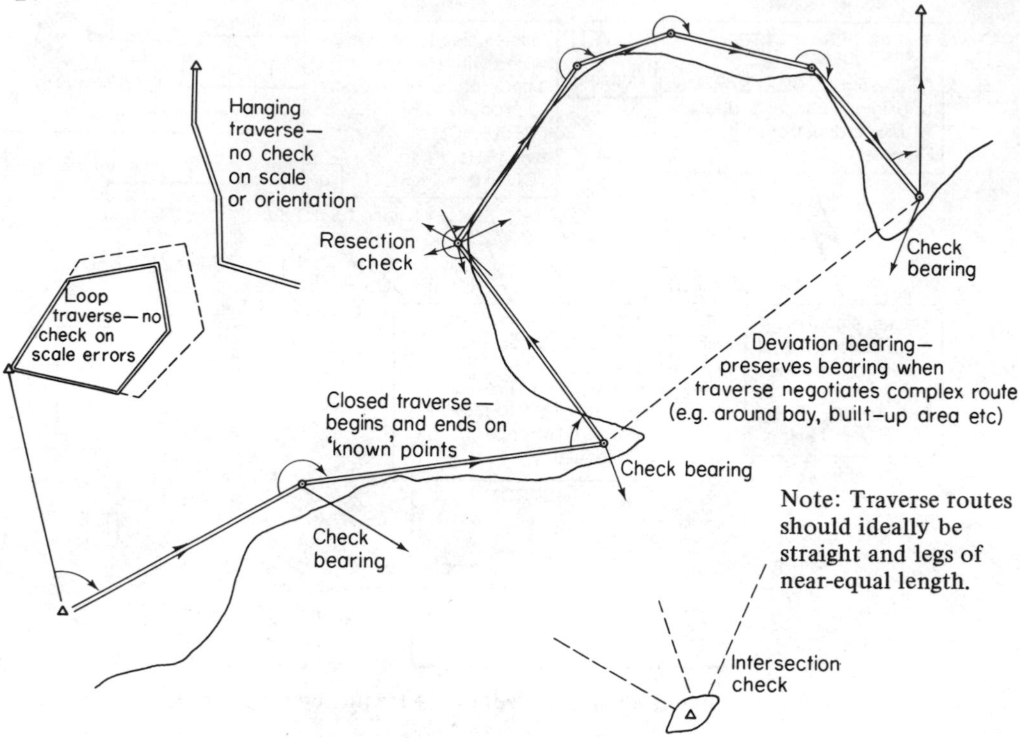
Hanging
traverse—
no check
on scale
or orientation
Loop
traverse—no
check on
scale errors
Resection
check
Check
bearing
Deviation bearing—
preserves bearing when
traverse negotiates complex route
(e.g. around bay, built-up area etc)
Closed traverse—
begins and ends on
'known' points
Check bearing
Check
bearing
Note: Traverse routes
should ideally be
straight and legs of
near-equal length.
Intersection
check

Figure 39. Features of traverses.

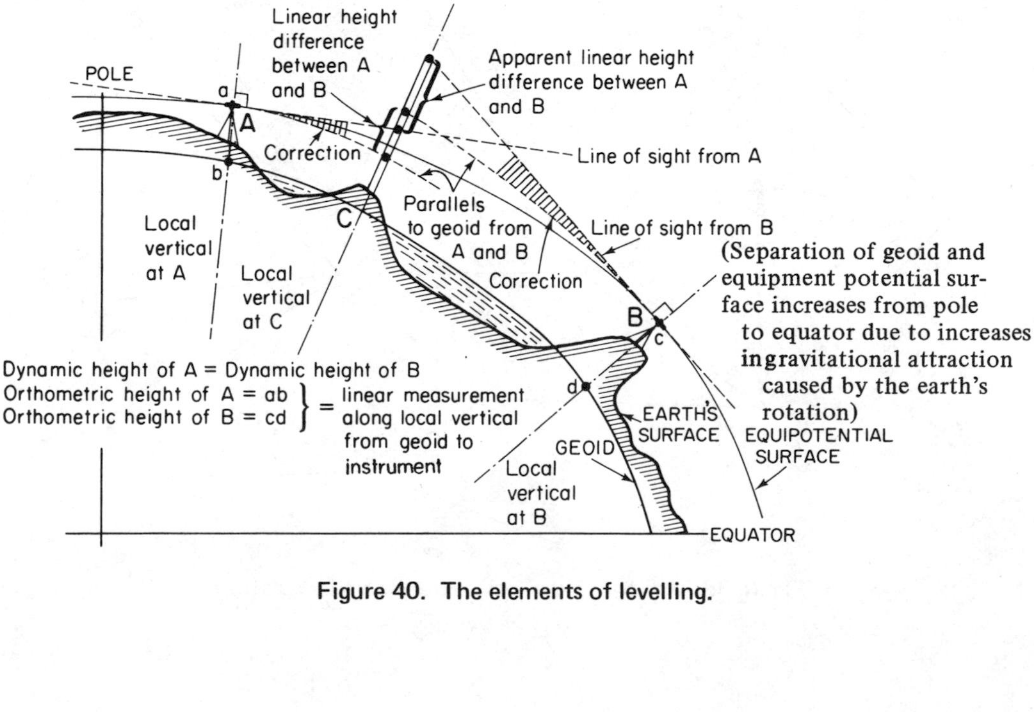
Linear height
difference
between A
and B
Apparent linear height
difference between A
and B
POLE
a
A
b
Correction
Line of sight from A
C
Parallels
to geoid from
A and B
Line of sight from B
Local
vertical
at A
Local
vertical
at C
Correction
(Separation of geoid and
equipment potential sur-
face increases from pole
to equator due to increases
ingravitational attraction
caused by the earth's
rotation)
B
c
d
Dynamic height of A = Dynamic height of B
Orthometric height of A = ab
Orthometric height of B = cd
= linear measurement
along local vertical
from geoid to
instrument
EARTH'S
SURFACE
EQUIPOTENTIAL
SURFACE
GEOID
Local
vertical
at B
EQUATOR

Figure 40. The elements of levelling.

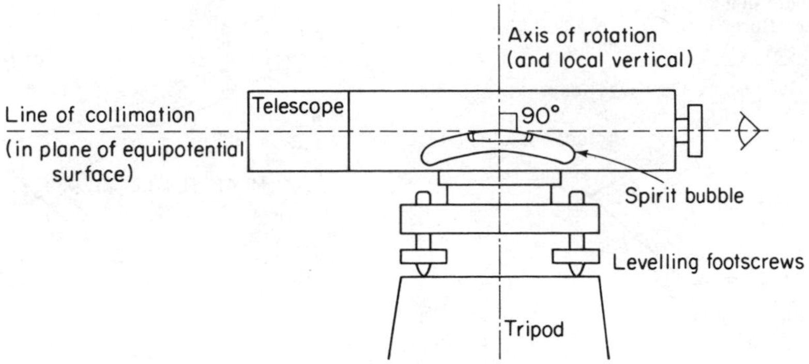

Figure 41. The simple level.

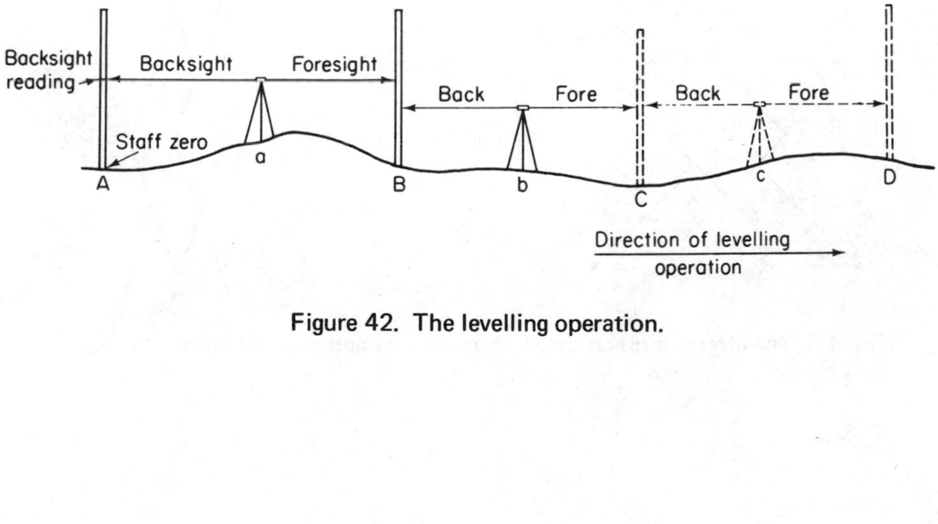

Figure 42. The levelling operation.

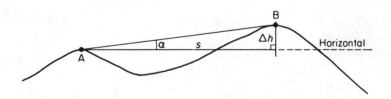

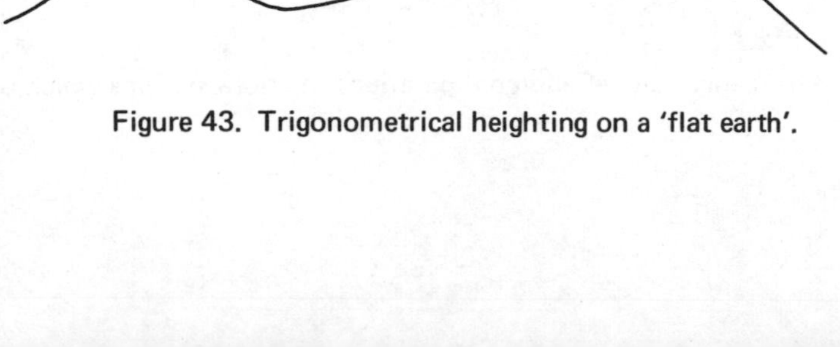

Figure 43. Trigonometrical heighting on a 'flat earth'.

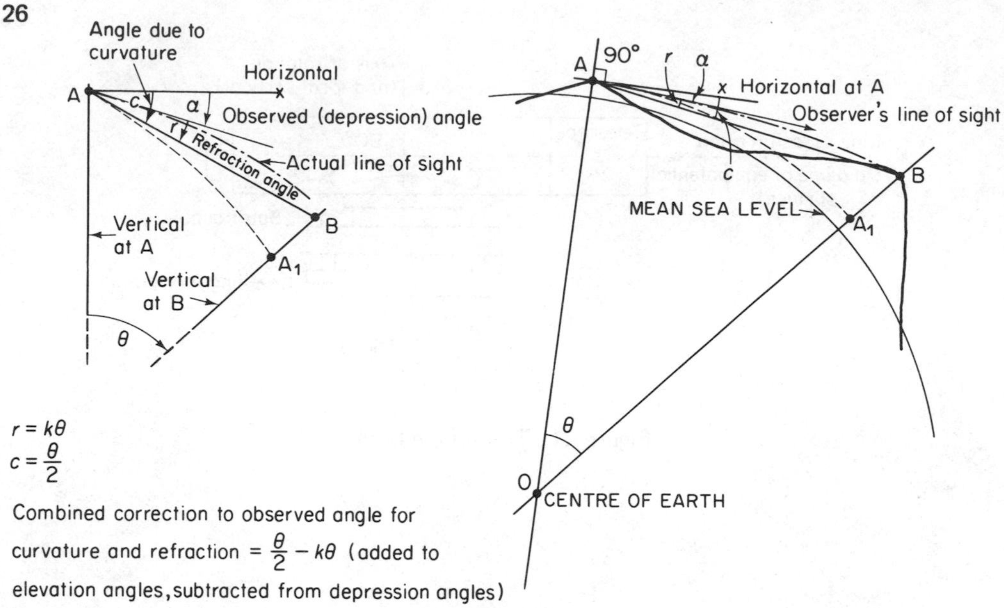

$r = k\theta$

$c = \dfrac{\theta}{2}$

Combined correction to observed angle for curvature and refraction $= \dfrac{\theta}{2} - k\theta$ (added to elevation angles, subtracted from depression angles)

Figure 44. The effects of curvature and refraction.

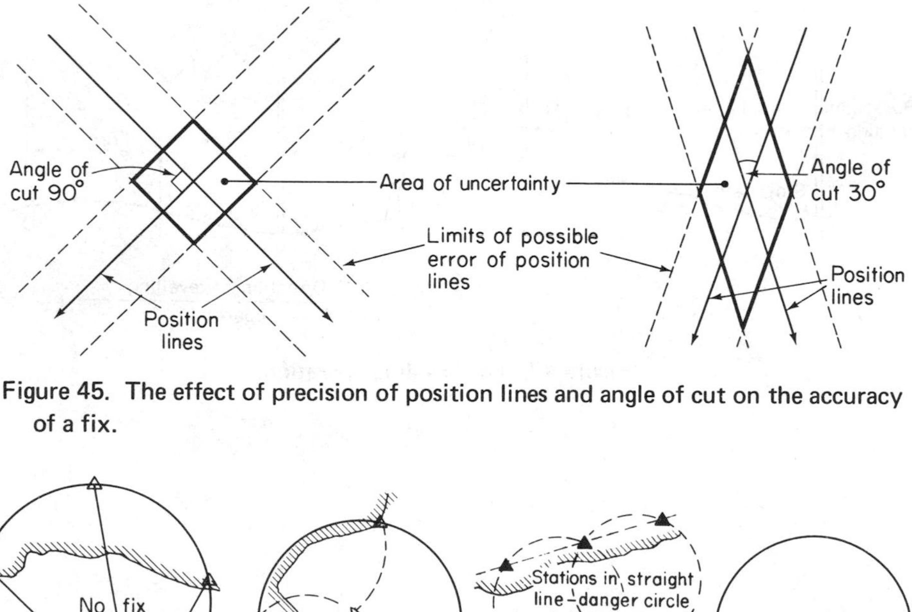

Figure 45. The effect of precision of position lines and angle of cut on the accuracy of a fix.

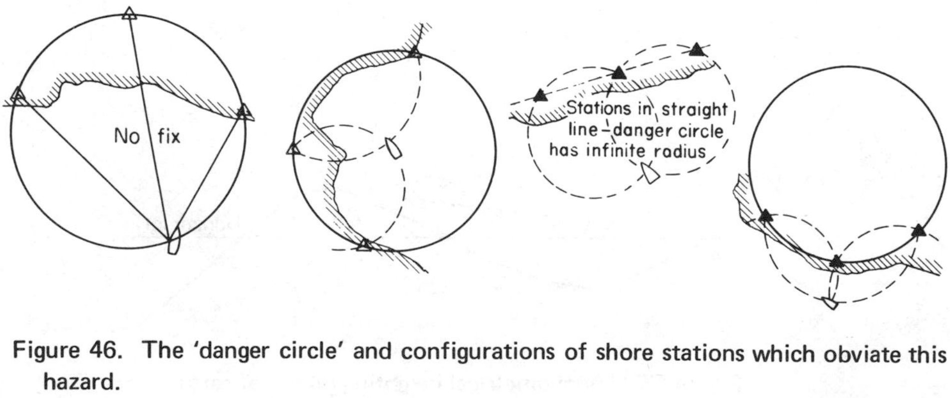

Figure 46. The 'danger circle' and configurations of shore stations which obviate this hazard.

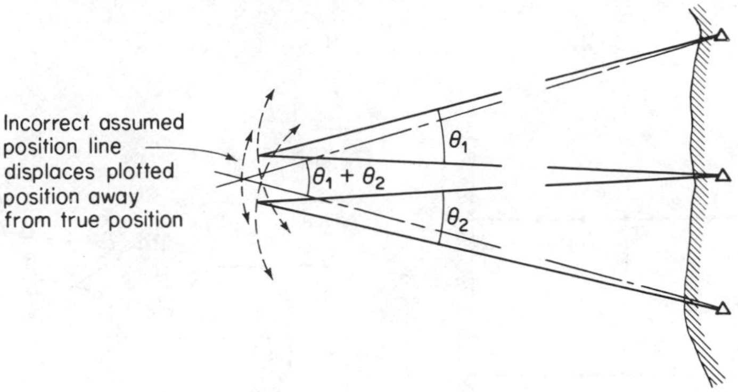

Figure 47. Parallax error due to separation of sextant anglers.

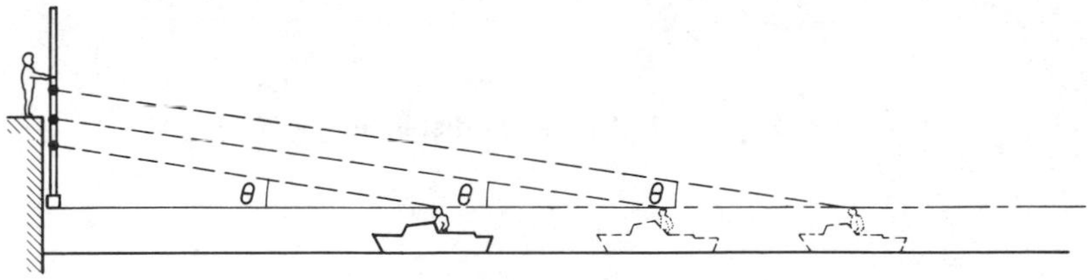

Figure 48. The subtense ranging technique.

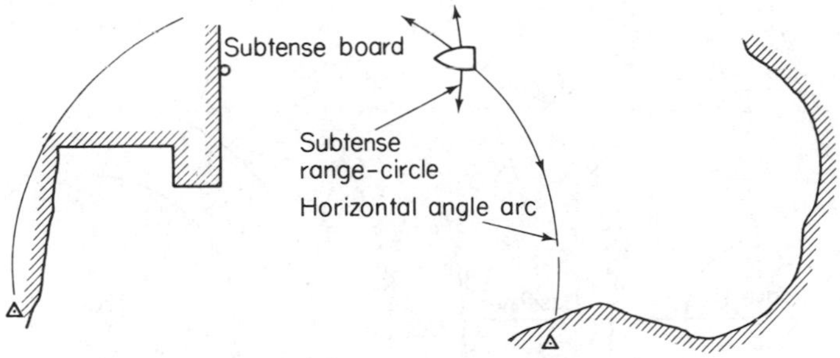

Figure 49. Fixing by subtense range and cut-off angle.

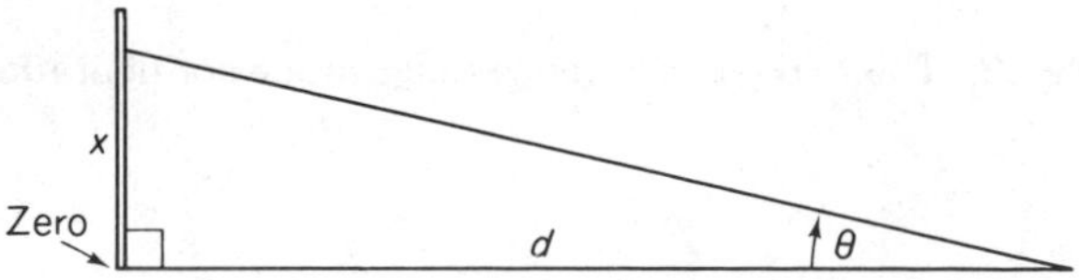

Figure 50. Marking a subtense board.

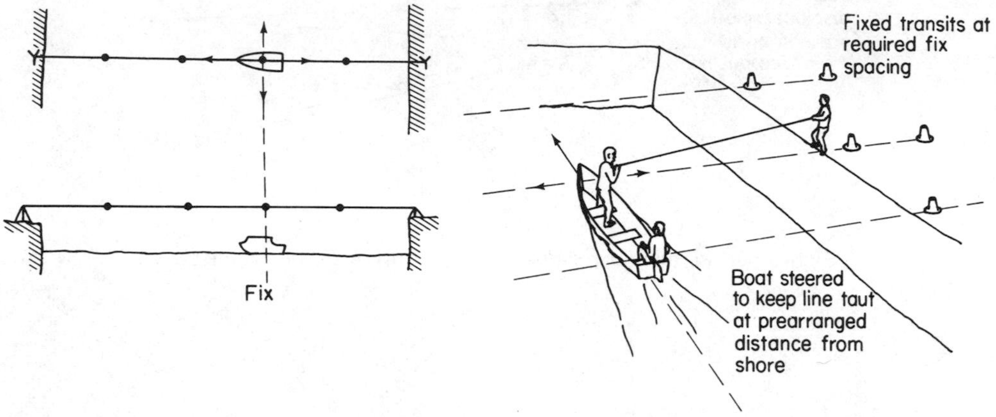

Figure 51. Fixing by distance line.

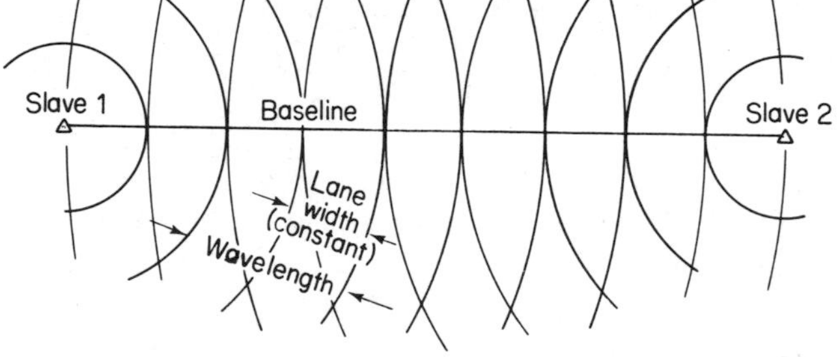

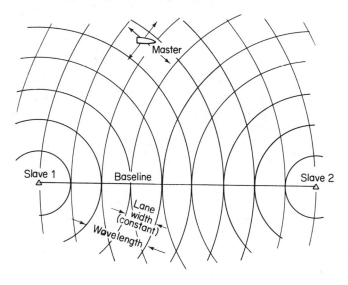

Figure 52. The two-range or range-range mode configuration.

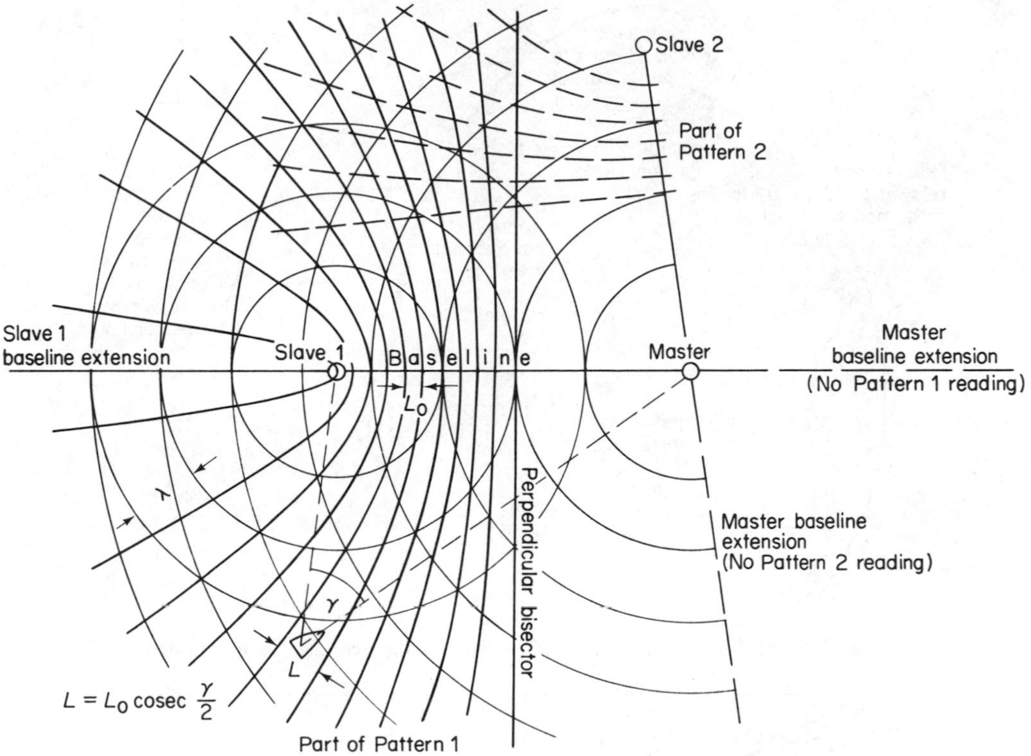

Figure 53. The hyperbolic mode configuration and principal features.

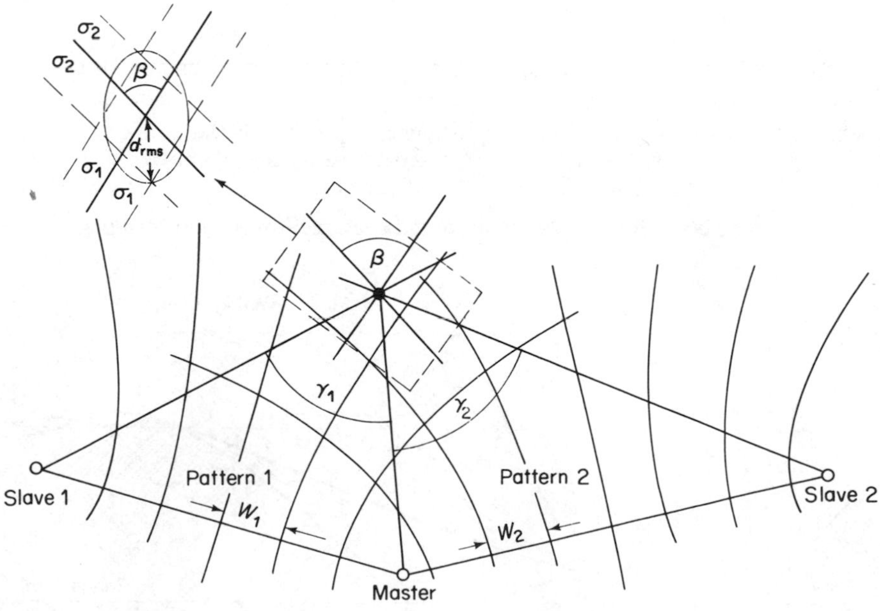

Figure 54. The root mean square error of an E.P.F. hyperbolic chain.

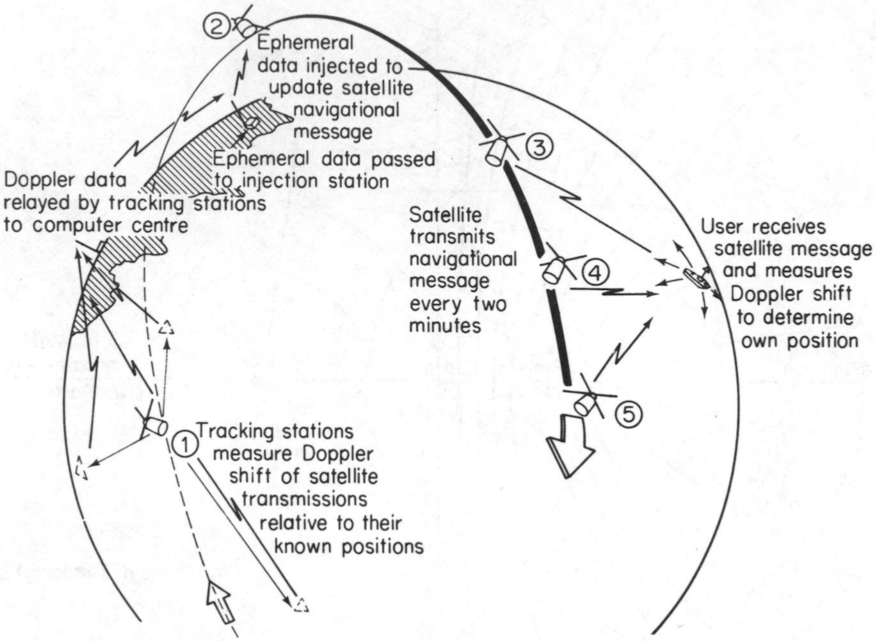

②
Ephemeral
data injected to
update satellite
navigational
message
Ephemeral data passed
to injection station
Doppler data
relayed by tracking stations
to computer centre
③
Satellite
transmits
navigational
message
every two
minutes
④
User receives
satellite message
and measures
Doppler shift
to determine
own position
⑤
①
Tracking stations
measure Doppler
shift of satellite
transmissions
relative to their
known positions

Figure 55. The TRANSIT system of position fixing by artificial satellite.

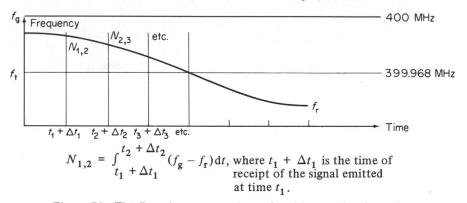

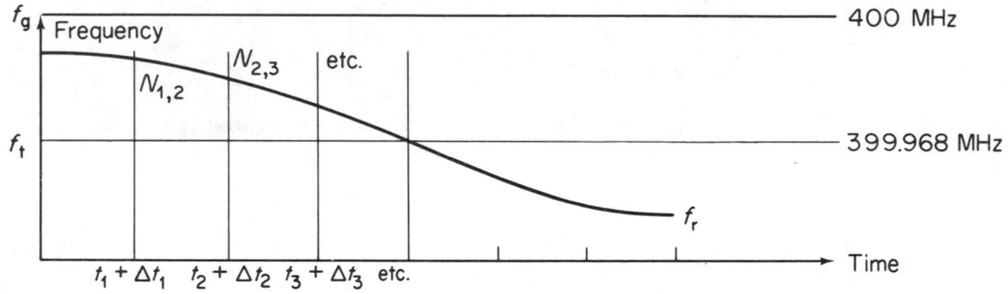

f_g
Frequency
400 MHz
$N_{1,2}$
$N_{2,3}$
etc.
f_t
399.968 MHz
f_r
Time
$t_1 + \Delta t_1$
$t_2 + \Delta t_2$
$t_3 + \Delta t_3$
etc.

$$N_{1,2} = \int_{t_1 + \Delta t_1}^{t_2 + \Delta t_2} (f_g - f_r)\,dt,$$ where $t_1 + \Delta t_1$ is the time of receipt of the signal emitted at time t_1.

Figure 56. The Doppler count and associated integration formula.

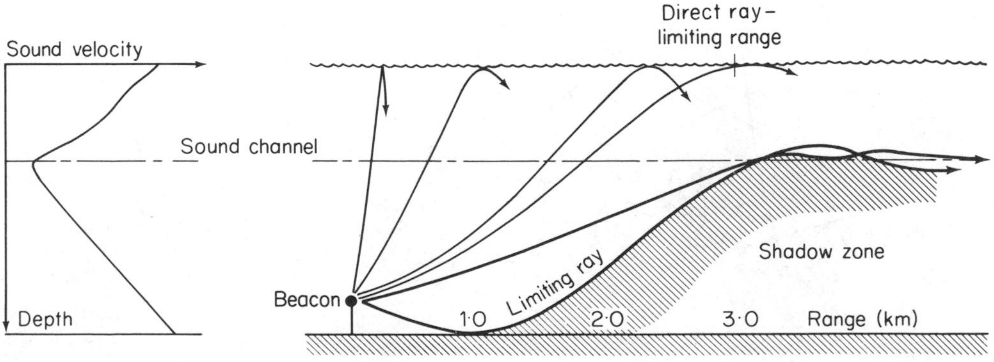

Sound velocity
Depth
Direct ray–
limiting range
Sound channel
Shadow zone
Limiting ray
Beacon
1·0
2·0
3·0
Range (km)

Figure 57. Characteristic ray paths in ocean depths.

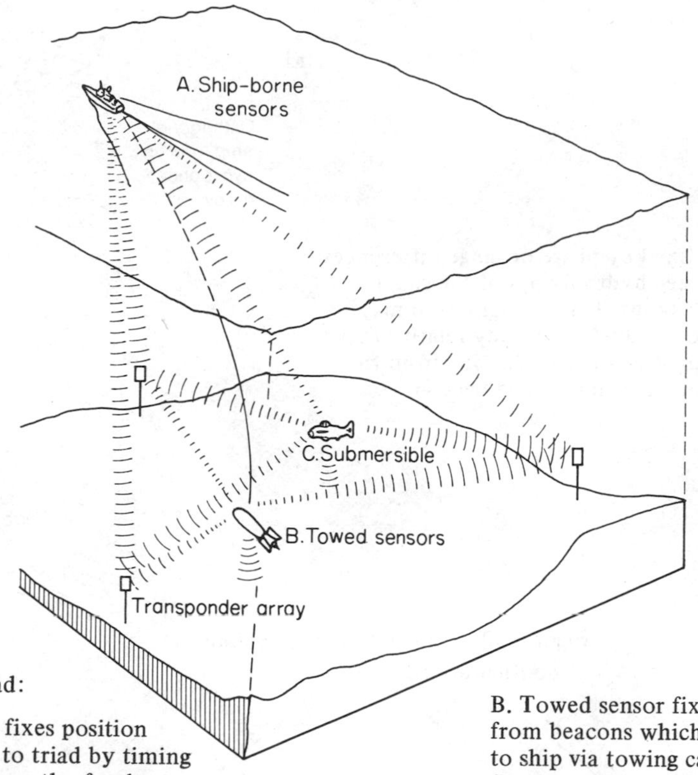

The triad:

A. Ship fixes position relative to triad by timing two-way path of pulse to beacons and return;

B. Towed sensor fixed by ranges from beacons which are telemetered to ship via towing cable – ship fixed by satellite, E.P.F. or as in A;

C. Submersible fixed as in A.

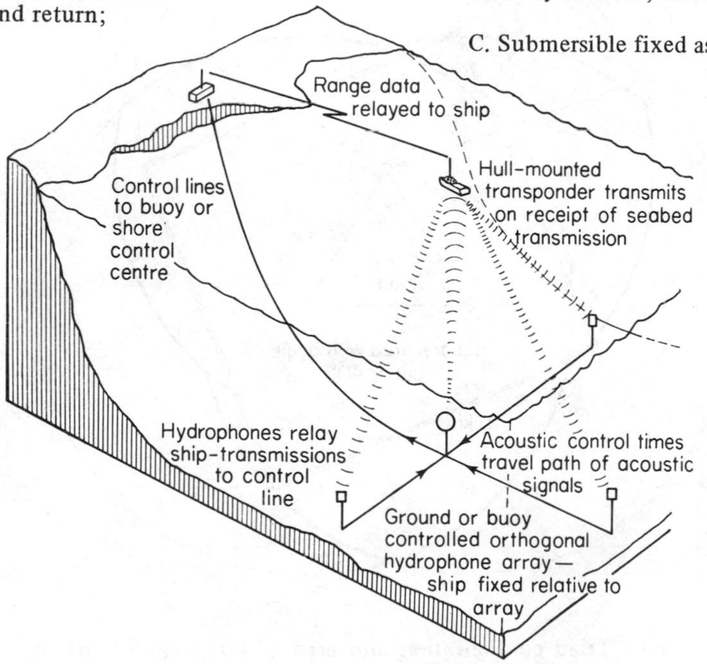

Figure 58. Long baseline system configurations.

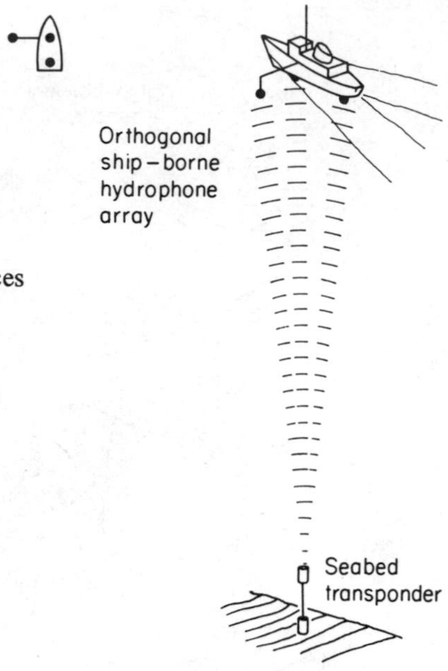

Ship position fixed by phase or range differences on arrival at three hydrophones of the single beacon transmissions. This configuration may also be used to fixed a towed body relative to the ship, the transmissions emanating from the towed body rather than a seabed beacon.

Figure 59. Short baseline system configuration.

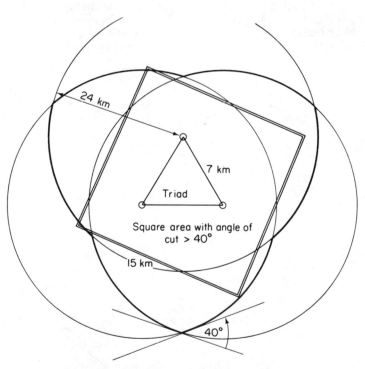

Figure 60. Triad configuration and area of 40° angle of cut or greater. Dimensions of the position-spheres and the area of best cut depend, of course, on depth of water.

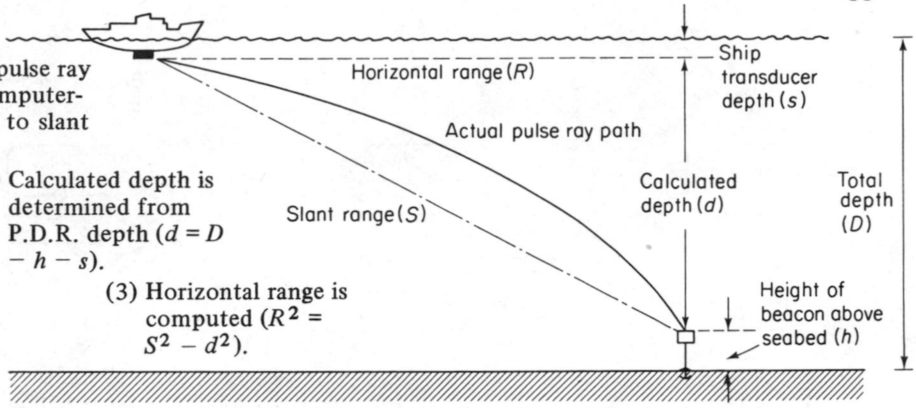

Figure 61. System geometry to obtain horizontal range from plotted position of beacon.

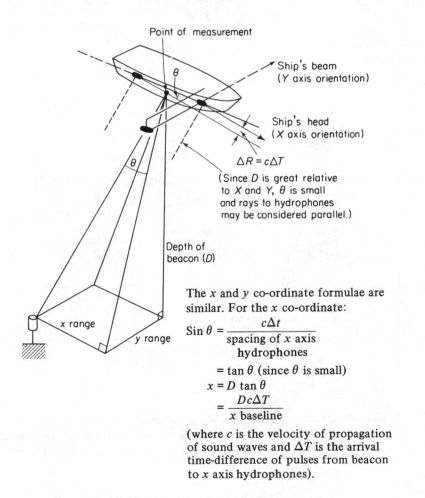

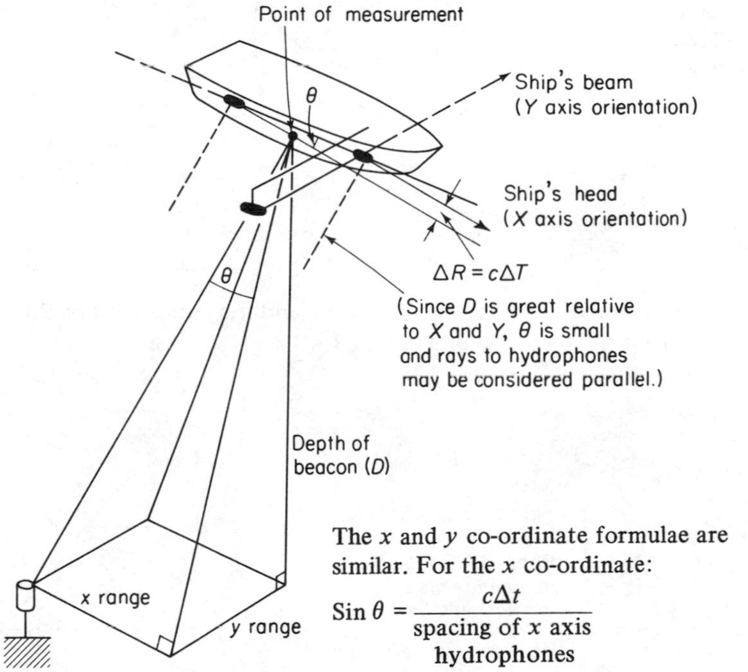

The x and y co-ordinate formulae are similar. For the x co-ordinate:

$$\text{Sin}\ \theta = \frac{c\Delta t}{\text{spacing of } x \text{ axis hydrophones}}$$

$$= \tan\theta \ \text{(since } \theta \text{ is small)}$$

$$x = D\tan\theta$$

$$= \frac{Dc\Delta T}{x \text{ baseline}}$$

(where c is the velocity of propagation of sound waves and ΔT is the arrival time-difference of pulses from beacon to x axis hydrophones).

Figure 62. Short baseline system geometry.

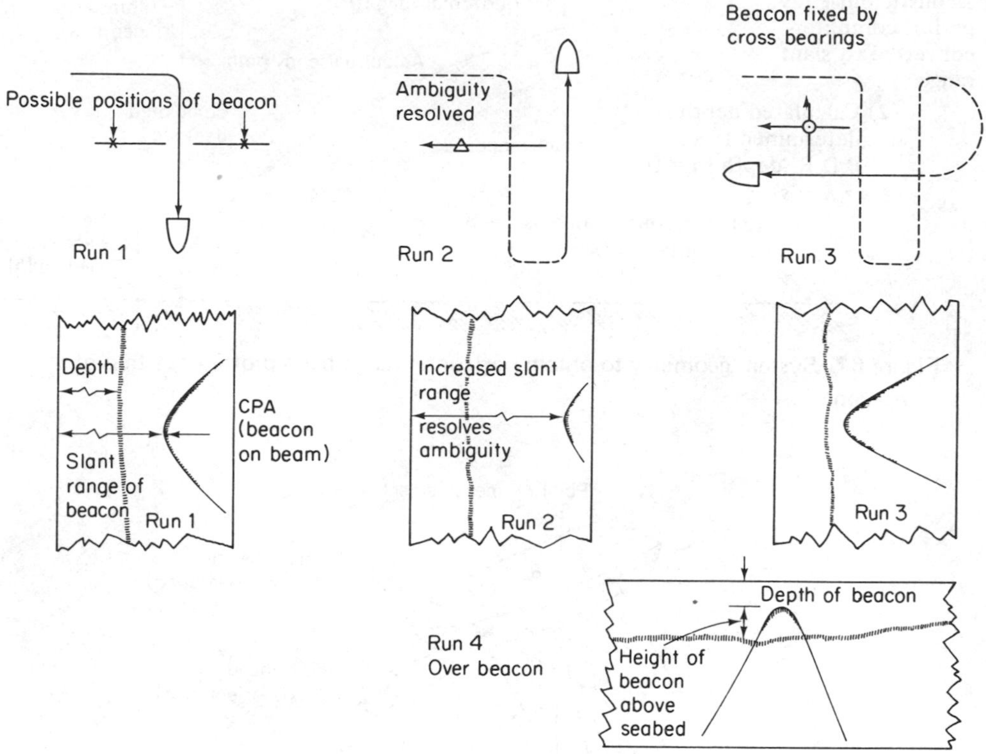

Figure 63. The manoeuvre to fix a single beacon and the corresponding P.D.R. traces (an expanded scale is shown in use on the P.D.R.).

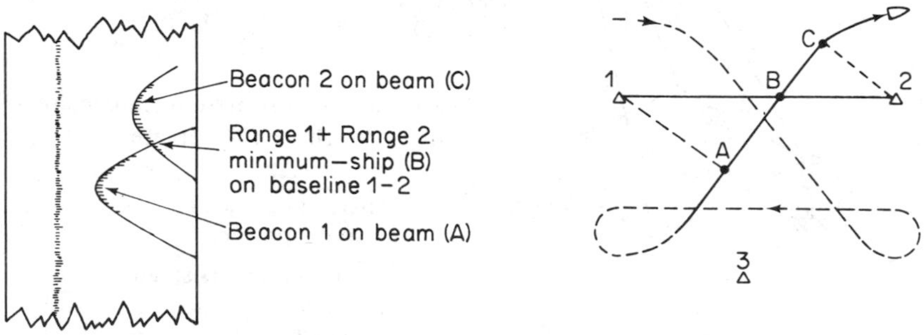

Figure 64. Determining the position of other beacons in a transponder array In each run between a pair of beacons the minimum sum of the two ranges indicates the point at which the ship crosses the baseline joining the two beacons.

(a) Decide where to go (e.g. for the methodical sounding of an area the surveyor might decide on parallel lines, closely spaced so as to cross the depth contours at right angles).

(b) Attempt to make good the chosen course (having planned the sounding lines the surveyor might erect transit marks on which the helmsman is to steer).

(c) Determine the actual track made good (the boat's position is fixed at intervals and the track followed is assumed to lie along the lines joining the fixes).

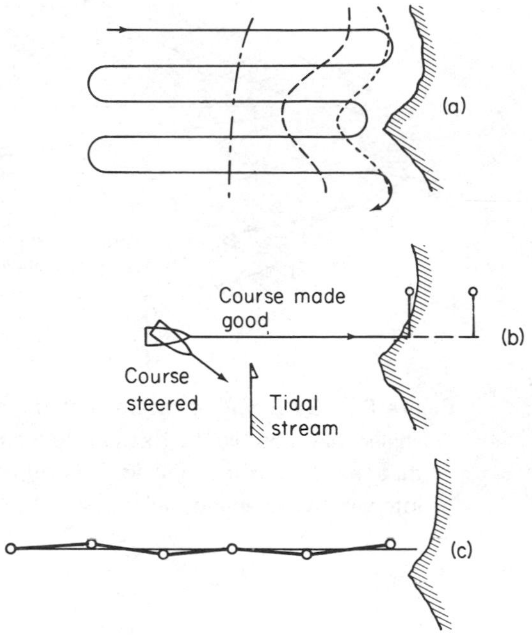

Figure 65. The control and determination of track.

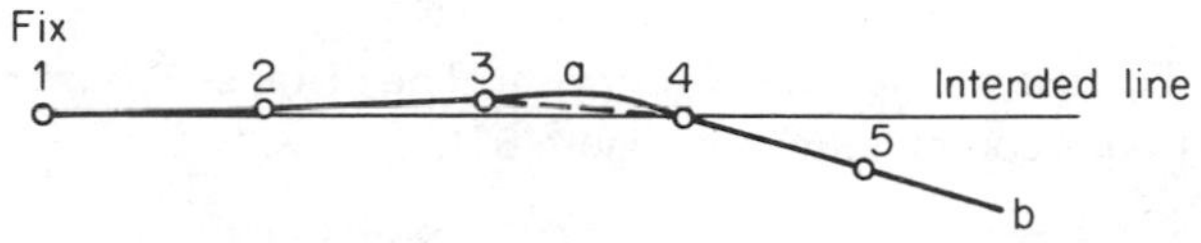

Figure 66. On plotting fix no. 3 the surveyor finds that he is north of the intended track. A course alteration is made and at fix no. 4 he finds that the line has been regained. The inexperienced surveyor might join fixes no. 3 and no. 4 by the dashed line. However, the course alteration could not have been made until fix no. 3 had been plotted, by which time the vessel would be at a point a. The new course is therefore not represented by the straight line 3-4, but by the line a-4, and by the time fix no. 5 is plotted the vessel will have reached point b, well off the line again. The course alteration was, of course, correct, but the surveyor should anticipate the situation at the next fix and, *at that fix,* bring the vessel back to the base course.

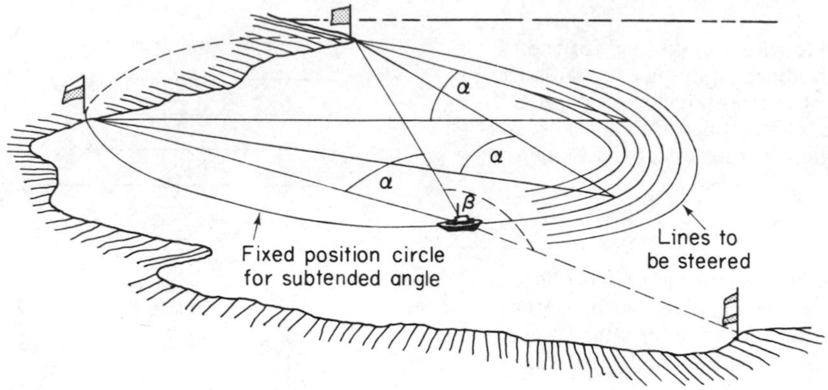

Figure 67. By planning coverage of the survey area by fixed-angle arcs, one angler may observe the fixed angle α constantly, directing the helmsman so that the angle is maintained. The other angler completes the fix at chosen intervals by observing angle β.

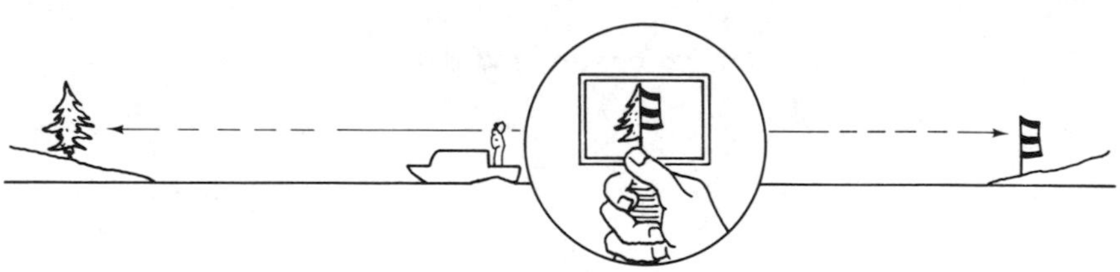

Figure 68. The use of the collimator prism. The distance line method deserves mention in this context also (see Figure 51).

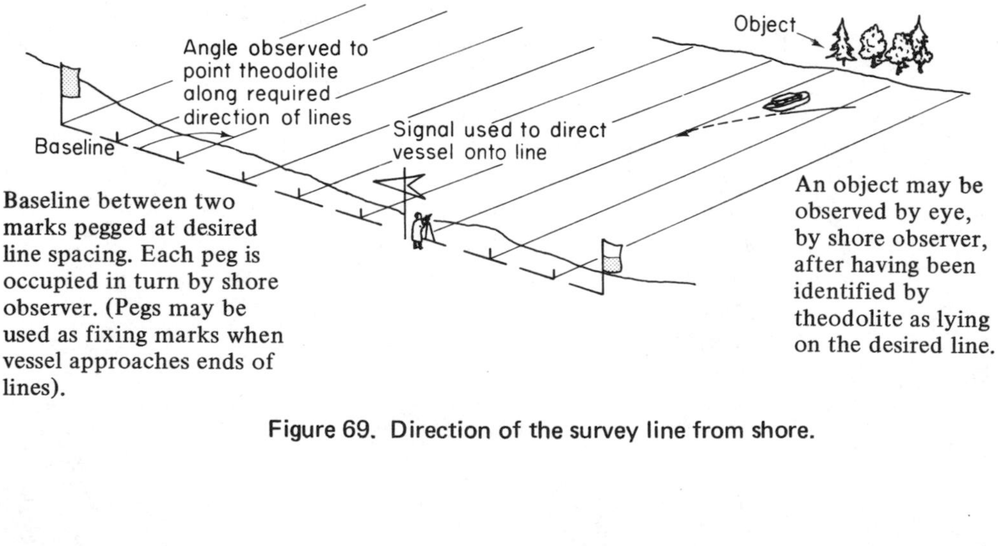

Figure 69. Direction of the survey line from shore.

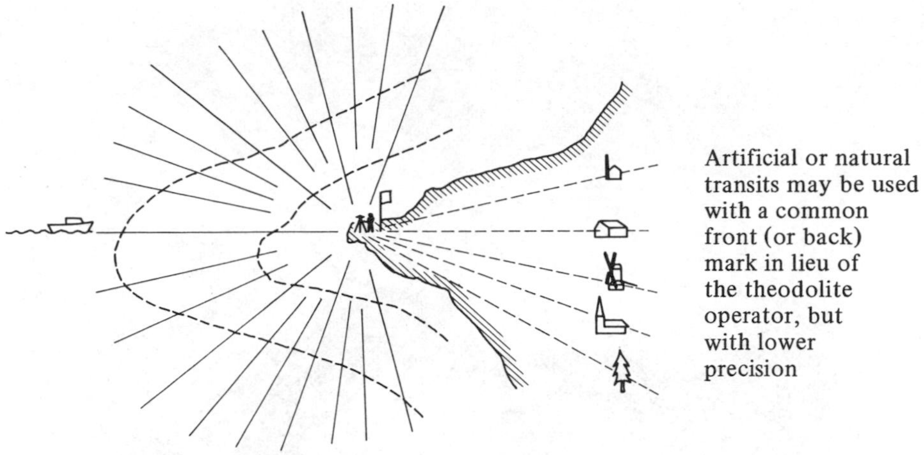

Figure 70. The 'starring' method of steering control.

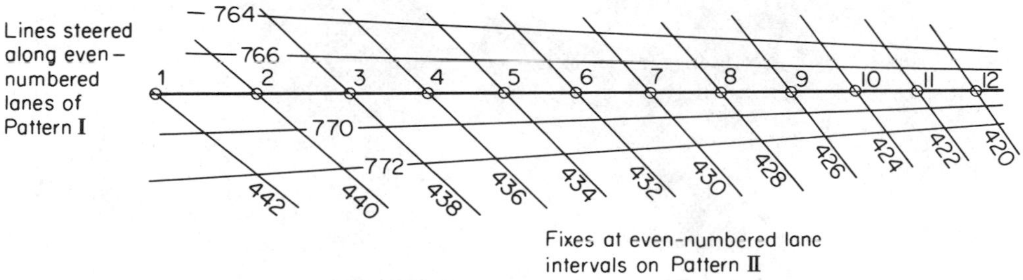

Figure 71. Steering survey lines planned along the lanes of one pattern of an E.P.F. system fixing at the intersections by the other pattern.

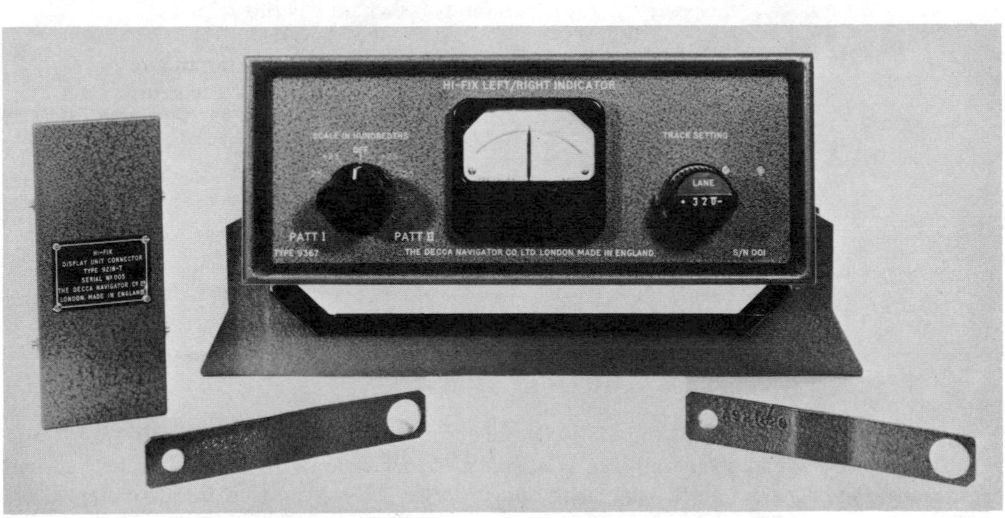

Figure 72. The Decca Hi-fix Left/Right Indicator.

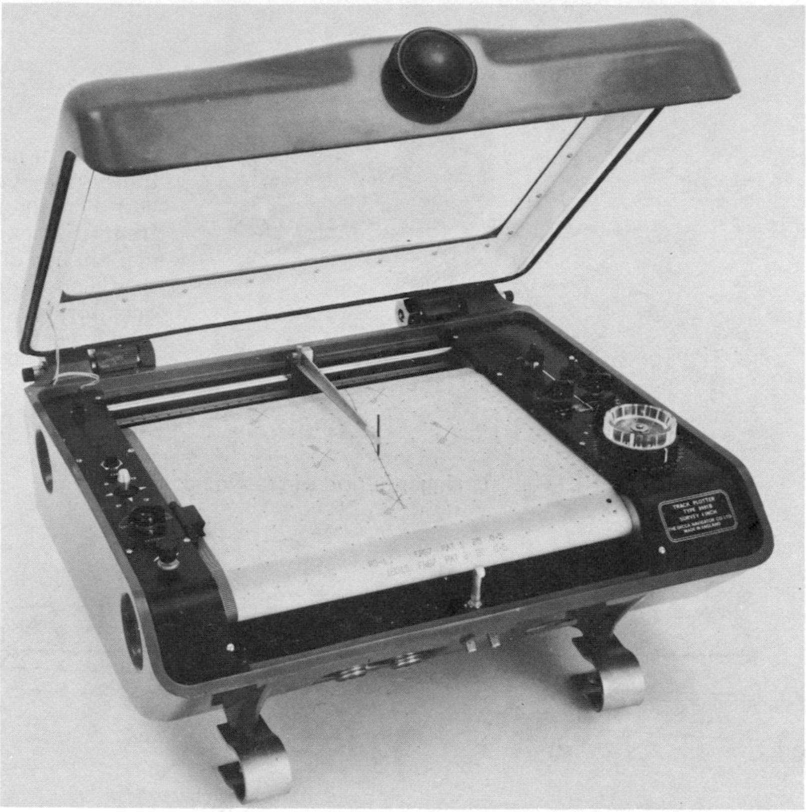

Figure 73. The Decca Survey Track Plotter.

$$\Delta f = f_t - f_r = \frac{2 f_t V \cos\theta}{c} \text{ (approximately)}$$

where c is the velocity of propagation of sound waves in seawater and $d\theta$ is negligible for the short time interval involved, and where V is small with respect to c.

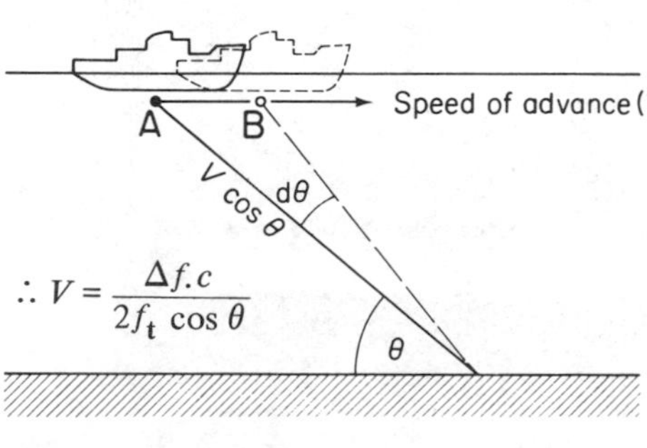

$$\therefore V = \frac{\Delta f . c}{2 f_t \cos\theta}$$

Figure 74. The geometry and theory of Doppler sonar systems. A ship at position A, transmitting a directional acoustic pulse at an angle of inclination θ to the horizontal, receives the reflected signal when it has reached position B. The transmitted frequency f_t differs from the received frequency f_r by an amount Δf, which is a measure of the ship's motion at velocity V relative to the seabed.

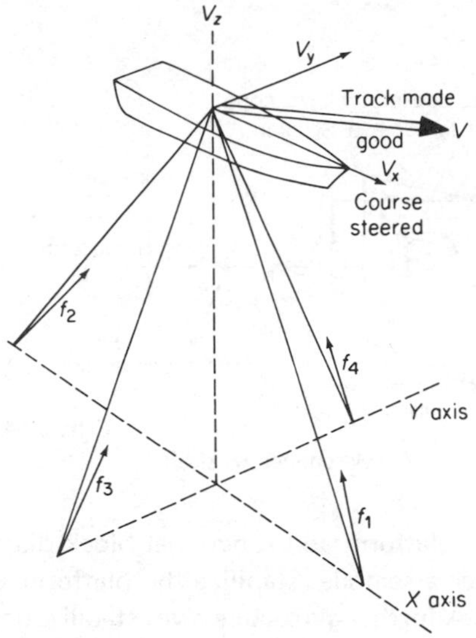

Figure 75. The geometry of the four-beam system. The difference-frequency for each pair of transmissions is $f_1 - f_2$ (and similarly $f_3 - f_4$), where $f_1 = f_t + \frac{2f_t}{c}(V_x \cos\theta - V_z \sin\theta)$ and $f_2 = f_t + \frac{2f_t}{c}(-V_x \cos\theta - V_z \sin\theta)$ $\therefore f_1 - f_2 = \frac{4f_t V_x \cos\theta}{c}$ and the effect of pitch is, in theory, cancelled. The equation for $f_3 - f_4$ similarly cancels out the effect of roll.

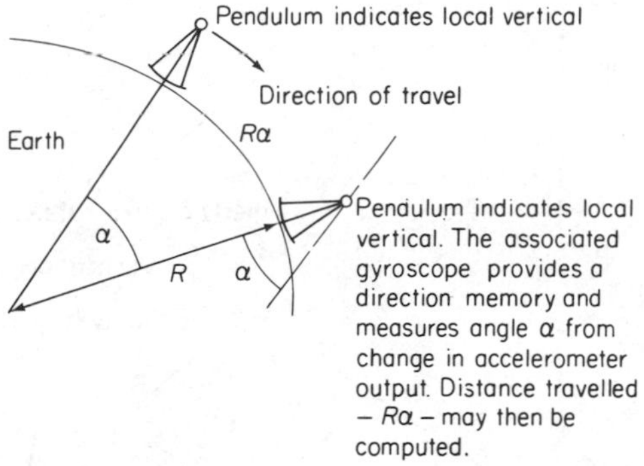

Figure 76. The Schuler pendulum principle, applied in accelerometers to define the local vertical. When the vessel accelerates, the accelerometer pendulum begins to swing. The local vertical rotates as the vessel moves across the earth's surface. If the pendulum period is 84 minutes, the movement due to the vessel's acceleration is equal to that required to keep the pendulum vertical. In order to produce the effect of an 84-minute pendulum in the miniature accelerometer, an associated gyroscope applies the necessary counteracting force as 'ordered' by the accelerometer read-out. The accelerometer is then said to be 'Schuler tuned'. In practice, two accelerometer/gyroscope arrangements are used to give stabilization in two directions at right angles and eliminate the effects of short-period ship accelerations.

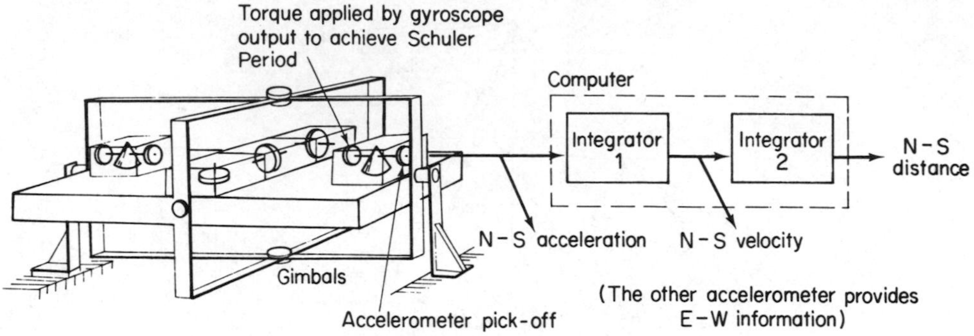

Figure 77. **The stabilized platform and functional block diagram of an inertial system. Gyroscope/accelerometer assemblies stabilize the platform to the vertical in east-west and north-south planes. A further gyroscope gives stabilization in azimuth.**

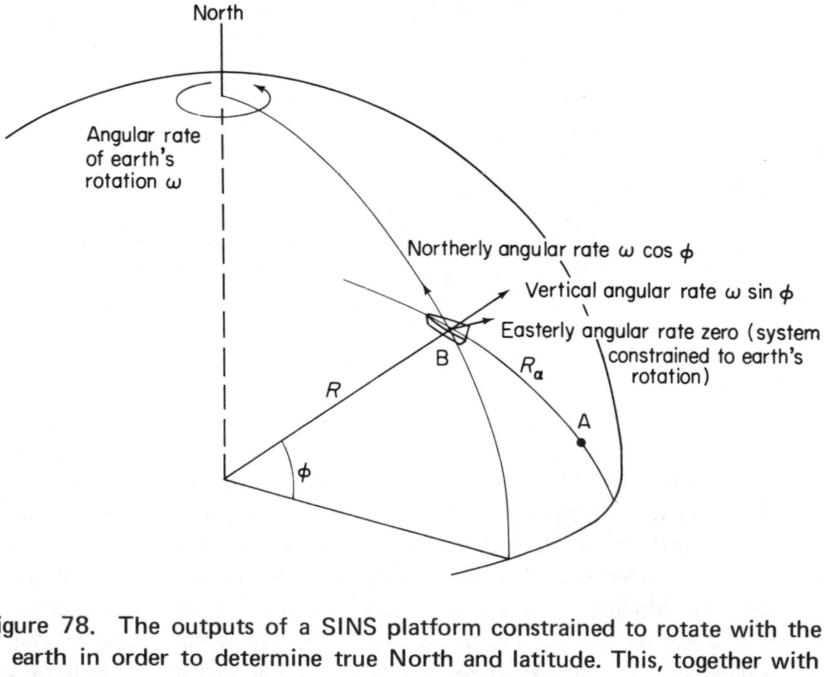

Figure 78. The outputs of a SINS platform constrained to rotate with the earth in order to determine true North and latitude. This, together with $R\alpha$, obtained as in Figure 76, defines the position of B relative to A.

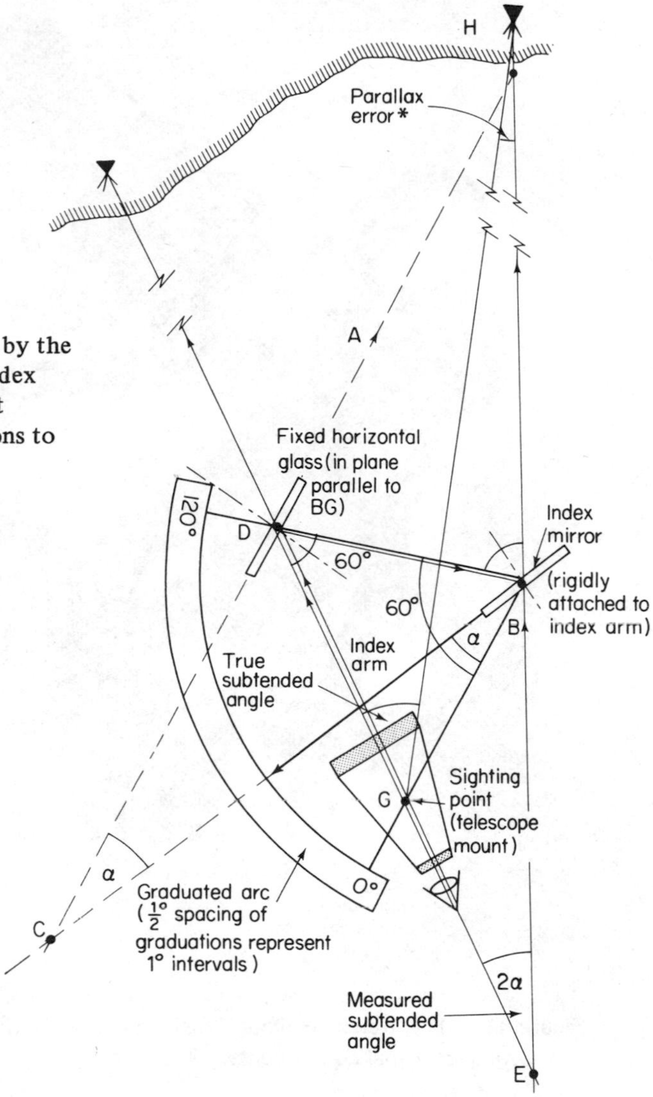

*Parallax error is caused by the physical separation of index mirror and sighting point (negligible for observations to distant marks,

when $\frac{GH}{BG}$ approaches infinity, but important in subtense ranging)

Let α be the angle through which the index mirror is moved in order to bring two observed marks into coincidence in the telescope field of view. The two marks subtend the angle at E as shown by the lines of sight in the diagram.

$D\hat{B}G = 60^\circ$ (thus 'sextant')
$D\hat{B}C = 60 - \alpha$
$H\hat{B}D = 180 - 2(60 - \alpha) = 60 + 2\alpha$
$\therefore H\hat{B}C = 60 + 2\alpha + 60 - \alpha = 120 + \alpha$
DC is parallel to BG $\therefore \hat{C} = \alpha$
$\therefore C\hat{A}B = 180 - (120 + \alpha) - \alpha = 60 - 2\alpha$
$A\hat{D}E = 120^\circ$
$\therefore \hat{E} = 180 - 120 - (60 - 2\alpha) = 2\alpha$

So, for a subtended angle between the observed marks of 2α, the index arm must be moved by an amount α to bring them to coincidence in the telescope.

Figure 79. The principle of angular measurement by sextant.

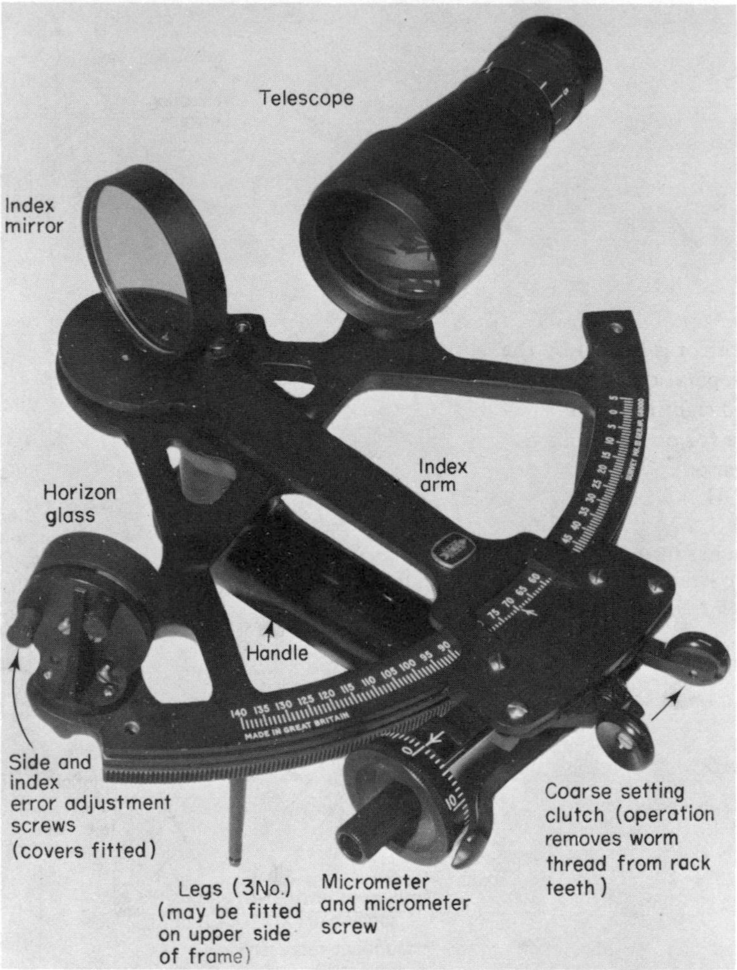

Figure 80. The Kelvin Hughes Challenger Mk. II Surveying Sextant, showing principal components.

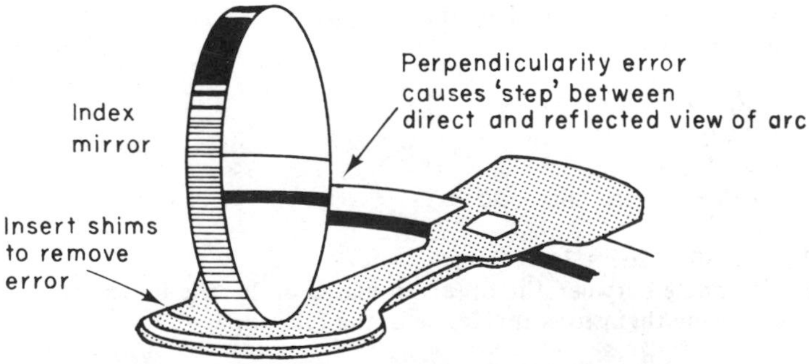

Figure 81. The method of detecting perpendicularity error.

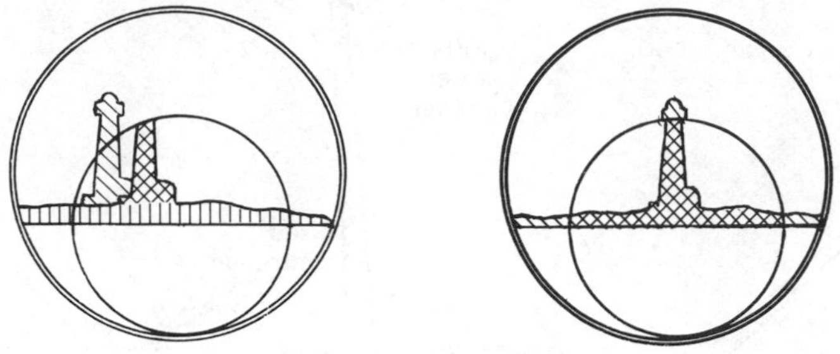

(a) (b)

Figure 82. Side error; the view through the telescope with side error present (a) and removed (b).

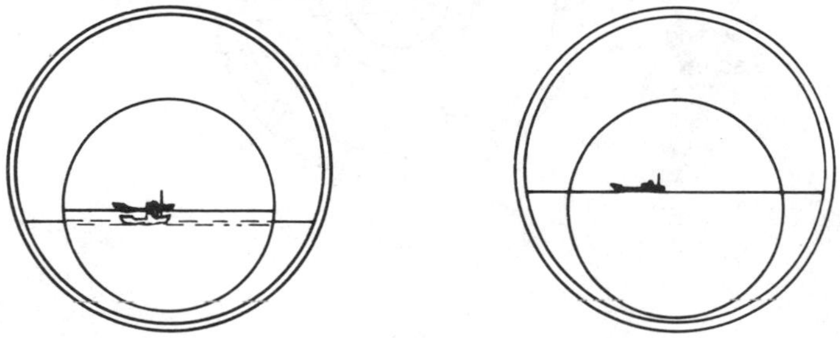

(a) (b)

Figure 83. Index error; the view through the telescope with index error present (a) and removed (b).

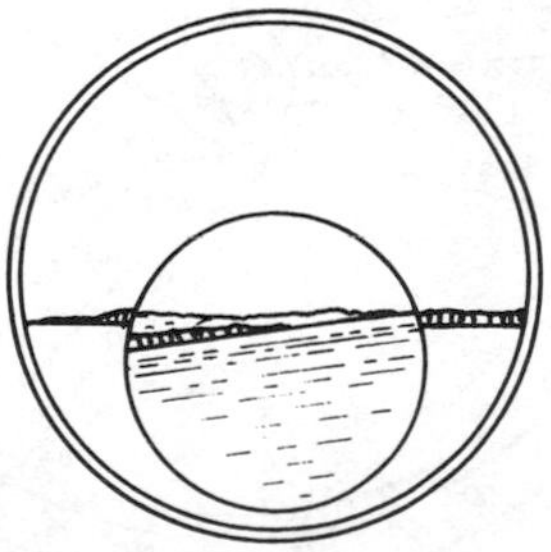

Figure 84. The view through the telescope when collimator error is present (most marked when a large angle is set on the instrument).

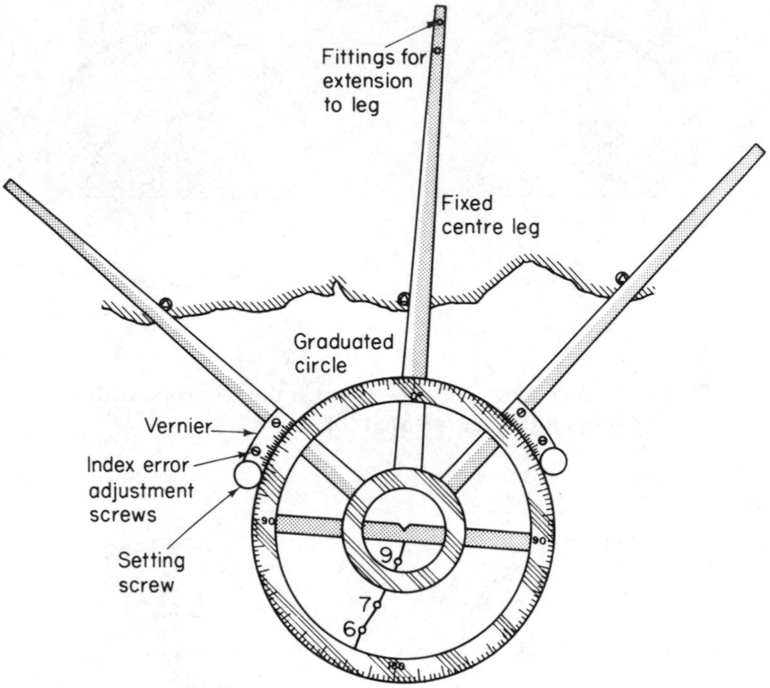

(a)

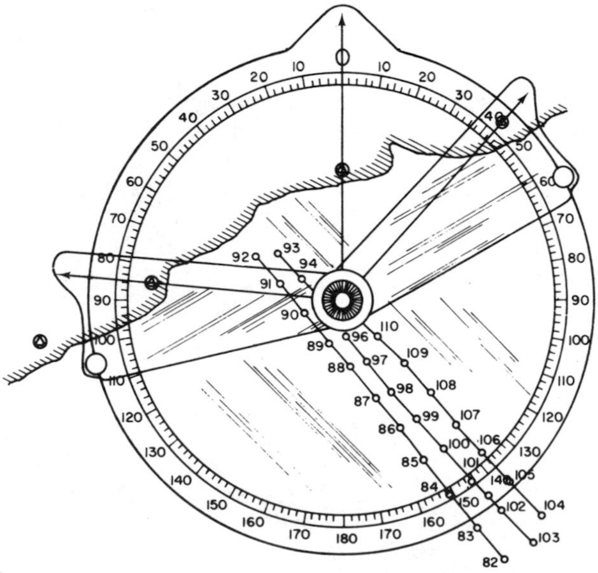

(b)

Figure 85. The station pointer (a) and three-arm protractor (b).

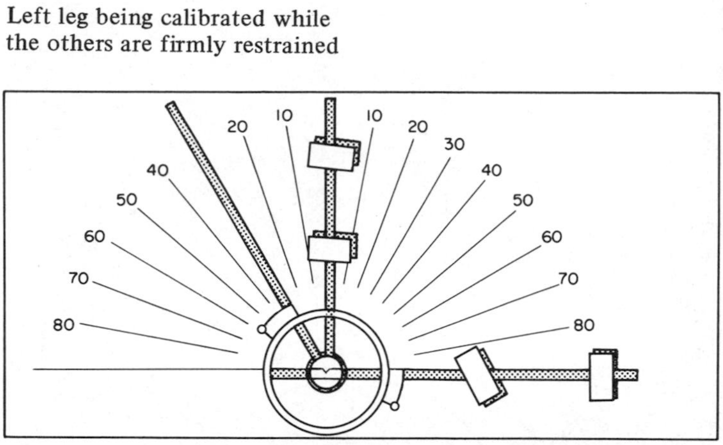

Figure 86. The station pointer corrector diagram.

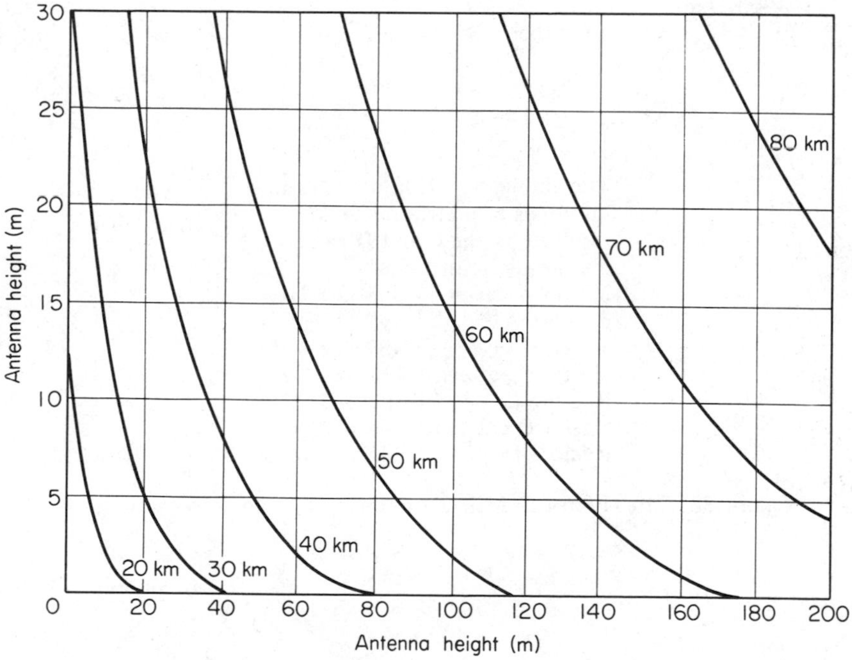

Figure 87. Radio line of sight ranges for given heights of antennae (microwave E.D.M.).

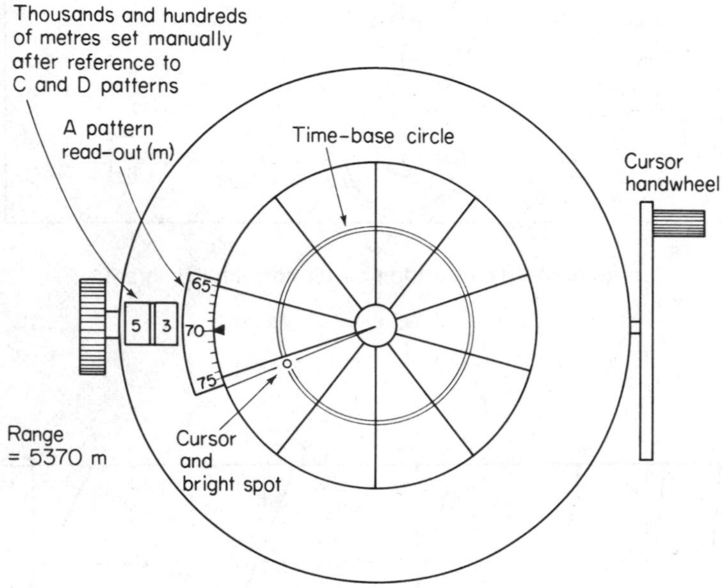

The ambiguity of thousands and hundreds of metres range is resolved by the C and D pattern measuring frequencies:

Master A pattern	1·498 468 MHz
Remote A–	1·499 468 MHz
A+	1·497 468 MHz
Master C pattern	1·483 483 MHz
Remote C	1·482 483 MHz
Master D pattern	1·348 621 MHz
Remote D	1·347 621 MHz

Figure 88. The Hydrodist MRB2—method of range presentation.

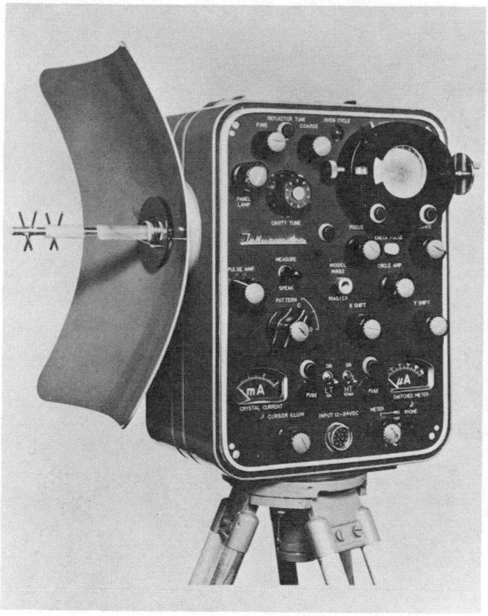

The Master instrument

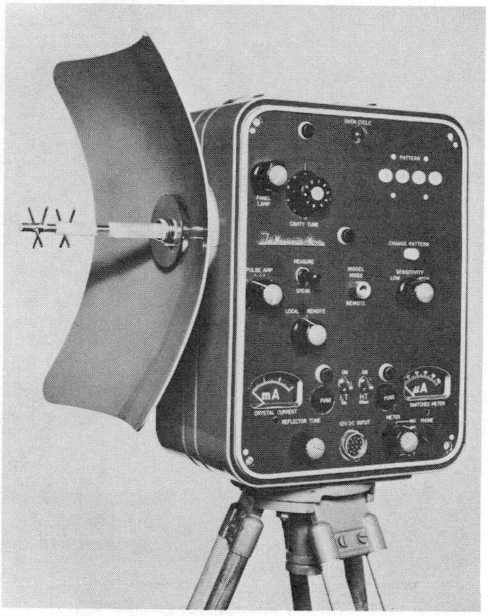

The Remote instrument

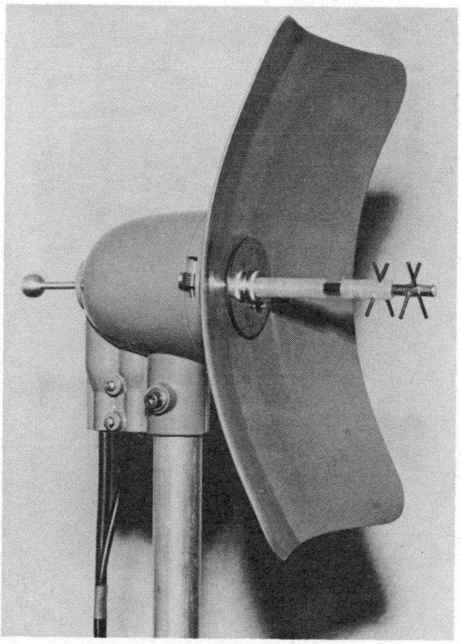

The Master unit may be installed in a sheltered position in the survey vessel, an aerial separator being used to isolate the dish and dipole antenna.

Figure 89. The Hydrodist MRB2 Master and Remote units.

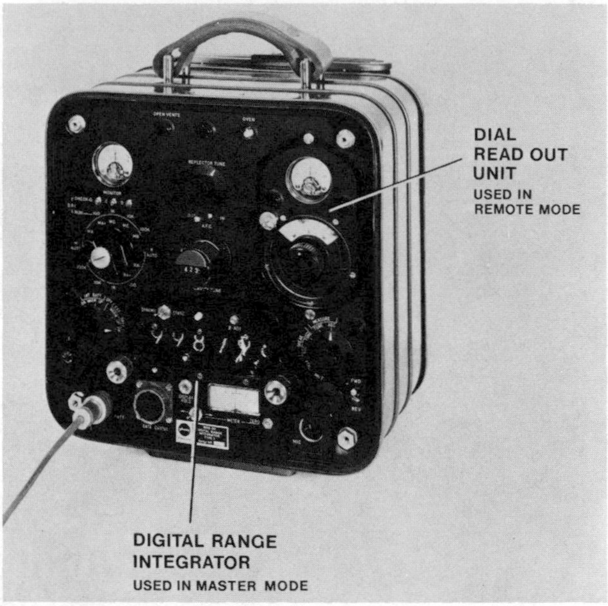

Figure 90. The Tellurometer MRB201 (blanking plates may be fitted to the D.R.O. or D.R.I. when not required for the Remote or Master function respectively).

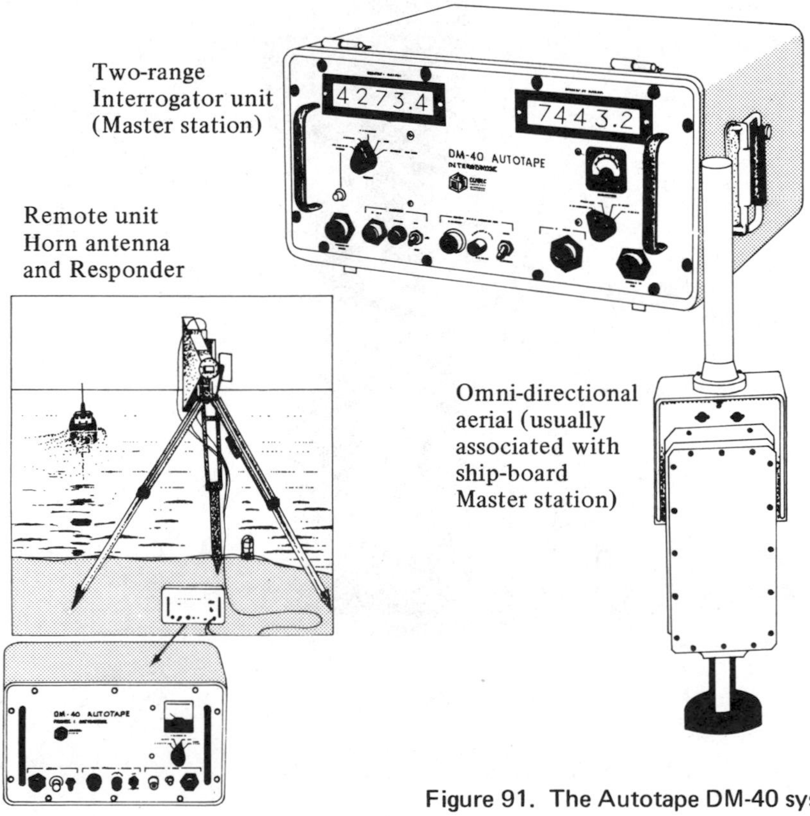

Figure 91. The Autotape DM-40 system components.

Figure 92a. Remote unit

Figure 92b. Base unit

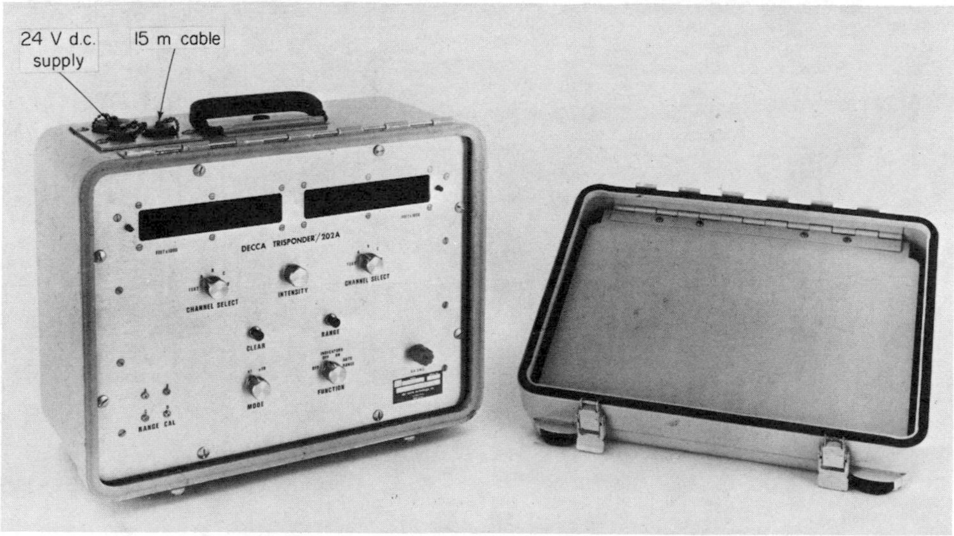

Figure 92c. Distance measuring unit

Figure 92. The Decca Trisponder system.

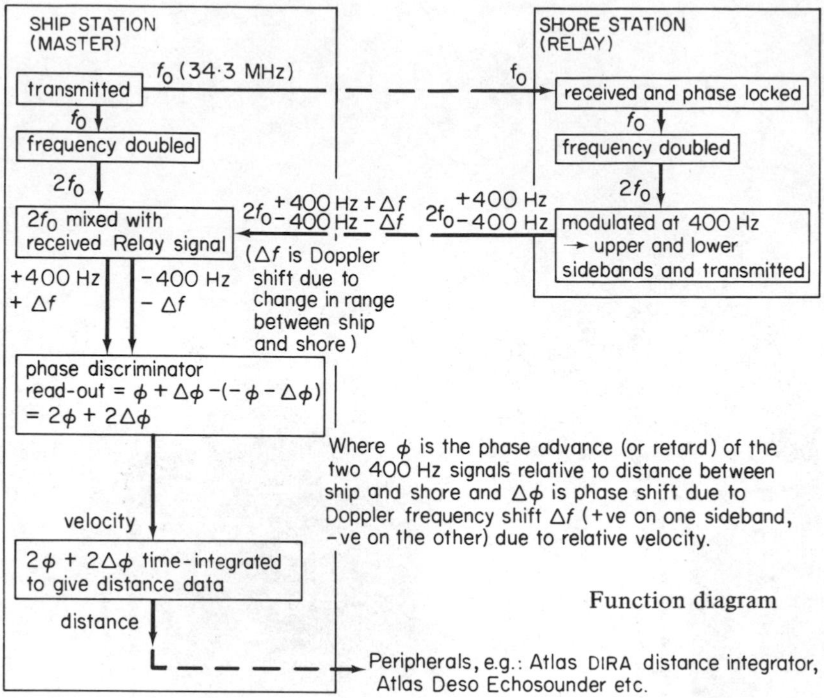
SHIP STATION
(MASTER)
transmitted
f_0 (34·3 MHz)
f_0
frequency doubled
$2f_0$
$2f_0$ mixed with received Relay signal
$2f_0 + 400$ Hz $+ \Delta f$
$2f_0 - 400$ Hz $- \Delta f$
(Δf is Doppler shift due to change in range between ship and shore)
$+400$ Hz $+ \Delta f$
-400 Hz $- \Delta f$
phase discriminator
read-out = $\phi + \Delta\phi - (-\phi - \Delta\phi)$
$= 2\phi + 2\Delta\phi$
velocity
$2\phi + 2\Delta\phi$ time-integrated to give distance data
distance
SHORE STATION
(RELAY)
f_0
received and phase locked
f_0
frequency doubled
$2f_0$
$2f_0 + 400$ Hz
$2f_0 - 400$ Hz
modulated at 400 Hz
→ upper and lower sidebands and transmitted
Where ϕ is the phase advance (or retard) of the two 400 Hz signals relative to distance between ship and shore and $\Delta\phi$ is phase shift due to Doppler frequency shift Δf (+ve on one sideband, −ve on the other) due to relative velocity.
Peripherals, e.g.: Atlas DIRA distance integrator, Atlas Deso Echosounder etc.

Function diagram

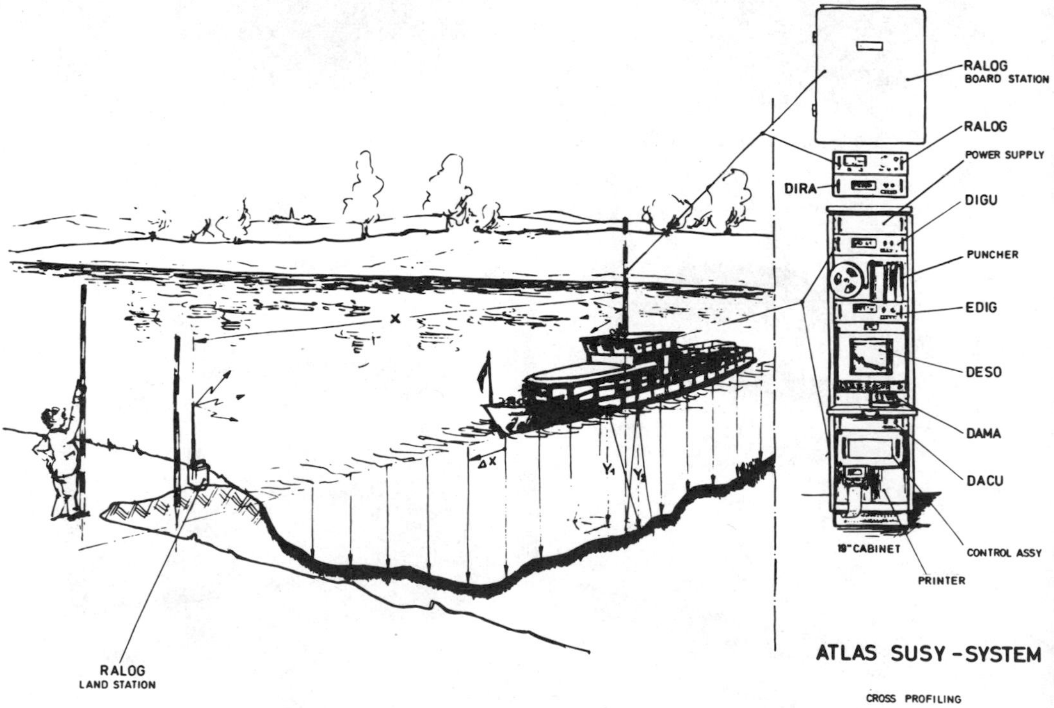
RALOG
BOARD STATION
RALOG
POWER SUPPLY
DIRA
DIGU
PUNCHER
EDIG
DESO
DAMA
DACU
CONTROL ASSY
19" CABINET
PRINTER
X
ΔX
RALOG
LAND STATION
ATLAS SUSY - SYSTEM
CROSS PROFILING

Ralog equipment in the automated sounding system 'Atlas Susy'.

Figure 93. Atlas-Ralog 10 function diagram and system equipment.

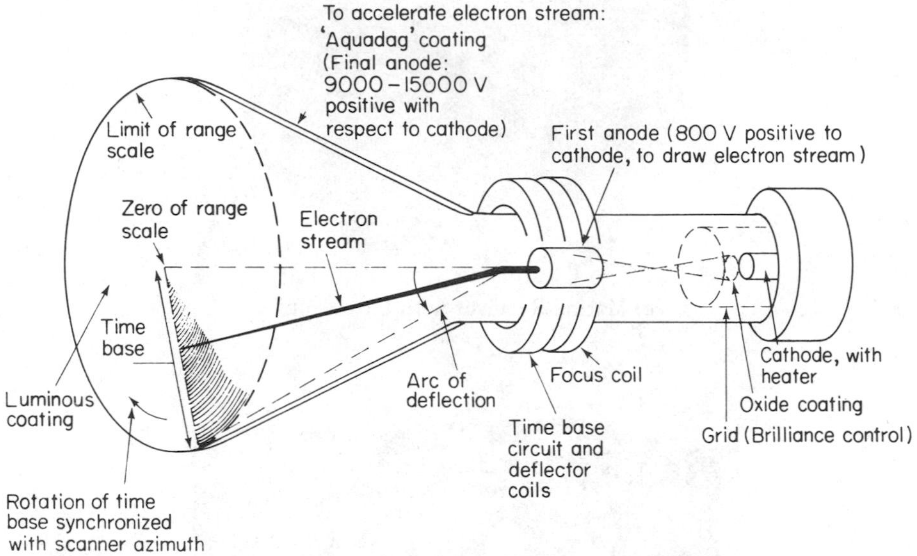

Figure 94. Cathode ray tube: Plan-Position Indicator (PPI) operation.

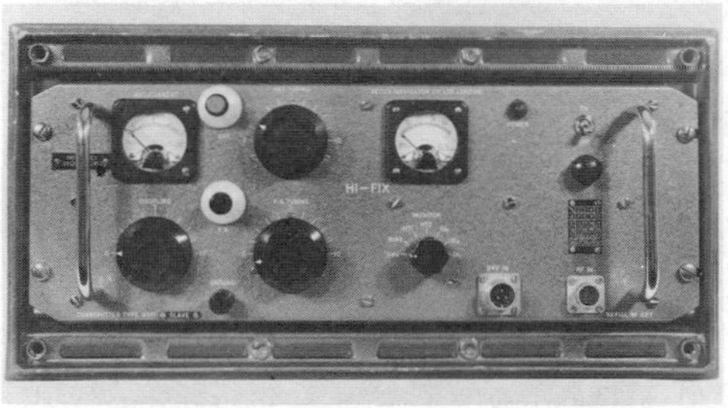

(a) Transmitter (Low Power)

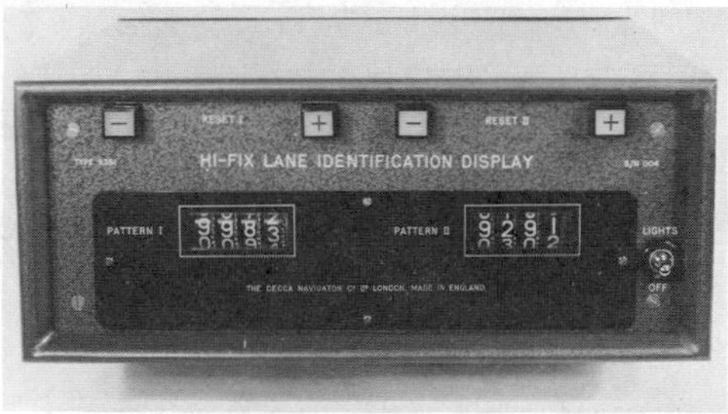

(b) Lane Identification Display

Figure 95. Hi-Fix equipment. A monitor unit is also available, driven by a Hi-Fix Marine Receiver, to monitor and record continuously at a stationary point as a check on the pattern stability of a chain. Any errors are revealed by the chart-paper record and enable corrections to be made to slave readings as necessary. (Continued overleaf.)

(c) Marine Receiver with blower unit

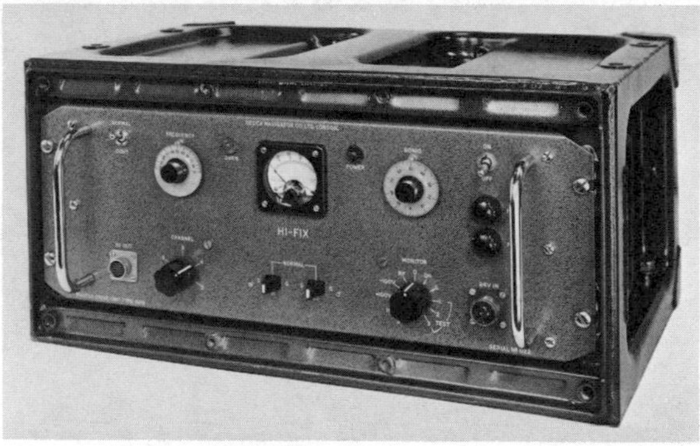

(d) Master Drive Unit

Figure 95 (contd.)

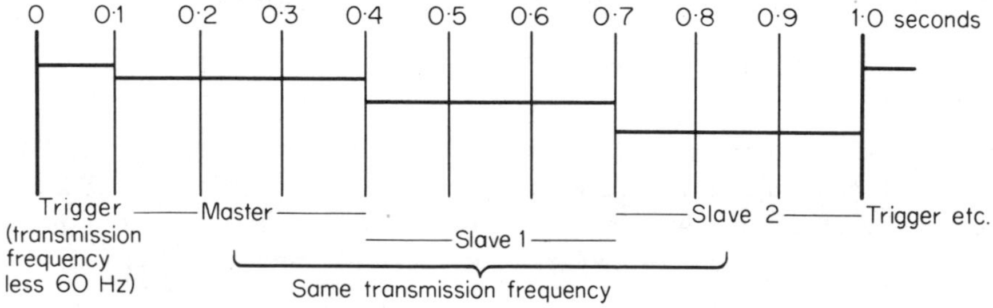

Figure 96. The characteristics of the Hi-Fix time-sharing transmission cycle.

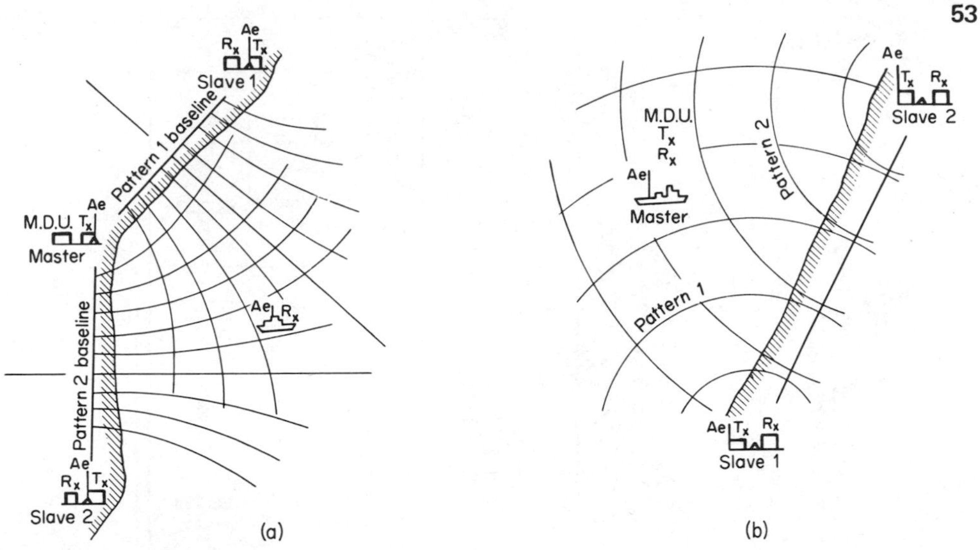

Figure 97. The Hi-Fix equipment requirements and disposition for hyperbolic (a) and two-range (b) working.

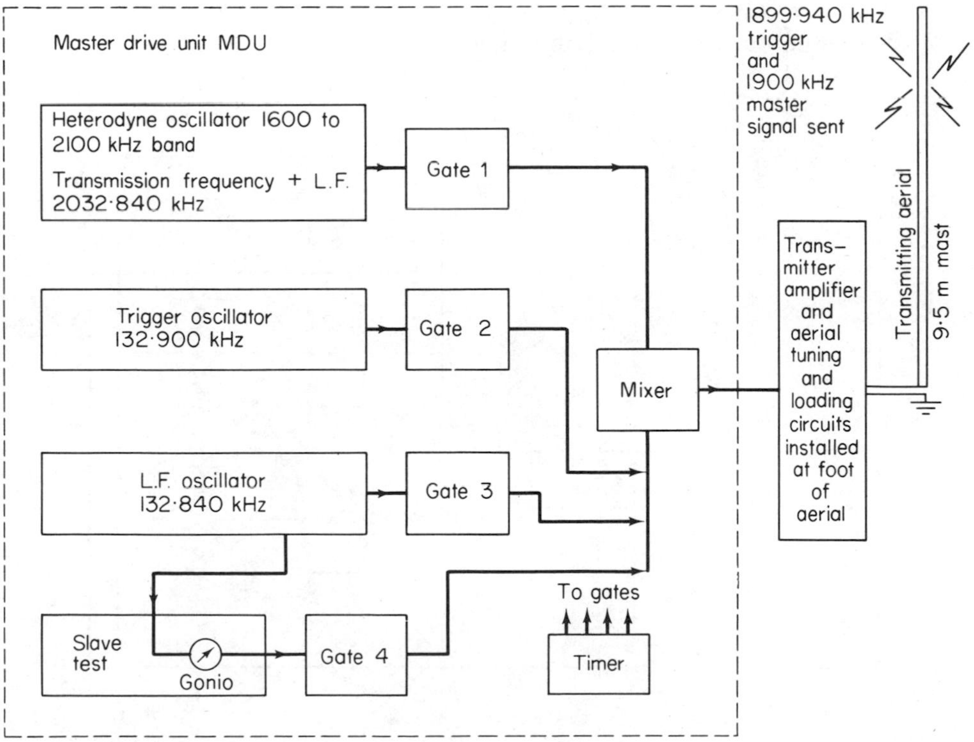

Figure 98. Block diagram of Master drive unit, transmitter and aerial assembly for Master station. A spot frequency of 1900 kHz is assumed.

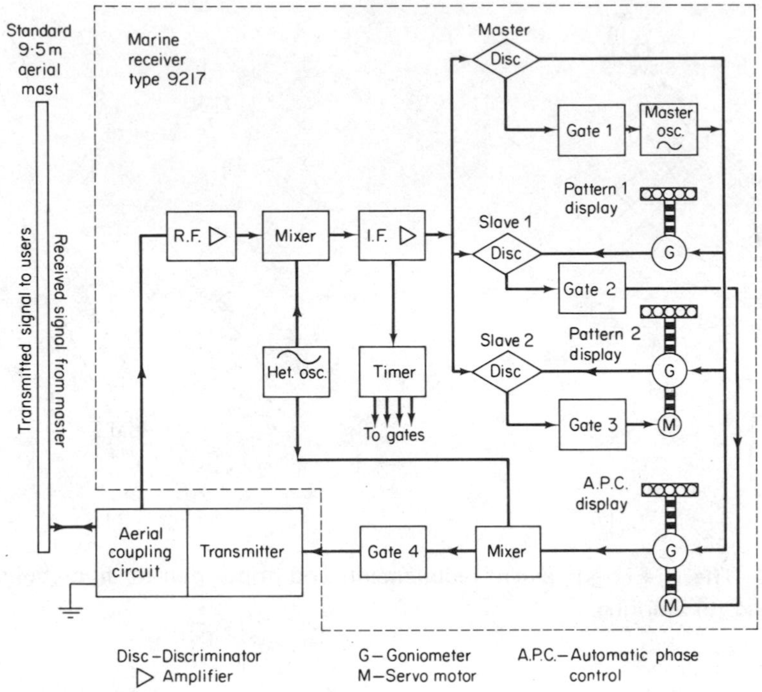

Figure 99. Block diagram of Hi-Fix Receiver in Slave Pattern 1 function—controls phase and initiates Slave transmission.

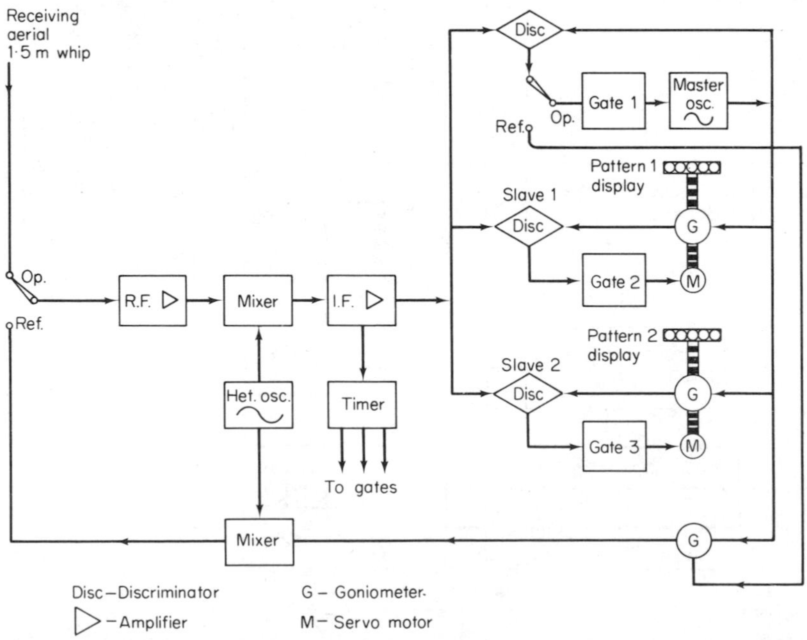

Figure 100. Block diagram of Hi-Fix Receiver in ship-borne phase measurement function.

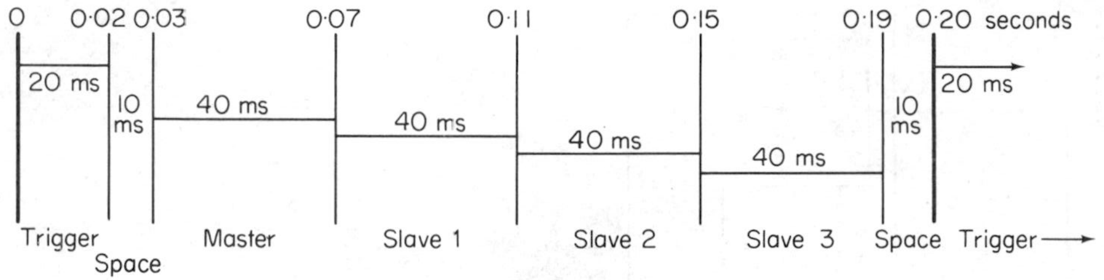

Figure 101. The Sea-Fix transmission cycle.

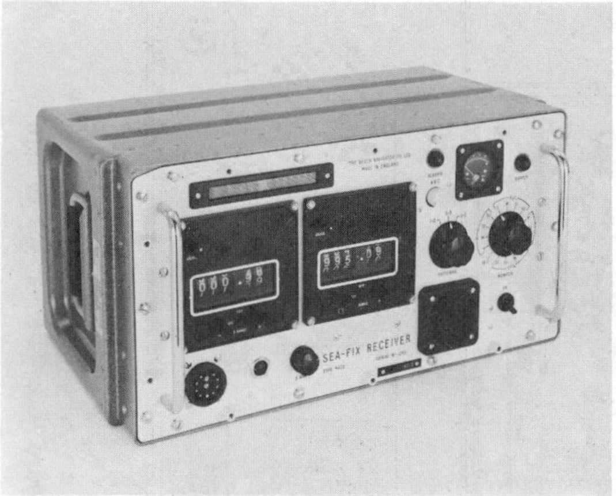

Marine positioning receiver

Slave components:
S.C.U.; Transmitter; T_x/R_x
Antenna with aerial matching coil
Master components:
M.C.U.; Transmitter; T_x/R_x
Antenna with aerial matching coil; plus Receiver if in two-range mode

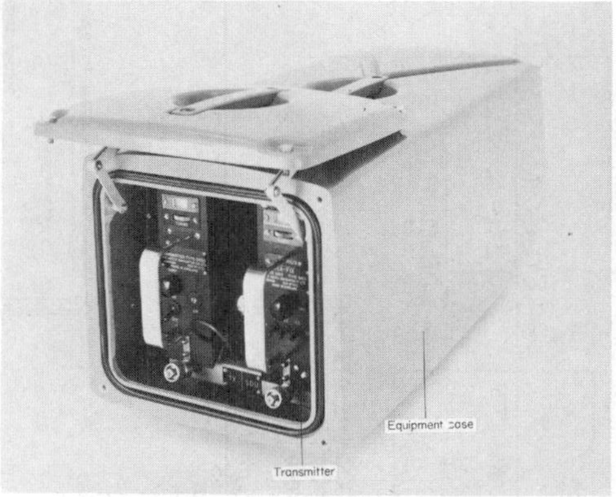

Slave control unit or Master control unit (according to function)

Master control unit

Figure 102. Sea-Fix equipment.

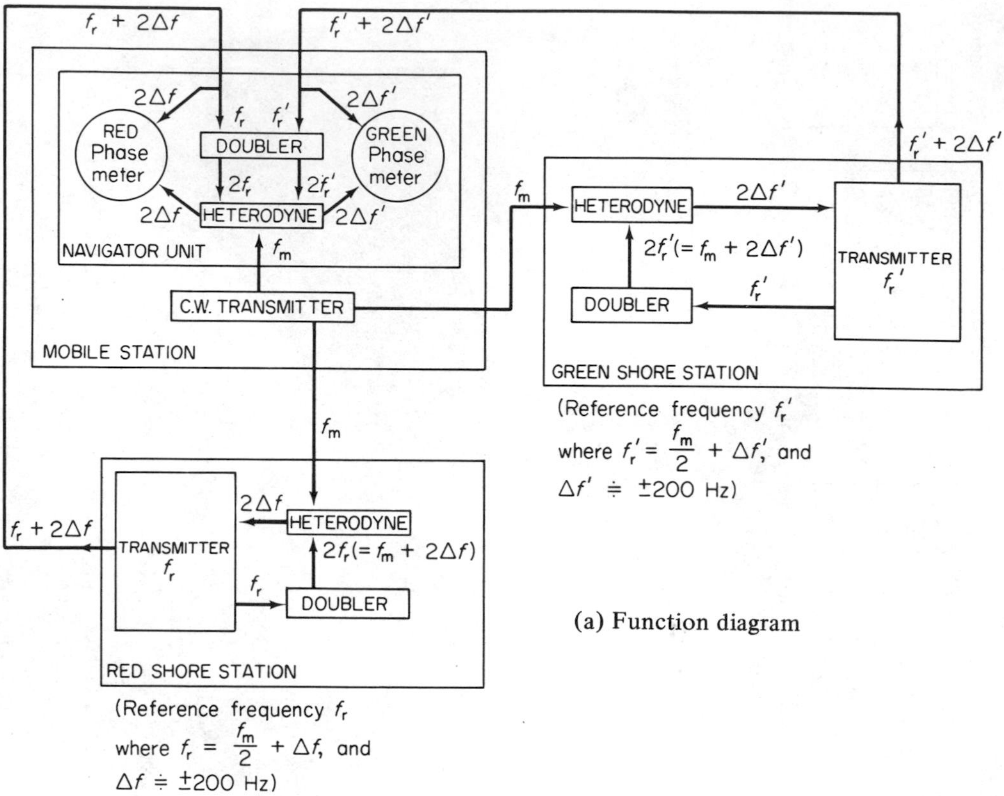

(a) Function diagram

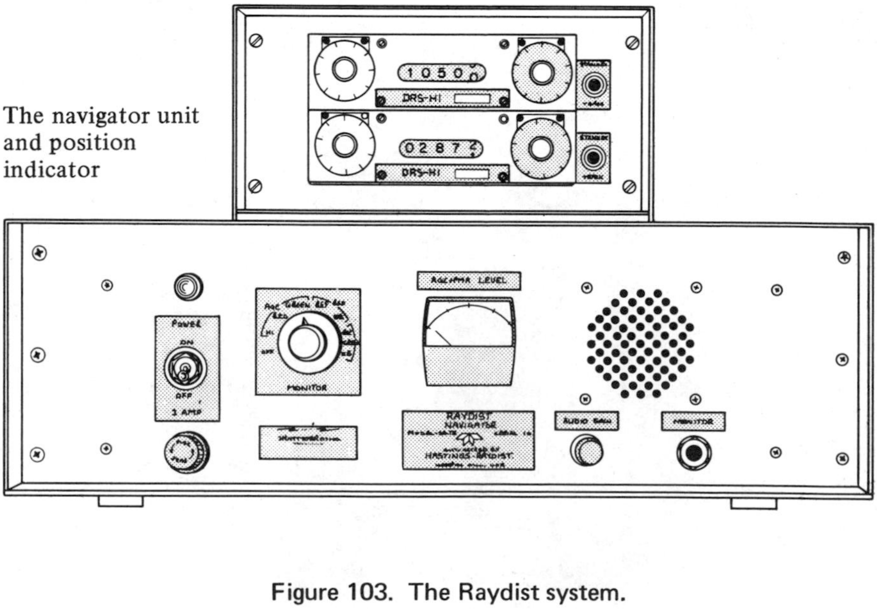

(b) The navigator unit and position indicator

Figure 103. The Raydist system.

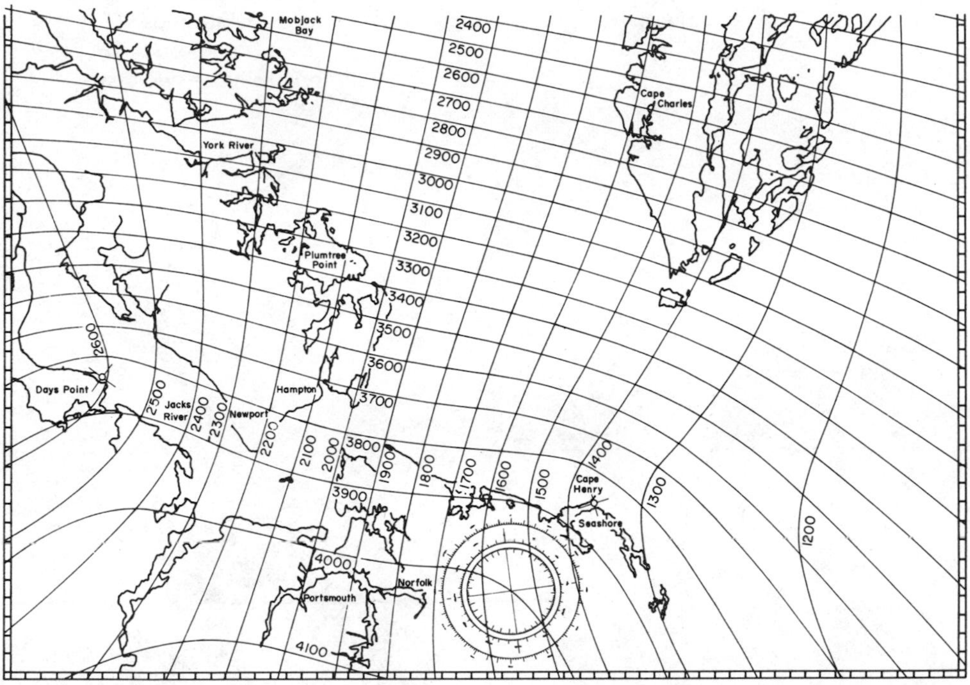

Figure 104. The Raydist T Halop geometry.

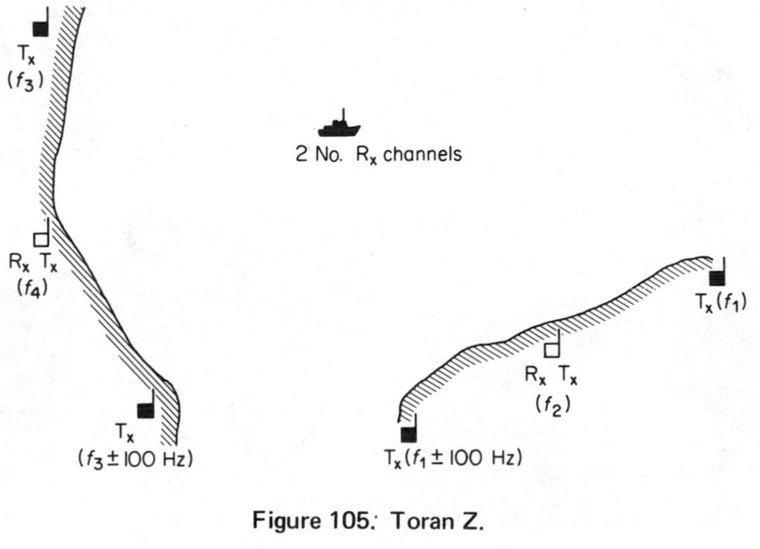

Figure 105. Toran Z.

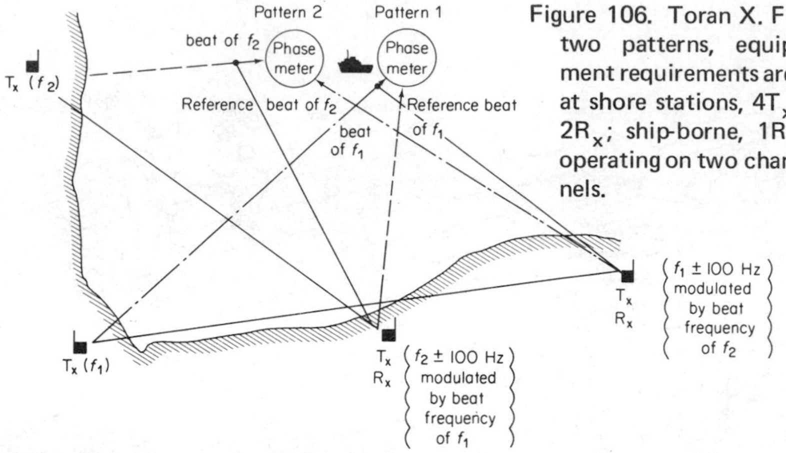

Figure 106. Toran X. For two patterns, equipment requirements are: at shore stations, $4T_x$, $2R_x$; ship-borne, $1R_x$ operating on two channels.

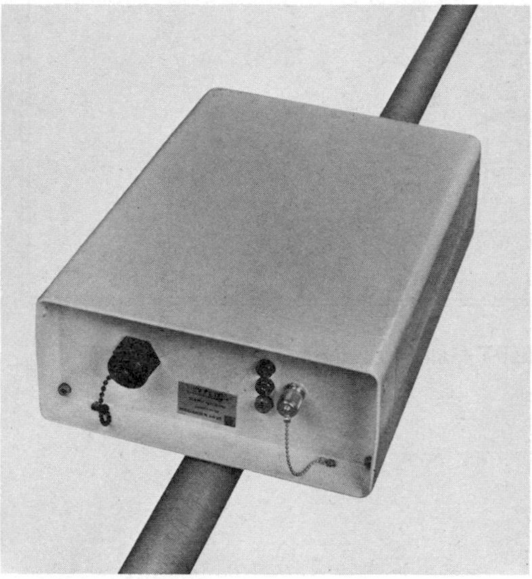

(a) Compensation receiver (modes X or Z)

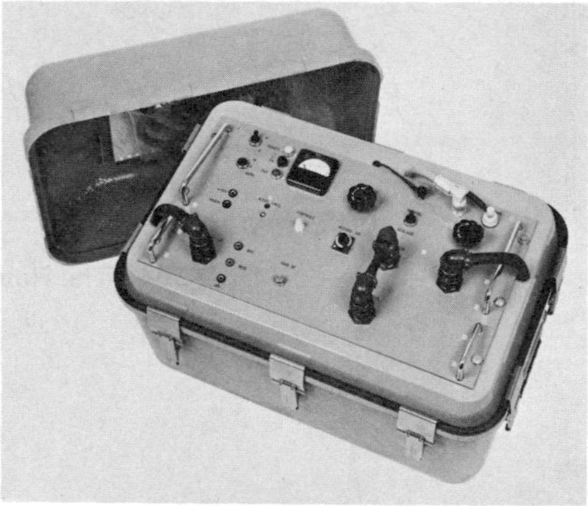

(b) P10 Non-modulated transmitter (short range)

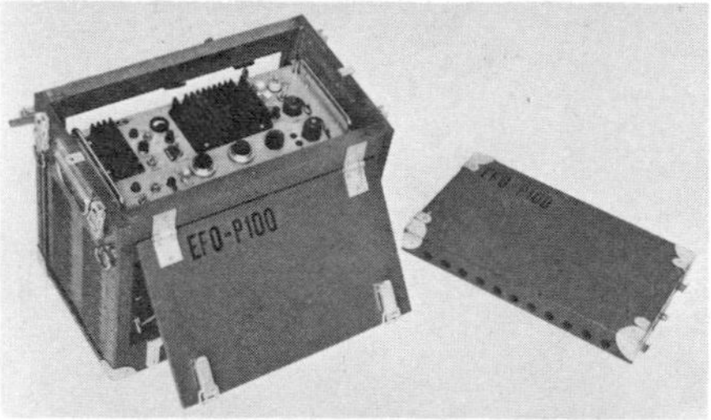

(c) P100 Transmitter (long range)

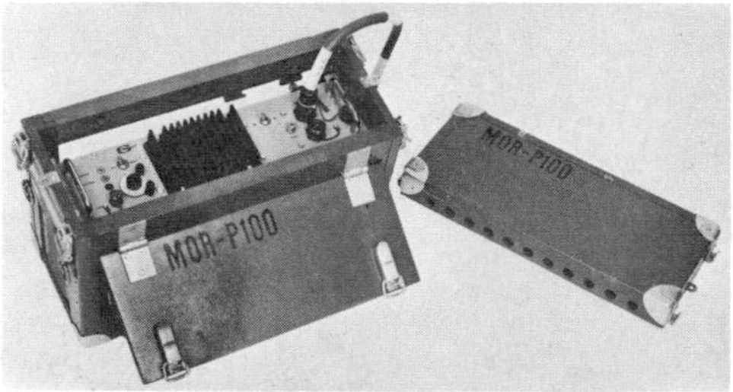

(d) P100 Modulator

Figure 107. Toran equipment. Two versions are available, the P100 transportable type, capable of long-range operation, and the highly portable P10 type with range limited to about 75 km. All equipment incorporates solid state, semiconductor circuitry and is thoroughly weatherproofed. The mobile and compensation receivers are common to both types.

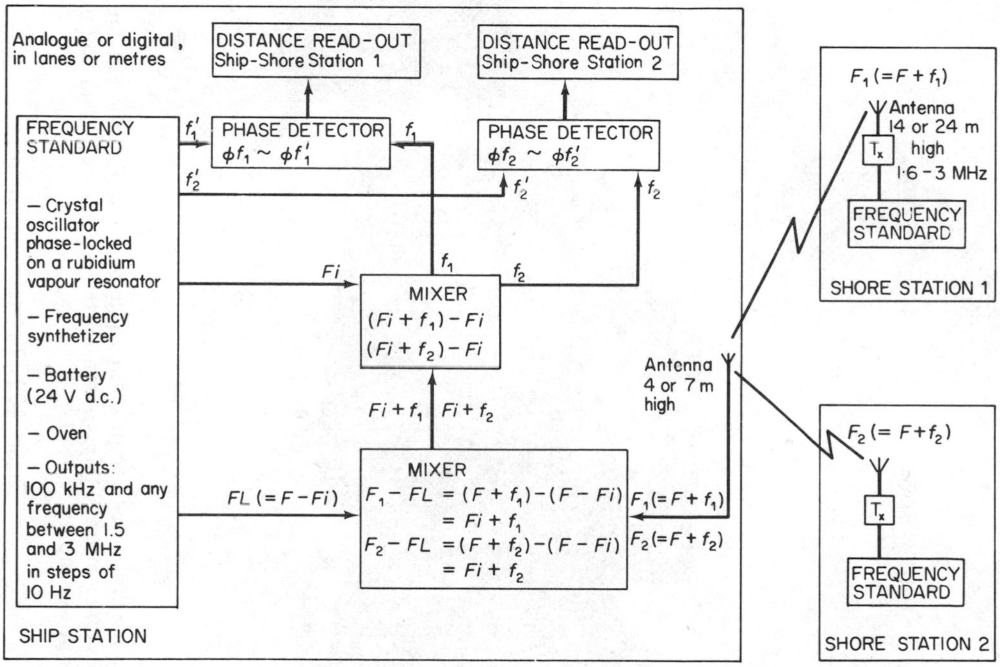

Figure 108. Function diagram of the Toran O system, showing two shore stations.

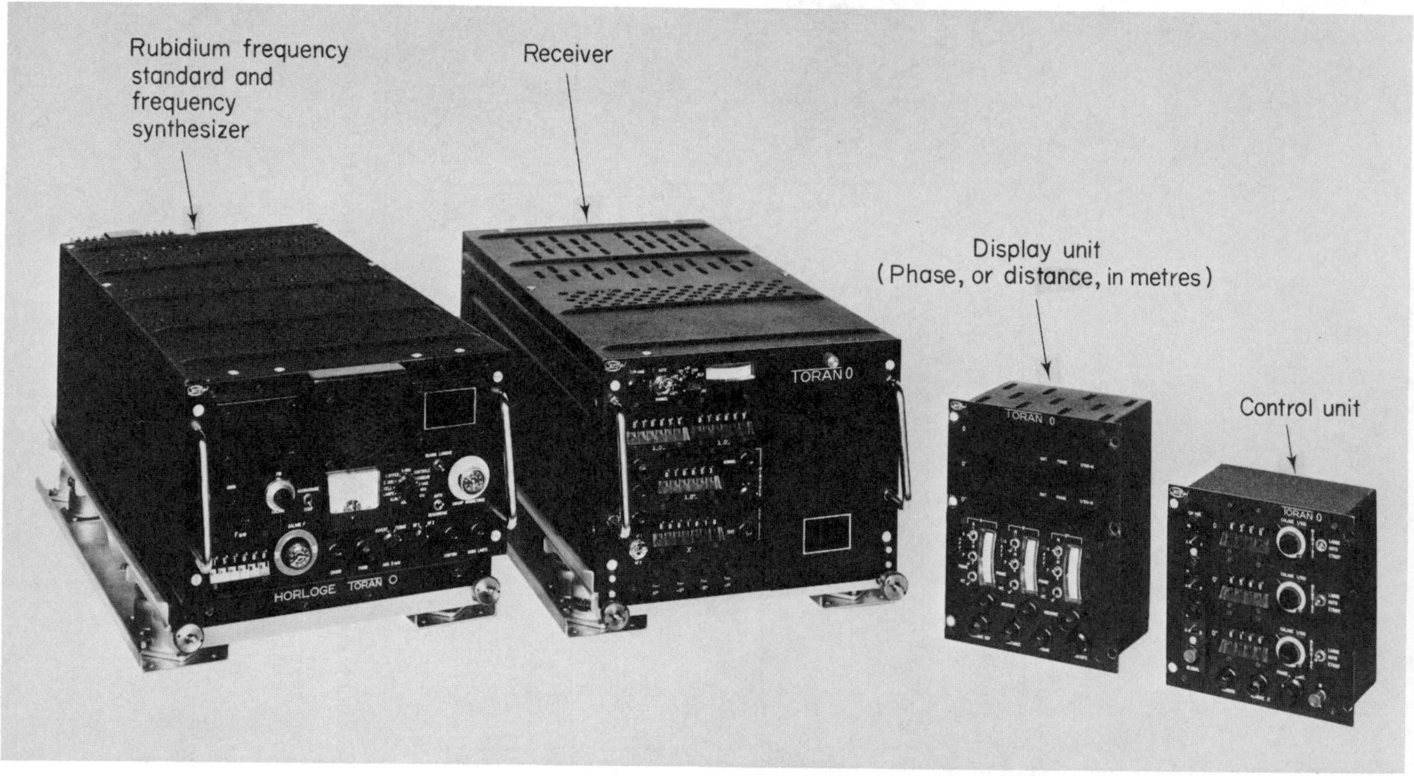

The units illustrated comprise the equipment required for the ship station. Each shore station requires a transmitter, a frequency standard (as above) and a power supply.

Figure 109. Toran O equipment.

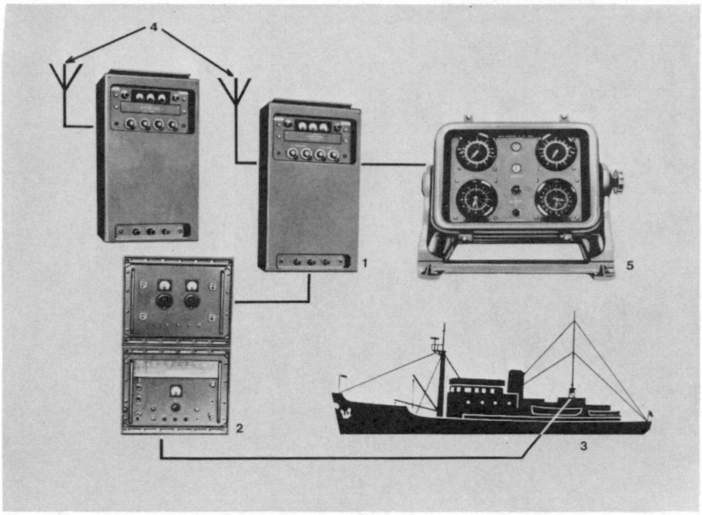

(a) Mobile Master Station.
1. Master Control Unit (M.C.U.) (2 No.) + power supply units.
2. Transmitter + Tank unit.
3. Transmitting Mast and Aerial Coil Unit.
4. Receiving aerials (2 No.).
5. Decometer Bowl.

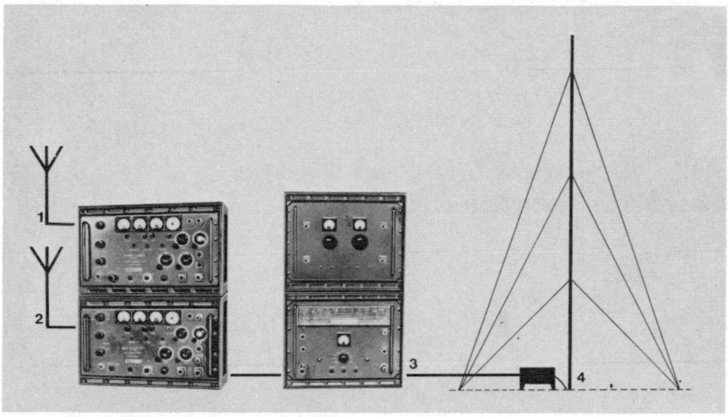

(Courtesy Decca Survey Ltd.)

(b) Shore (Slave) Station.
1. Receiving aerials (2 No.).
2. Slave Control Units (2 No.) (S.C.U.) + power supply units.
3. Transmitter + Tank unit (as for Master units).
4. Transmitting Mast (with two transmitters) and Aerial Coil unit (as for Master unit).

Note: The second Control Unit at Slave and Master Stations is installed as a standby.

Figure 110. Lambda equipment.

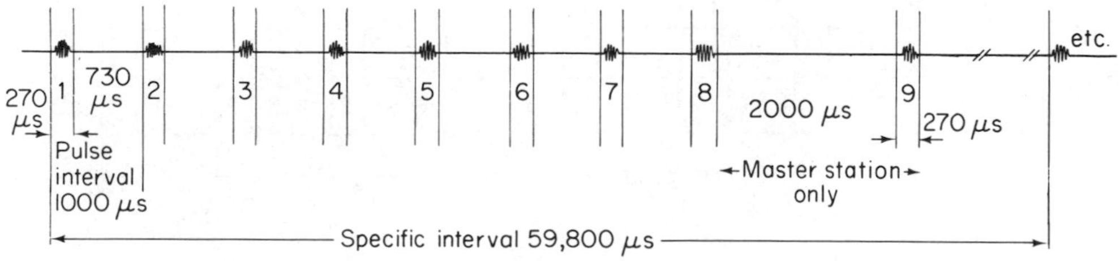

Figure 111. The time scale of a typical pulse-group.

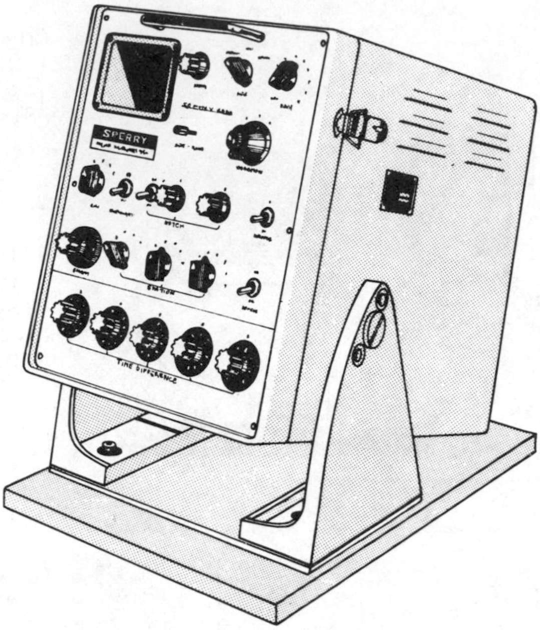

Figure 112. An example of a Loran C receiver (courtesy Sperry Rand Corporation).

Station	Frequencies (kHz)																
A	10·2		13·6		11·333												
B			10·2		13·6		11·333										
C					10·2		13·6		11·333								
D							10·2		13·6		11·333						
E ↓ etc									10·2		13·6		11·333				
Segment	A		B		C		D		E		F		G		H		A
Duration (seconds)	0·9	·2	1·0	·2	1·1	·2	1·2	·2	1·1	·2	0·9	·2	1·2	·2	1·0	·2	0·9

← 10 second interval →

Figure 113. The Omega transmission format.

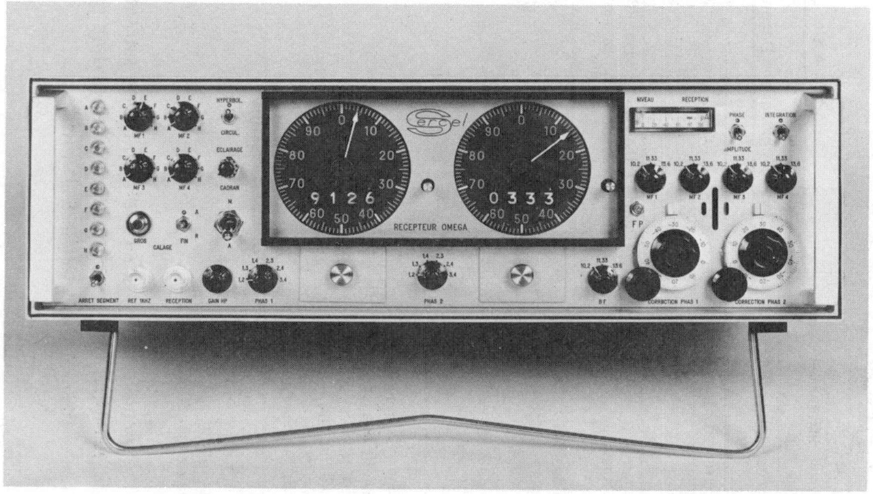

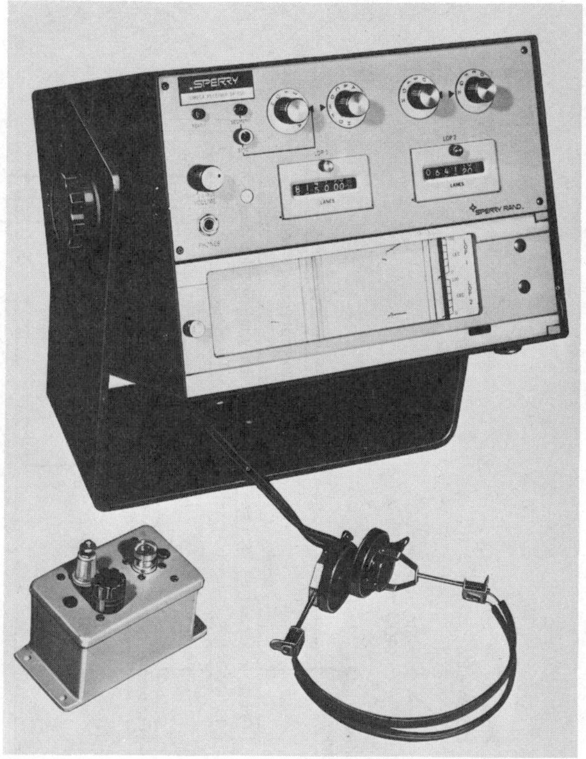

Figure 114. Omega Receivers, (a) with dial read-out and (b) with digital read-out and strip-chart record (courtesy Messrs. Sercel and Sperry Rand).

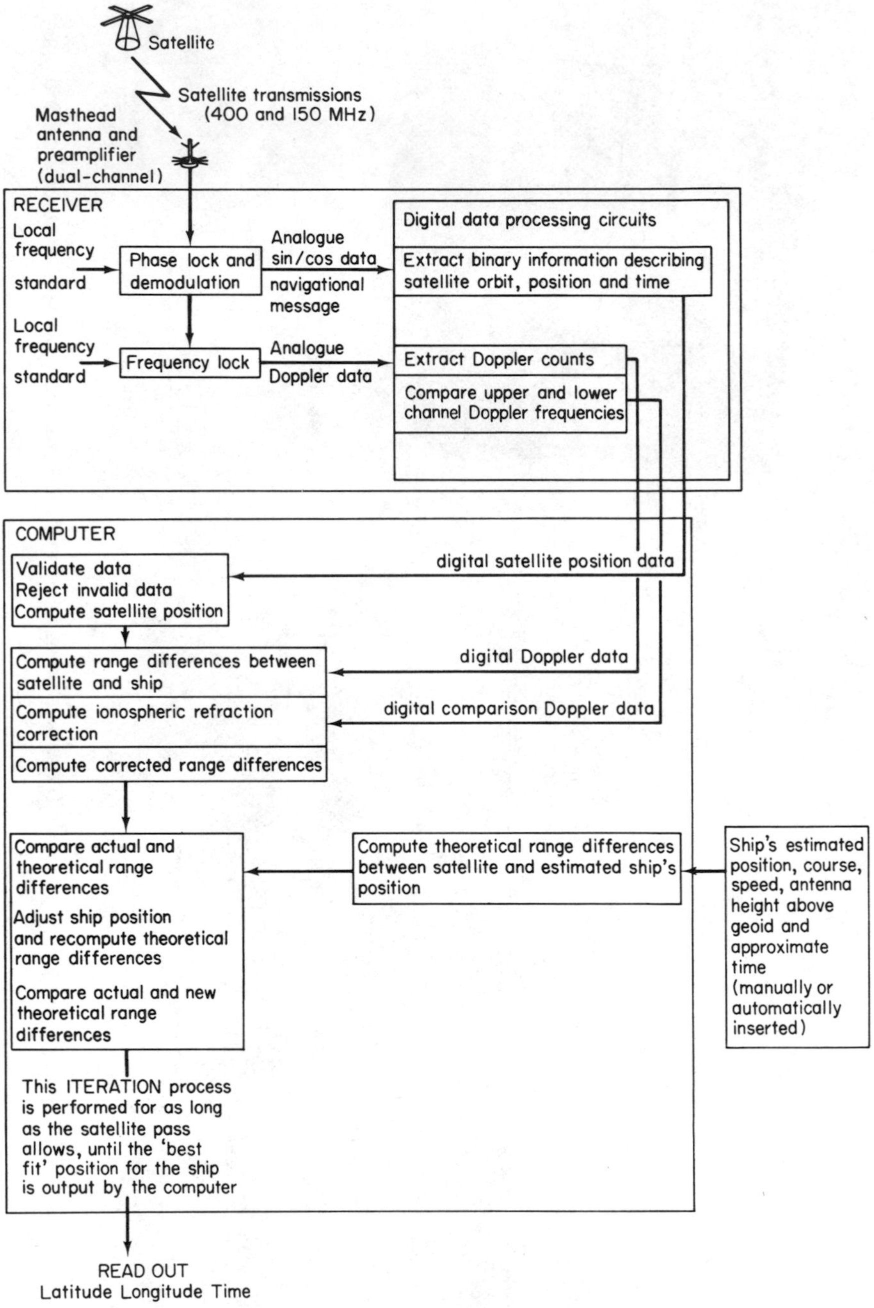

Figure 115. Function diagram of a ship-borne satellite position fixing system. It should be noted that the computer receives data from the receiver in respect of both channels of the satellite transmissions.

MX-702A Receiver Specifications:

Carrier frequencies: 399.968 MHz ± 10 kHz
149.980 MHz + 3.750 kHz

Sensitivity: 145 dBm for 3 dB s/n

Selectivity:

Pre-amplifier	3 dB	30 dB
High channel	1.0 MHz	2.6 MHz
Low channel	1.5 MHz	3.1 MHz
Equivalent final noise		
Bandwidth		
High channel	40 Hz (20 Hz loop noise bandwidth)	
Low channel	20 Hz (10 Hz loop noise bandwidth)	

Dynamic range: – 90 dBm to – 145 dBm
Intermodulation products: – 55 dB
Phase tracking error: 15° maximum
AGC: Coherent with less than ± 5% variation over entire dynamic range
Frequency stability: 2 parts in 10^{11} per 2 minutes
Signal acquisition: Search–Automatic programmed for satellite Doppler at horizon (manual or optional computer override available)

Lock–Automatic when satellite signal is in loop passband
Single channel reacquire: Automatic slaved to locked channel (manual override available)
Dual channel reacquire: Automatic saw-tooth search in direction of Doppler rise (manual override available)
Signal rejection: Automatic if doublet synchronization is not achieved in 30 seconds to prevent false lock on undesired signal
Receiver output logic levels: 0 volts = logical one + 5 volts = logical zero
Doppler integration: 4.6 second readout synchronized with satellite message allows a combination of 23 second Doppler periods to be used for position determination
Message recovery: Automatic from either locked channel

Memory: 8192 words, 16 bits per word expandable to 32,768 words in main frame 0.980 microsecond cycle time
Input/Output: 14 individually buffered 1/0 channels: one for Teleprinter, 13 available for sensor interface cards for automatic transfer of sensor data
Software (binary tapes): Navigation program (including position fix, dead-reckoning, automatic update, alerts) and hardware diagnostics
Overall dimensions of electronics unit: 1156 x 584 x 787 mm Weight 169 kg
Power requirement: 115 V a.c., 1000 watts nominal

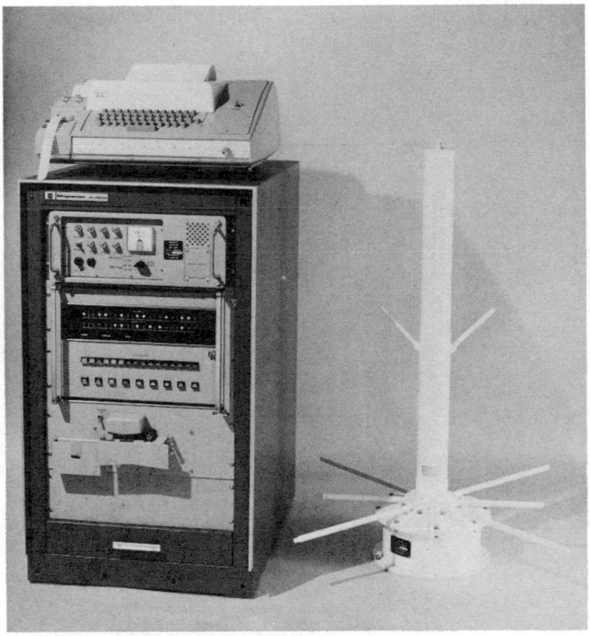

(Courtesy The Magnavox Company)

Figure 116. The Magnavox MX/702A Satellite Navigation System.

```
DAY       TIME       LAT          LON           ANT       HDG      SPD
Ø224      Ø935       Ø51 4Ø.ØØØ   ØØØ ØØ.ØØ     15Ø.Ø     .ØØØØ    .ØØØØ
                                                 Initial information passed to
                                                 computer
WO: VLØ CH

   NO.  4ØØ-CH     15Ø-CH     COUNT
                              (Refraction)

   Ø1   ØØØØØØØ    ØØØØØØØ    ØØ.
   Ø2   Ø6Ø84Ø3    Ø6Ø8372    3Ø.    Five 23-second Doppler counts
   Ø3   Ø599535    Ø5995Ø5    3Ø.    in 2-minute message period (up to
   Ø4   Ø588959    Ø588932    26.    40 counts are possible during a pass)
   Ø5   Ø699195    Ø699164    3Ø.
   Ø6   Ø55819Ø    Ø558167    21.
   Ø7   Ø54Ø625    Ø54Ø6Ø8    17.
                .
                .
                .
                .

   Ø2982237
   83668953
   825Ø5278
   89Ø2Ø34Ø
   8ØØØ6613
   8Ø746236
   82784989
   9ØØØØØ26   Fixed orbital
   8ØØØØ247   parameters
   8Ø343157
   82Ø2Ø137
   8114Ø224
   81ØØØØØØ
   8Ø129ØØØ                               Satellite navigation message
   ØØØØØØØØ
   ØØØØØØØØ
   2Ø16Ø118
   2Ø27Ø226
   2Ø37Ø4Ø7
   2Ø45Ø646   Variable orbital
   2Ø51Ø916   parameters
   2Ø541226
   2Ø551543
   2Ø531856
   6Ø47214Ø
   6Ø392395
   ØØØØØØØØ
   ØØØØØØØØ
   ØØØØØØØØ

   Ø938  Ø4
   Ø1    7Ø.283756      .ØØØ19557     -.ØØ655919
   Ø2    7Ø.2128Ø2      .ØØØØØ512     -.ØØØ13Ø32     3 iterations
   Ø3    7Ø.218Ø95     -.ØØØØØØ39     -.ØØØØØØ61
   Ø1111111111111111111111111ØØ11ØØØØØØØØØØ           Indicator line
                  ↑↑
                                       (arrows indicate closest approach)
   MXØ9-D-72ØØ4
```

DAY	TIME	LAT	LON	ANT	HDG	SPD
Ø224	Ø942	Ø51 4Ø.689	-ØØØ 22.999	15Ø.Ø	.ØØØØ	.ØØØØ

ITER	ELEV	
Ø3	6Ø	Final position fix information

(Courtesy S.G. Brown Ltd.)

Figure 117. Extracts from an MX/702 read-out.

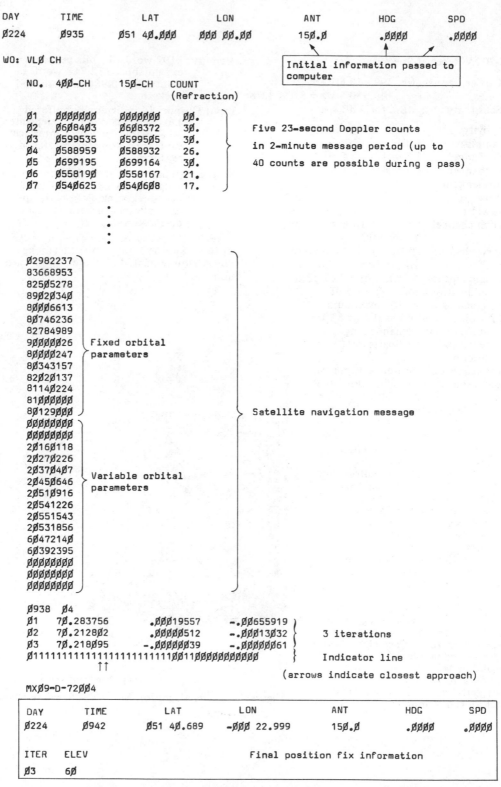

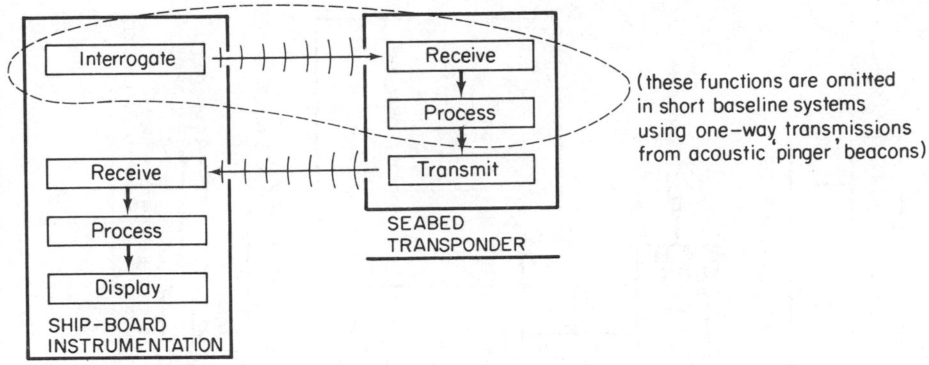

Figure 118. Functional diagram of an acoustic position fixing system.

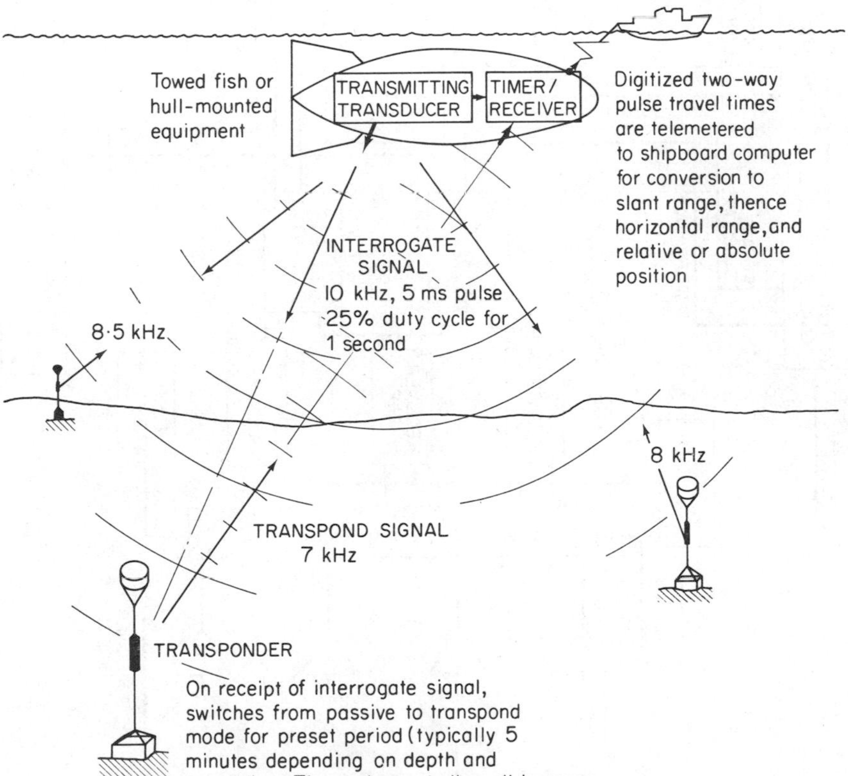

Figure 119. A typical long baseline system.

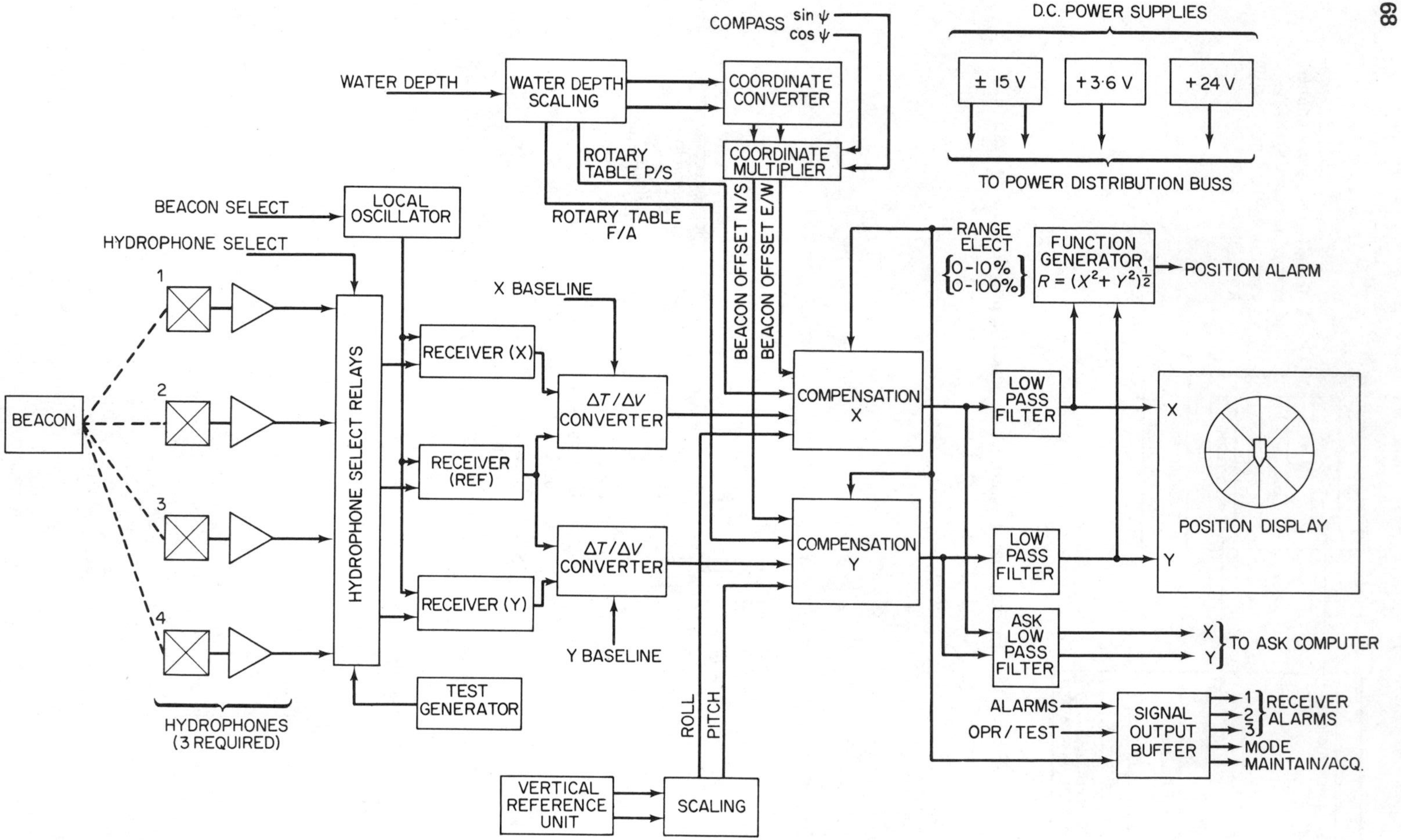

(Courtesy Honeywell Inc.)

Figure 120. The Honeywell RS-5 short baseline system function diagram.

ORE Transponder model 460A.
T_x and R_x frequencies 3.5-20 kHz.
Pulse length 15 ms.
Source level 85 dB.
Maximum interrogation rate
1.2 per second
Battery life 3 years or 3 x 10^6 pings.
Maximum operating depth 8000 m.
Dimensions:
Beacon 610 x 915 mm,
Battery (mooring) 508 x 787 x 686 mm.

Figure 121. An example of an acoustic beacon with mooring

The Honeywell RS505 acoustic position indicator—operator's display and control cabinet (signal processor has additional cabinet)

(Courtesy Honeywell Inc.)

Figure 122. An example of signal processing and display instrumentation.

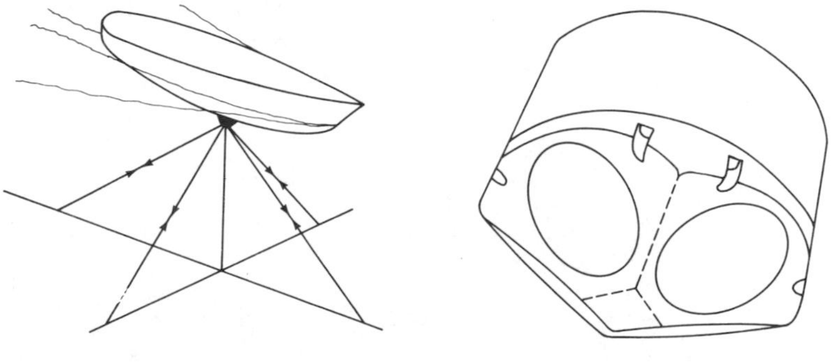

Figure 123. The 'Janus' transducer configuration.

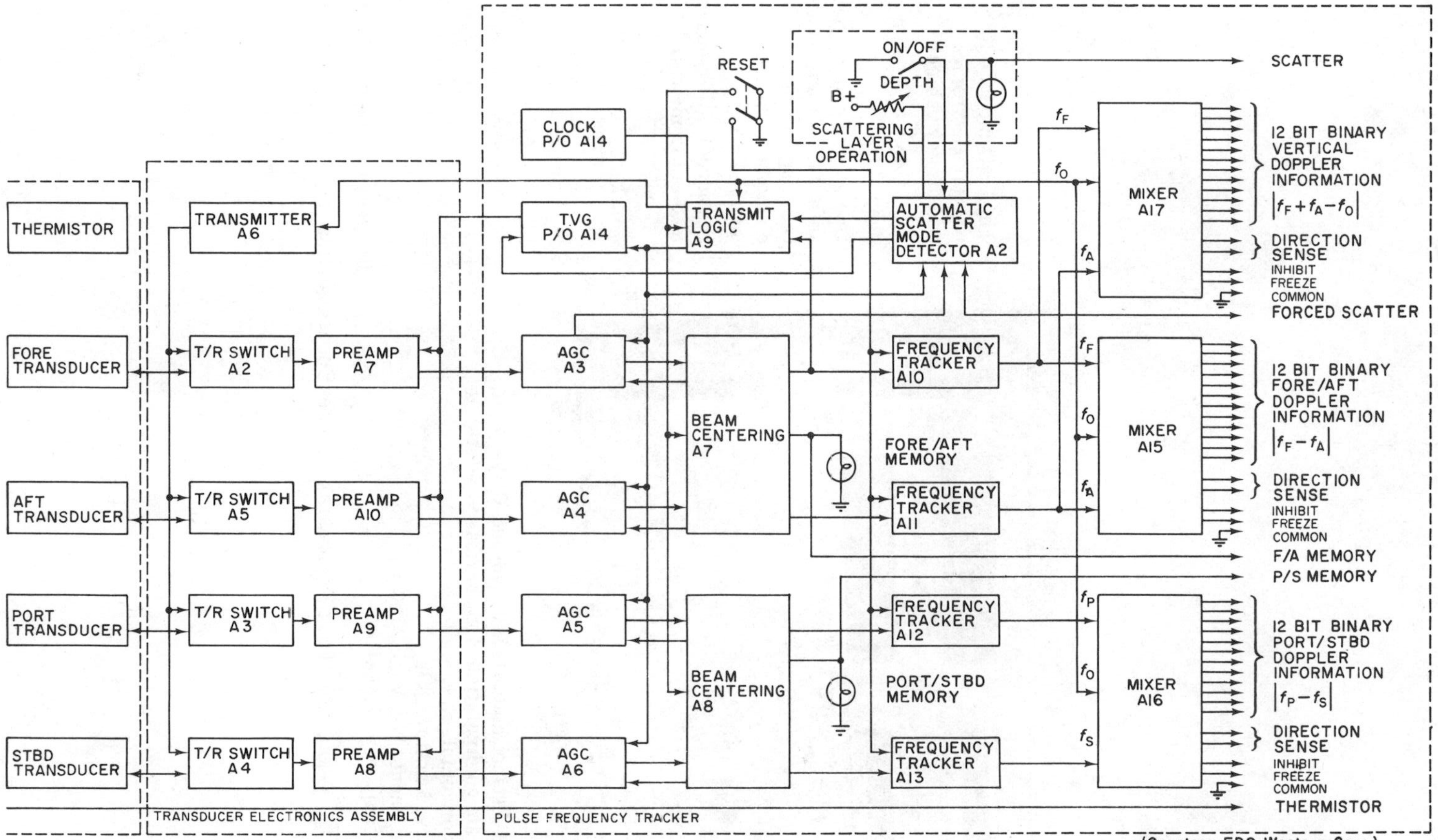

(Courtesy EDO Western Corp.)

Figure 124. Functional block diagram of the EDO Navtrak system.

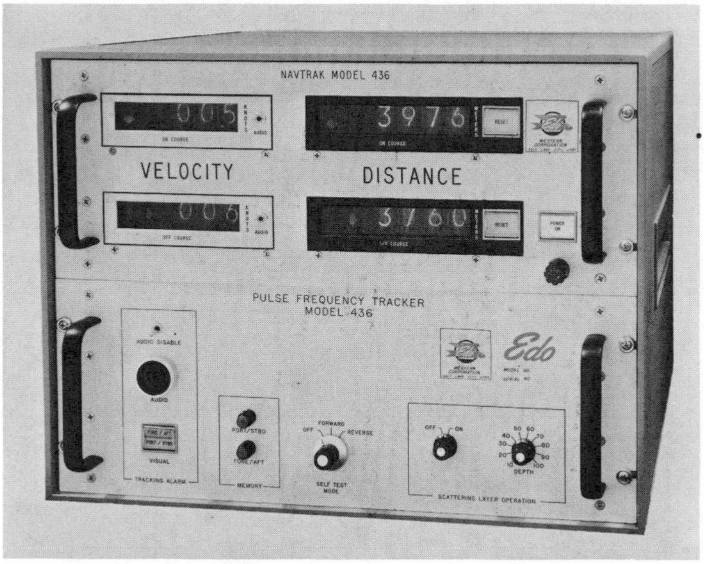

The EDO model 436 Navtrak electronics and display cabinet

(356 x 483 x 381 mm deep; 38 kg).

Operated in conjunction with 'Janus' transducer (housing 356 mm diameter x 254 mm high; 54.4 kg).

Distance on/off course read-out in metres.

Accuracy of distance and velocity read-out, 0.5%.
Acoustic beam configuration: four beams (5° width) at 60° to the horizontal.
Drift stability; 3 m/h for 8 h.
Power requirement: 115 V a.c., 120 W.
Transducer frequency: 150 kHz.

(Courtesy EDO Western Corp.)

Figure 125. The EDO Navtrak Doppler Sonar equipment.

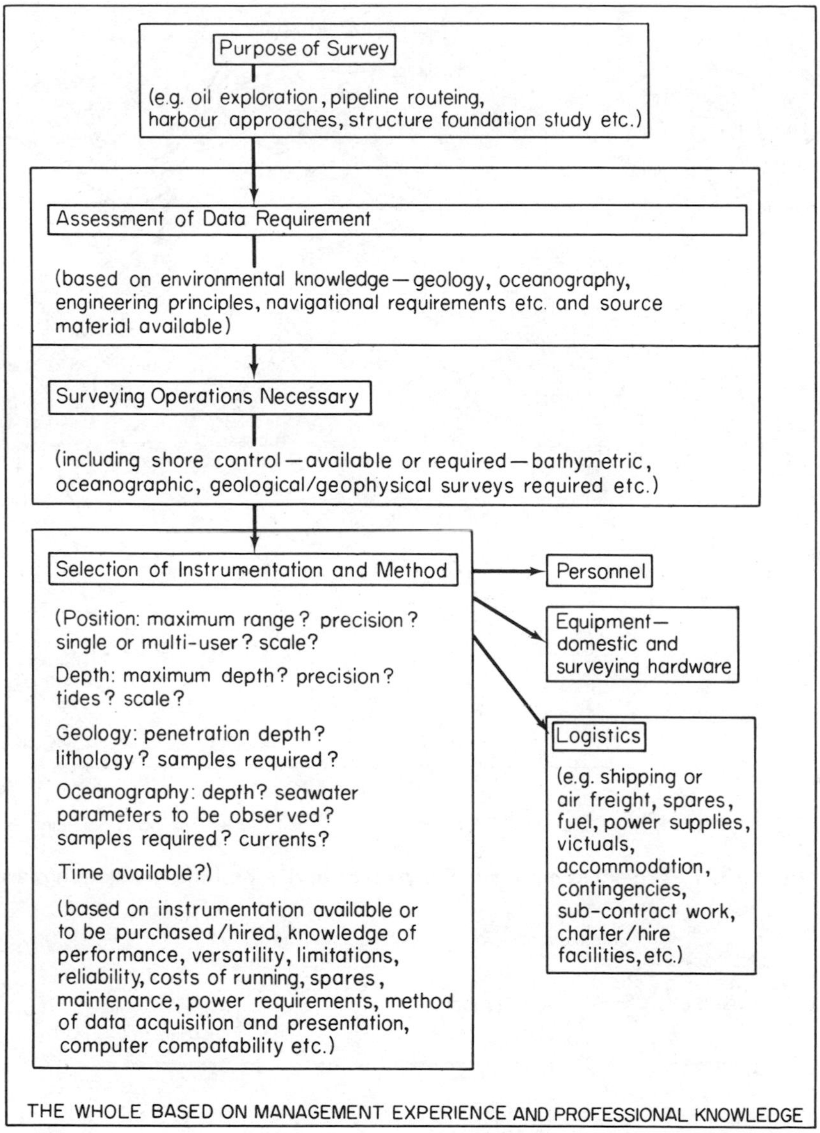
Purpose of Survey
(e.g. oil exploration, pipeline routeing, harbour approaches, structure foundation study etc.)
Assessment of Data Requirement
(based on environmental knowledge—geology, oceanography, engineering principles, navigational requirements etc. and source material available)
Surveying Operations Necessary
(including shore control—available or required—bathymetric, oceanographic, geological/geophysical surveys required etc.)
Selection of Instrumentation and Method
(Position: maximum range? precision? single or multi-user? scale?
Depth: maximum depth? precision? tides? scale?
Geology: penetration depth? lithology? samples required?
Oceanography: depth? seawater parameters to be observed? samples required? currents?
Time available?)
(based on instrumentation available or to be purchased/hired, knowledge of performance, versatility, limitations, reliability, costs of running, spares, maintenance, power requirements, method of data acquisition and presentation, computer compatability etc.)
Personnel
Equipment—domestic and surveying hardware
Logistics
(e.g. shipping or air freight, spares, fuel, power supplies, victuals, accommodation, contingencies, sub-contract work, charter/hire facilities, etc.)
THE WHOLE BASED ON MANAGEMENT EXPERIENCE AND PROFESSIONAL KNOWLEDGE

Figure 126. The job evaluation and practical considerations summarized. (This figure is based upon the premise that the instrument and method selection will determine the personnel etc. requirements. The reverse might be the case, the method being dictated by personnel and logistical availability.)

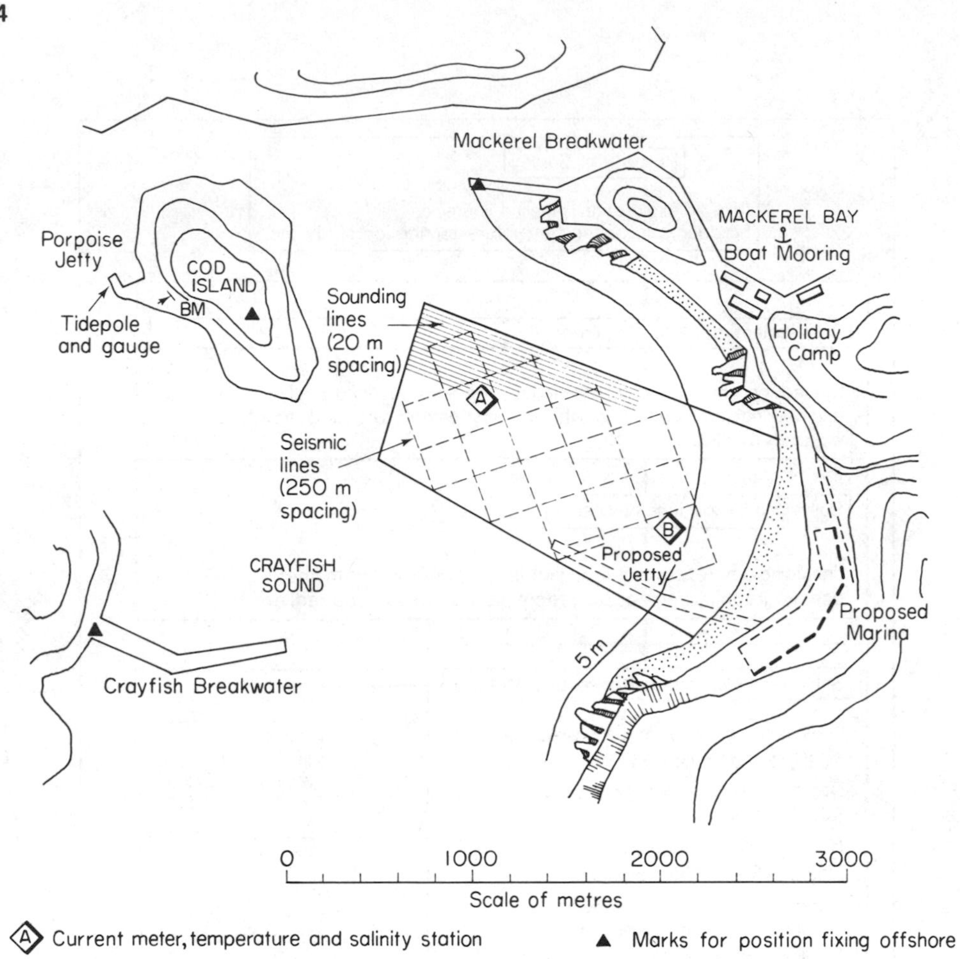

Figure 127. Projected marina in Crayfish Sound — preliminary sketch plan.

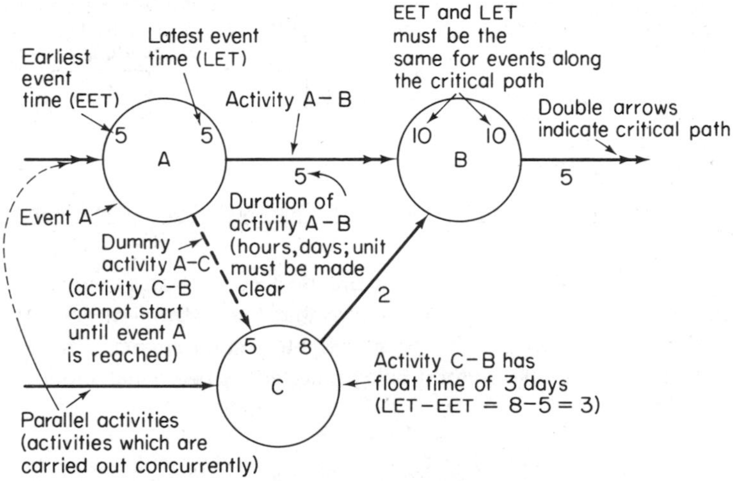

Figure 128. Suggested symbols for use in the arrow diagram.

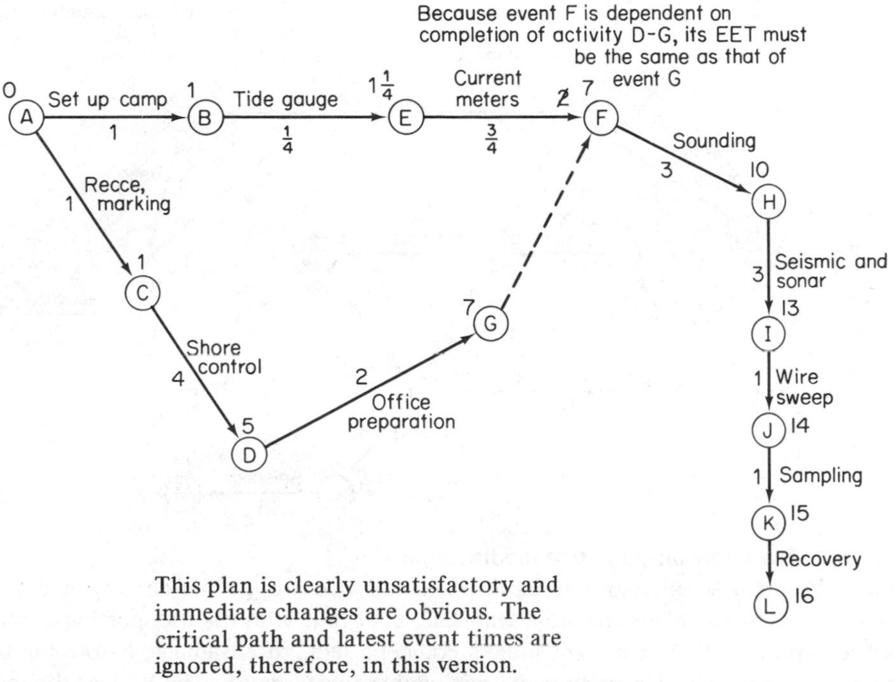

Figure 129. The first arrow diagram.

Notes. It is envisaged that activities A-B and A-C may be run concurrently, each requiring three men. Because sounding and other boat operations cannot start until positional control is achieved, a dummy arrow joins events G and F, therefore giving event F an EET of 7 days, instead of 2 days as indicated by the path A-B-E-F. The shore control and office preparation is therefore reconsidered and a revised arrow diagram is constructed (Figure 130).

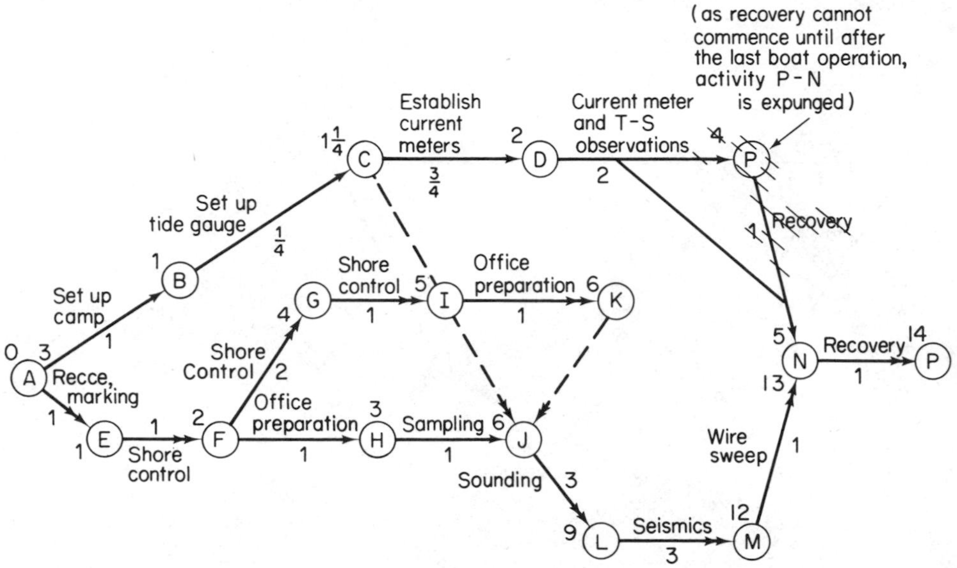

Figure 130. The arrow diagram, first modification.

Notes. The boat is required until day 2 for landing control parties and laying current meters etc. The sampling operation will share boat time with the transporting of shore parties until day 4. The current meters could be laid, to advantage, before the tide gauge is set up since tide readings are not required until day 5. The EET of the terminal event (P) is seen to be 14 days. Thus, 2 days have been saved over the initial arrow diagram by commencing the less precise boat operation before completion of preparation and shore control work. The critical path has been identified as A-E-F-G-I-K-J-L-M-N-P. It is pointless to change any activity not lying along the critical path if further shortening of the project time is desired. The critical path may be shortened by either crashing the shore control and boat operations or by cutting down on the degree of work accomplished. Preparation time can be halved if a fixed-angle plot is not required. If the boat operations are to be crashed, this can only be done by supplying a second craft and crew and not by increasing personnel alone. If these measures may be taken the following arrow diagram (Figure 131) can be drawn up.

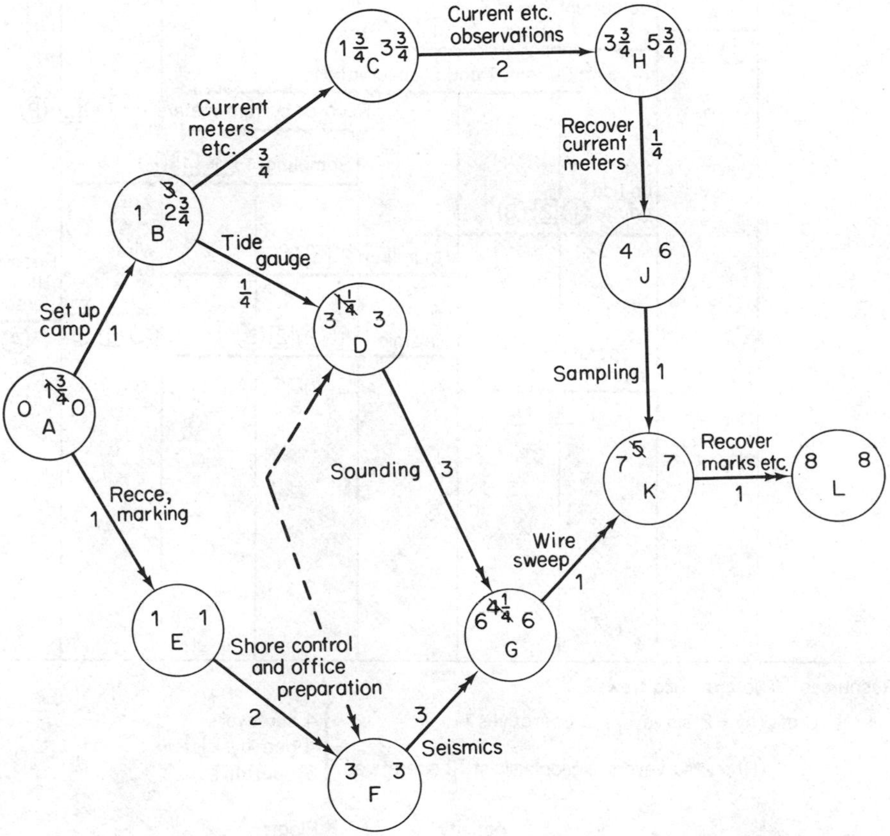

Figure 131. The arrow diagram, second modification.

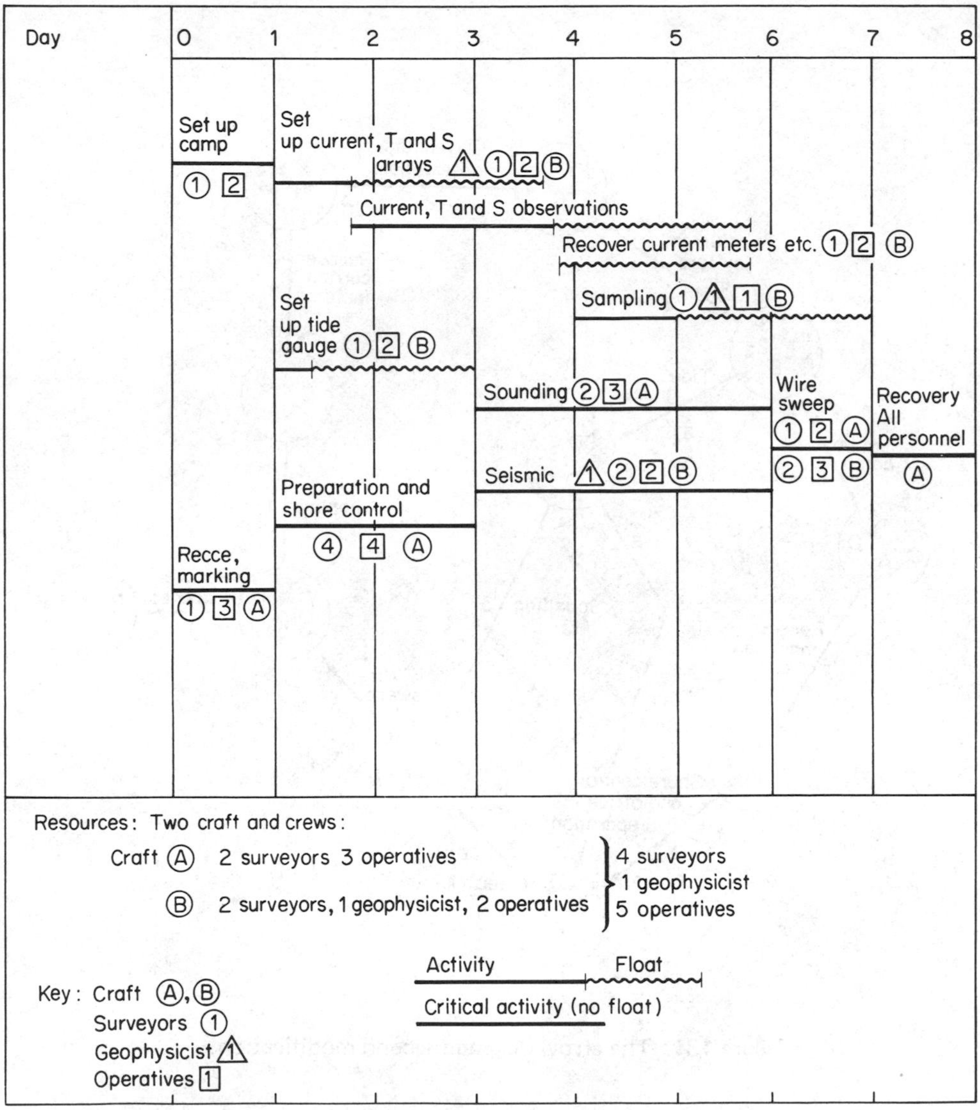

Day
0
1
2
3
4
5
6
7
8
Set up camp
① [2]
Set up current, T and S arrays △1 ① [2] Ⓑ
Current, T and S observations
Recover current meters etc. ① [2] Ⓑ
Sampling ① △1 [1] Ⓑ
Set up tide gauge ① [2] Ⓑ
Sounding ② [3] Ⓐ
Wire sweep
① [2] Ⓐ
② [3] Ⓑ
Recovery All personnel
Ⓐ
Seismic △1 ② [2] Ⓑ
Preparation and shore control
④ [4] Ⓐ
Recce, marking
① [3] Ⓐ
Resources: Two craft and crews:
Craft Ⓐ 2 surveyors 3 operatives
Ⓑ 2 surveyors, 1 geophysicist, 2 operatives
4 surveyors
1 geophysicist
5 operatives
Activity
Float
Critical activity (no float)
Key: Craft Ⓐ, Ⓑ
Surveyors ①
Geophysicist △1
Operatives [1]

Figure 132. The time-chart before levelling.

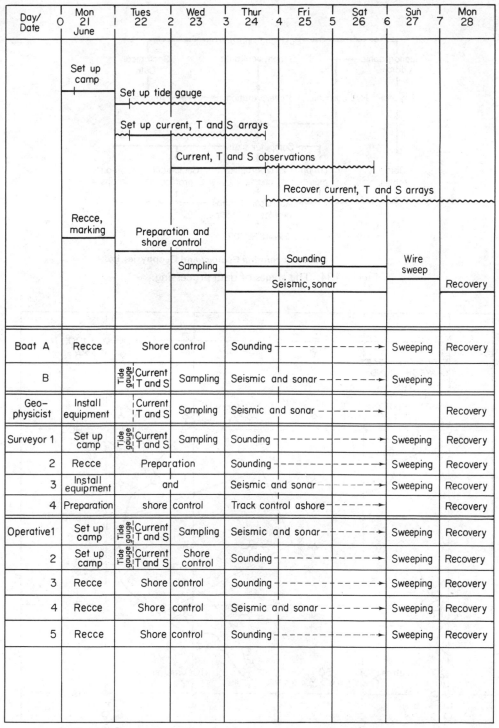

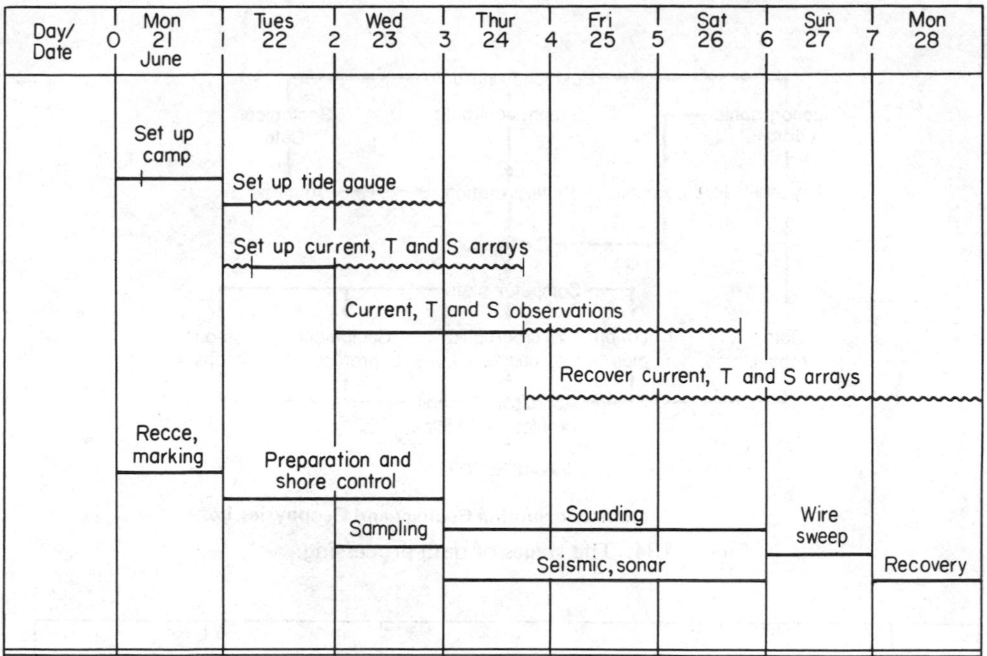

Day/Date	Mon 21 June	Tues 22	Wed 23	Thur 24	Fri 25	Sat 26	Sun 27	Mon 28
Boat A	Recce	Shore	control	Sounding ------	------	------→	Sweeping	Recovery
B		Tide gauge / Current T and S	Sampling	Seismic and sonar --	------	------→	Sweeping	
Geo-physicist	Install equipment	Current T and S	Sampling	Seismic and sonar --	------	------→		Recovery
Surveyor 1	Set up camp	Tide gauge / Current T and S	Sampling	Sounding ------	------	------→	Sweeping	Recovery
2	Recce	Prepar	ation	Sounding ------	------	------→	Sweeping	Recovery
3	Install equipment	and		Seismic and sonar --	------	------→	Sweeping	Recovery
4	Preparation	shore	control	Track control ashore	------	------→		Recovery
Operative 1	Set up camp	Tide gauge / Current T and S	Sampling	Seismic and sonar -	------	------→	Sweeping	Recovery
2	Set up camp	Tide gauge / Current T and S	Shore control	Sounding ------	------	------→	Sweeping	Recovery
3	Recce	Shore	control	Sounding ------	------	------→	Sweeping	Recovery
4	Recce	Shore	control	Seismic and sonar -	------	------→	Sweeping	Recovery
5	Recce	Shore	control	Sounding ------	------	------→	Sweeping	Recovery

Notes. (a) The sampling activity has been moved to an earlier time, the precision of the positional control being less essential.
(b) Surveyor 4 and the geophysicist are free to process data and progress the administrative work on one day each.

Figure 133. The time and resource charts after levelling and dating the schedule.

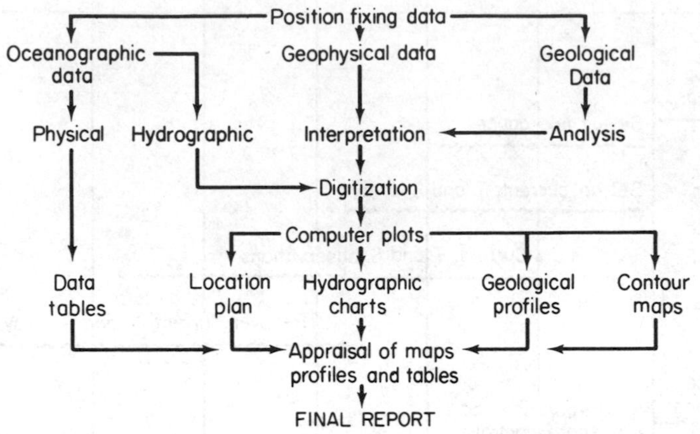

(Courtesy Hunting Geology and Geophysics Ltd.)

Figure 134. The stages of data processing.

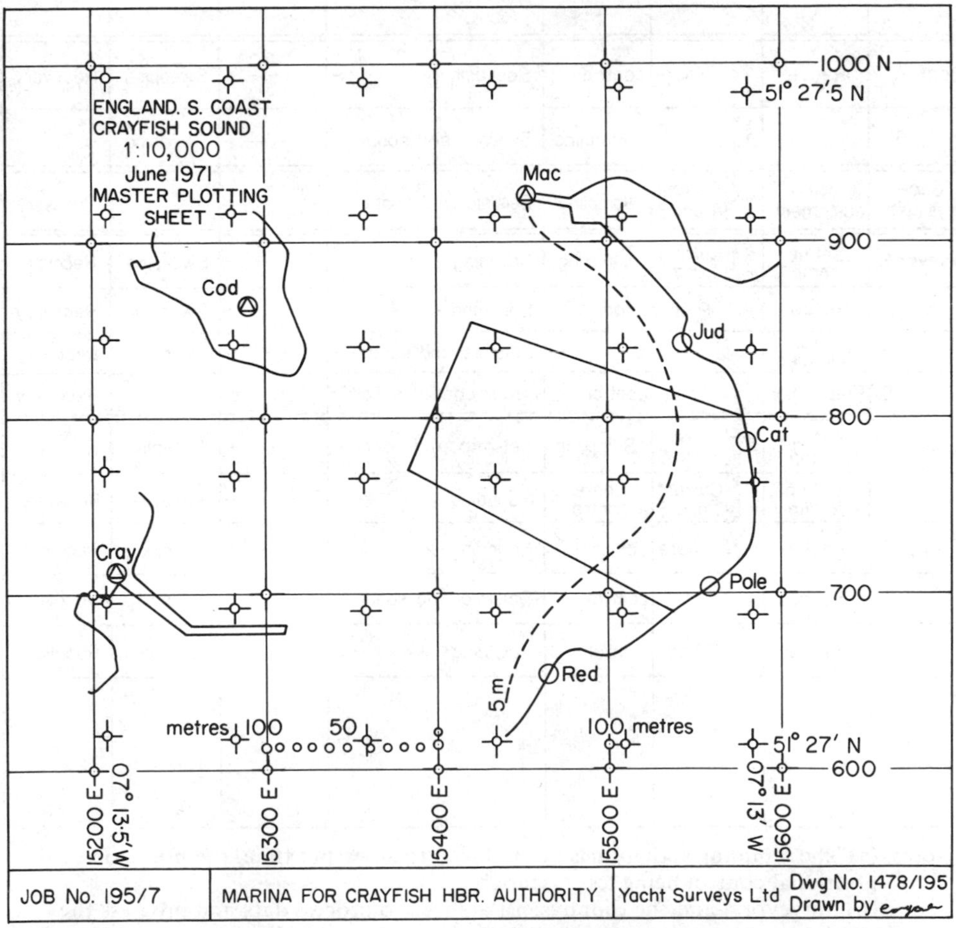

Figure 135. An example of a completed master plotting sheet or prick through.

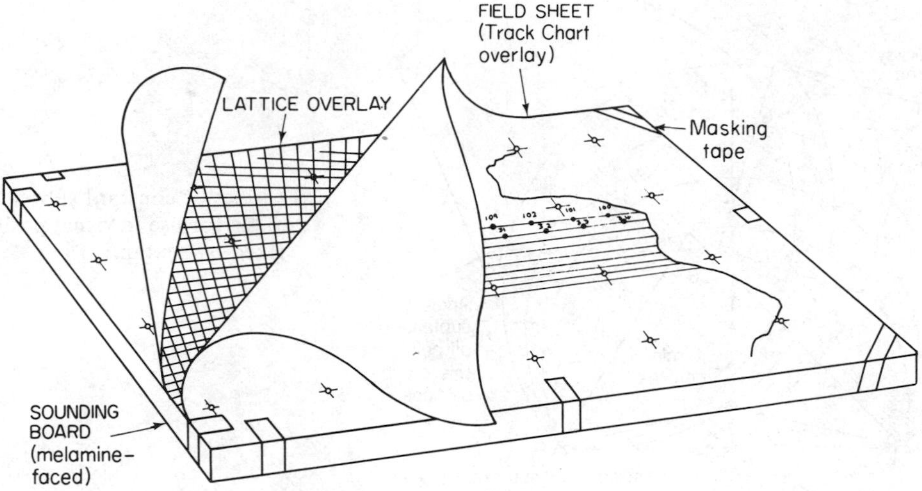

Data acquisition—e.g. overlays positioned over sounding board whilst carrying out hydrographic operation

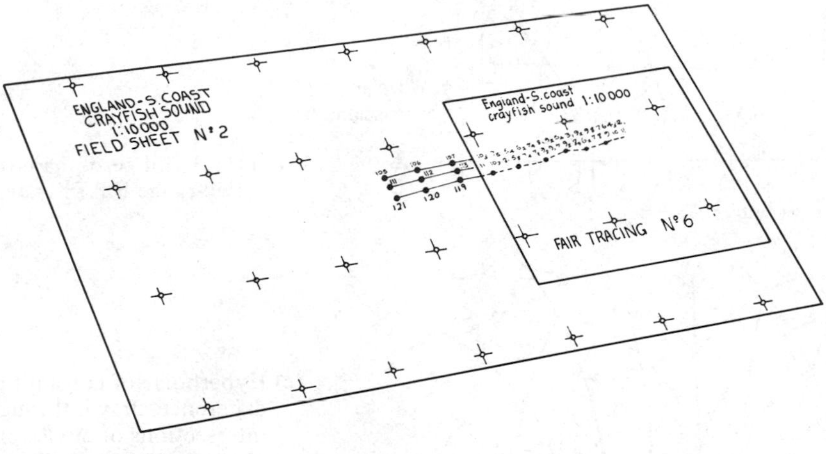

Data processing—e.g. fair tracing used to record reduced soundings by tracing direct over track plot on field sheet

Figure 136. The use of overlays and fair tracings.

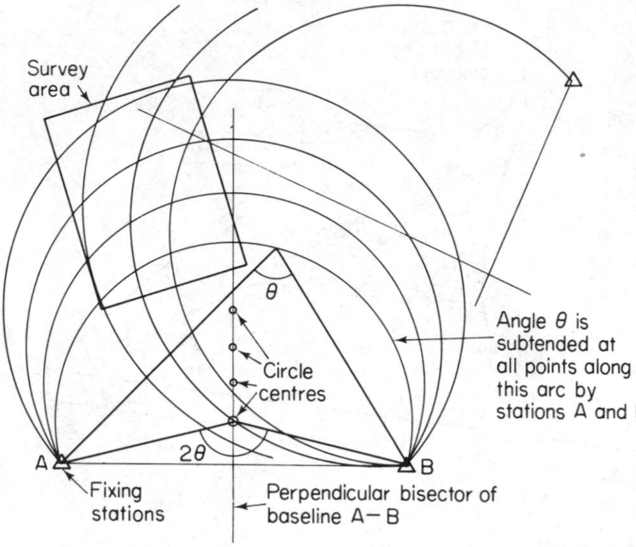
Survey area
θ
Angle θ is subtended at all points along this arc by stations A and B
Circle centres
2θ
A
B
Fixing stations
Perpendicular bisector of baseline A–B

(a) Circles of constant subtended angle for use in visual position fixed by sextant.

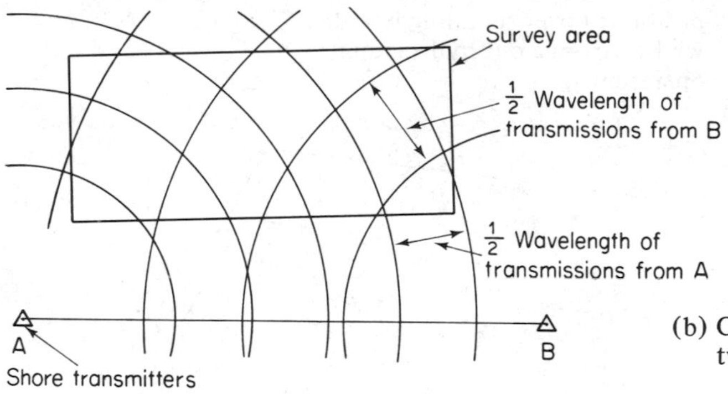
Survey area
1/2 Wavelength of transmissions from B
1/2 Wavelength of transmissions from A
A
B
Shore transmitters

(b) Circles of zero-phase for a two-range E.P.F. system.

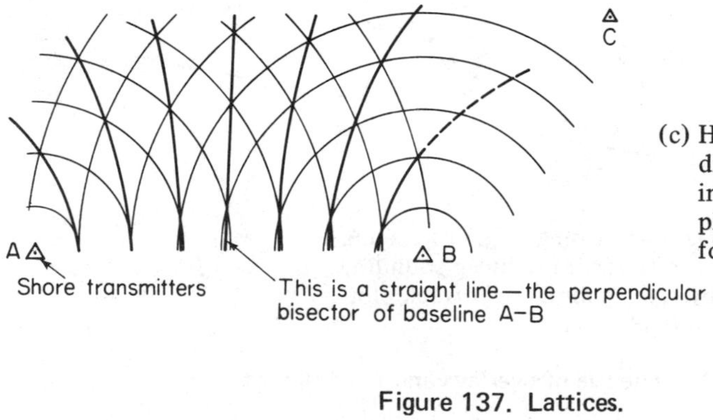
C
A
B
Shore transmitters
This is a straight line—the perpendicular bisector of baseline A–B

(c) Hyperbolae of constant phase-difference, drawn through intersections of circles of zero phase (plotted as in Figure 108) for a hyperbolic E.P.F. system.

Figure 137. Lattices.

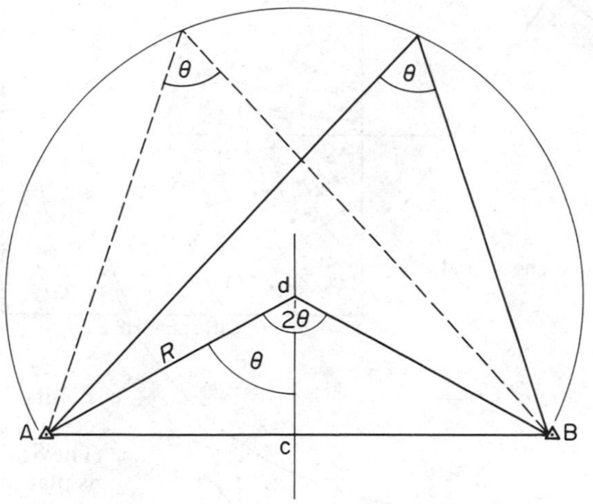

Figure 138. The geometry of constant subtended angle lattice construction.

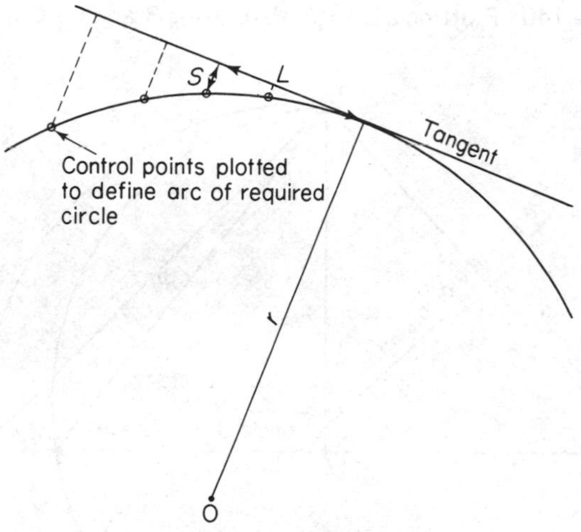

Figure 139. The geometry of plotting arcs by offsets from the tangent.

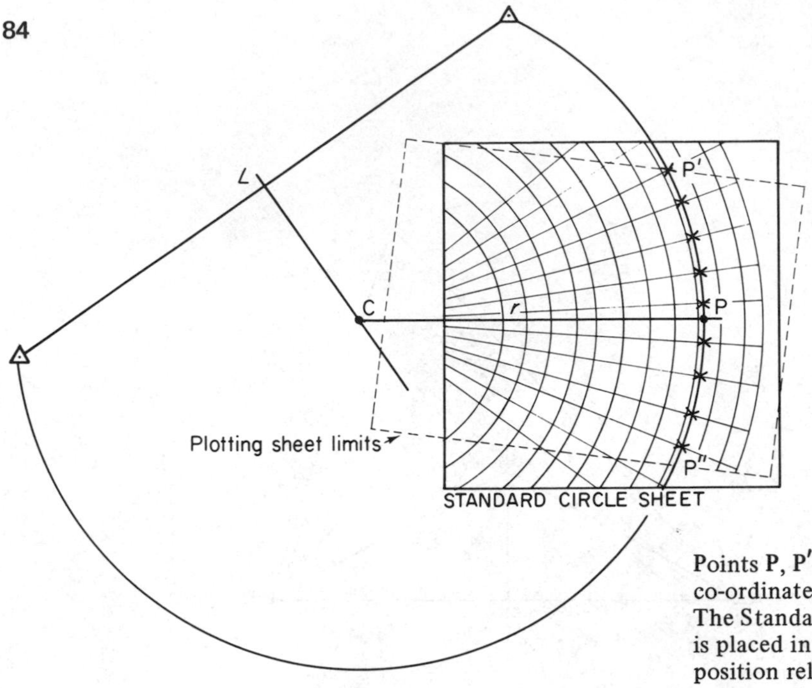

Points **P**, **P′** and **P″** are co-ordinated and plotted. The Standard Circle Sheet is placed in the correct position relative to these points and the radius (*r*) and control points are pricked through to the plotting sheet for placing of the splines (X)

Figure 140. Plotting a position arc using Standard Circle Sheets.

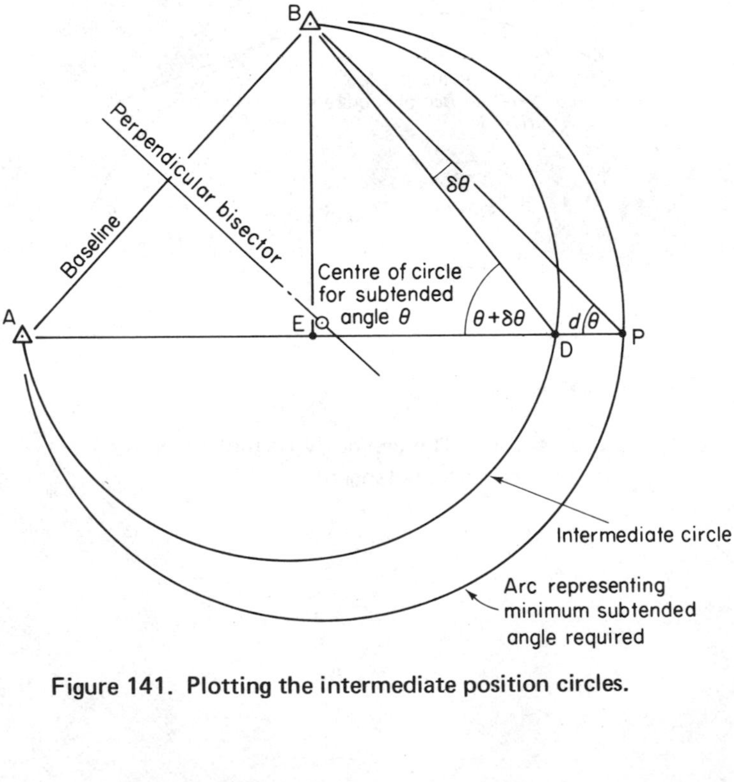

Figure 141. Plotting the intermediate position circles.

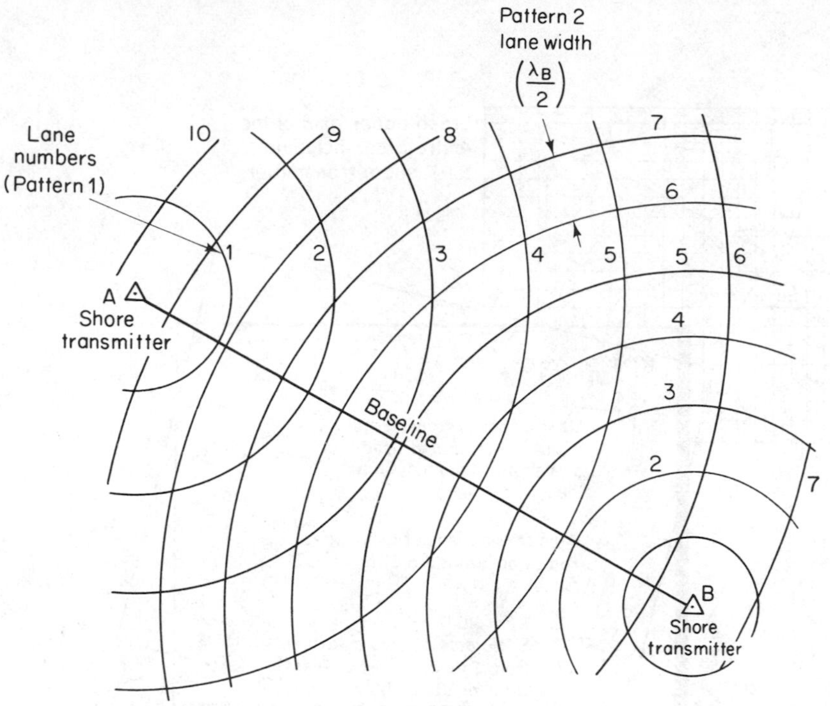

Figure 142. The two-range lattice.

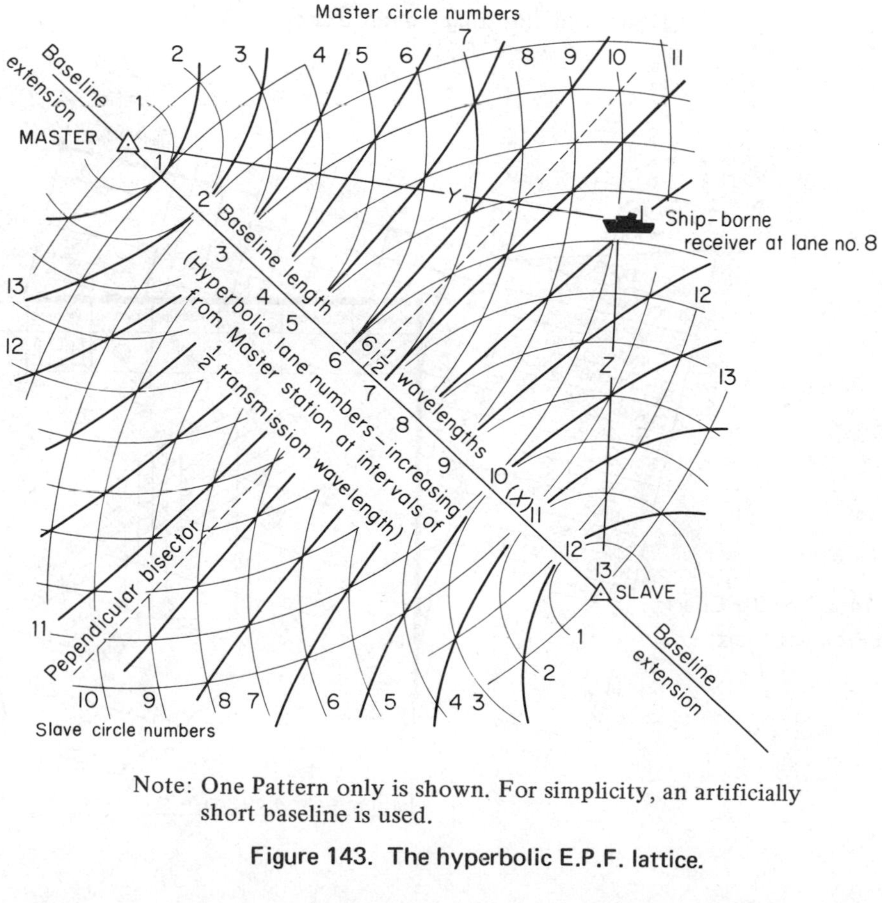

Note: One Pattern only is shown. For simplicity, an artificially short baseline is used.

Figure 143. The hyperbolic E.P.F. lattice.

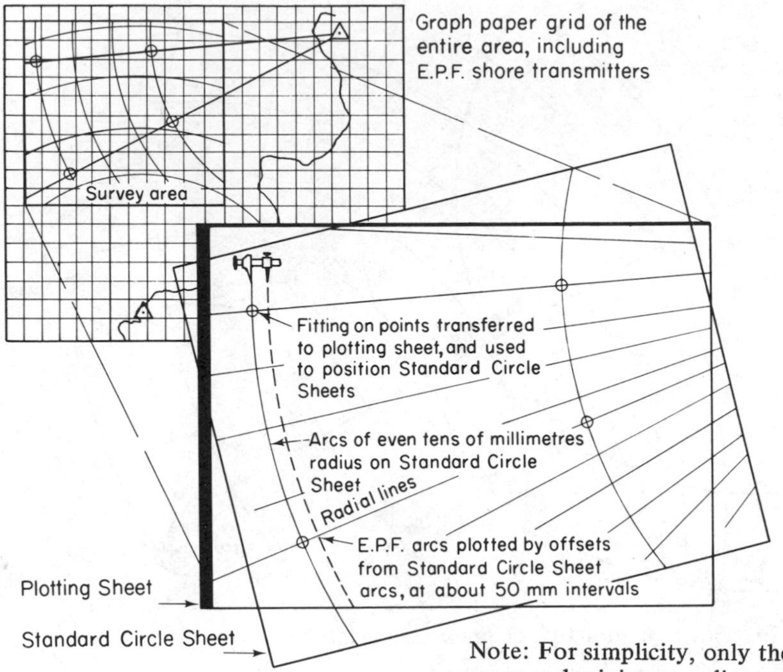

Note: For simplicity, only the maximum and minimum radius arcs are shown.

Figure 144. Plotting the range circles.

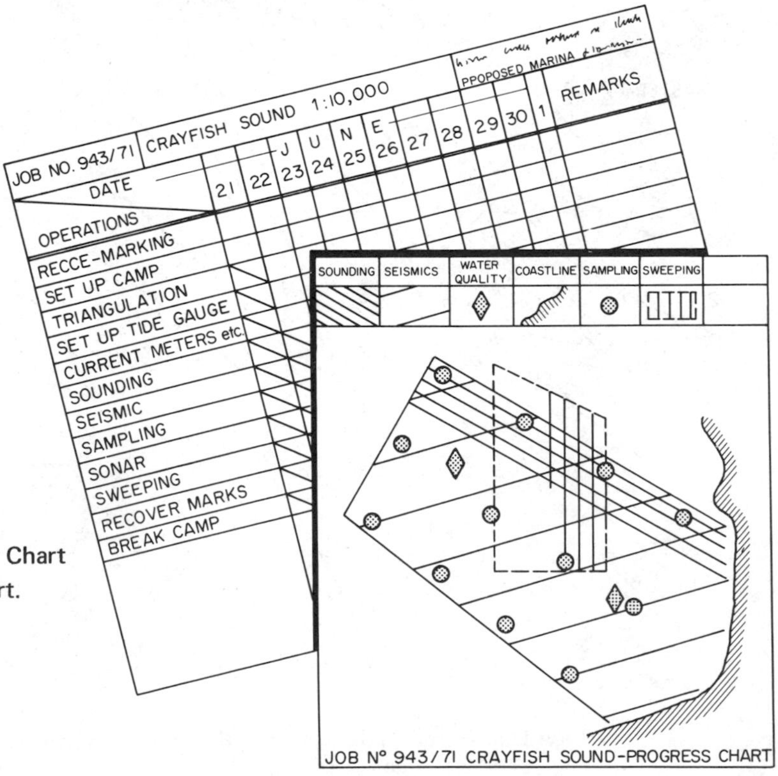

Figure 145. The Bar Chart and Progress Chart.

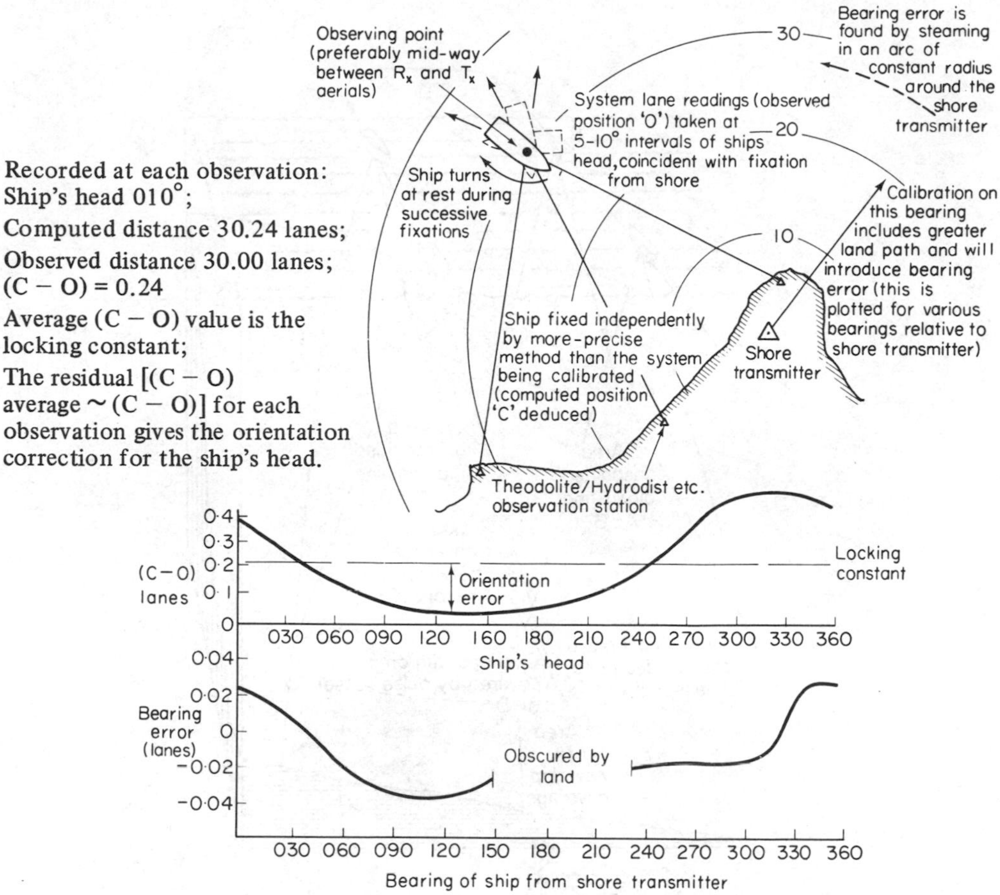
Observing point (preferably mid-way between R_x and T_x aerials)
System lane readings (observed position 'O') taken at 5-10° intervals of ships head, coincident with fixation from shore
Bearing error is found by steaming in an arc of constant radius around the shore transmitter
30
20
10
Recorded at each observation:
Ship's head 010°;
Computed distance 30.24 lanes;
Observed distance 30.00 lanes;
(C − O) = 0.24
Average (C − O) value is the locking constant;
The residual [(C − O) average ~ (C − O)] for each observation gives the orientation correction for the ship's head.
Ship turns at rest during successive fixations
Calibration on this bearing includes greater land path and will introduce bearing error (this is plotted for various bearings relative to shore transmitter)
Ship fixed independently by more-precise method than the system being calibrated (computed position 'C' deduced)
Shore transmitter
Theodolite/Hydrodist etc. observation station
(C−O) lanes
0·4
0·3
0·2
0·1
0
Orientation error
Locking constant
030 060 090 120 160 180 210 240 270 300 330 360
Ship's head
Bearing error (lanes)
0·04
0·02
0
−0·02
−0·04
Obscured by land
030 060 090 120 150 180 210 240 270 300 330 360
Bearing of ship from shore transmitter

Figure 146. A two-range chain calibration (calibration of one pattern only is shown).

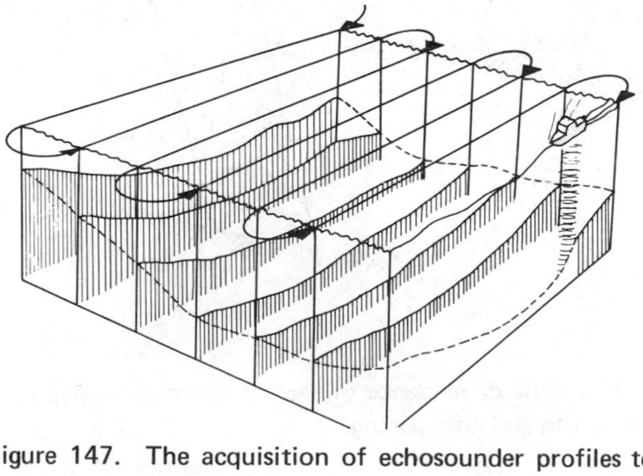

Figure 147. The acquisition of echosounder profiles to portray seabed relief.

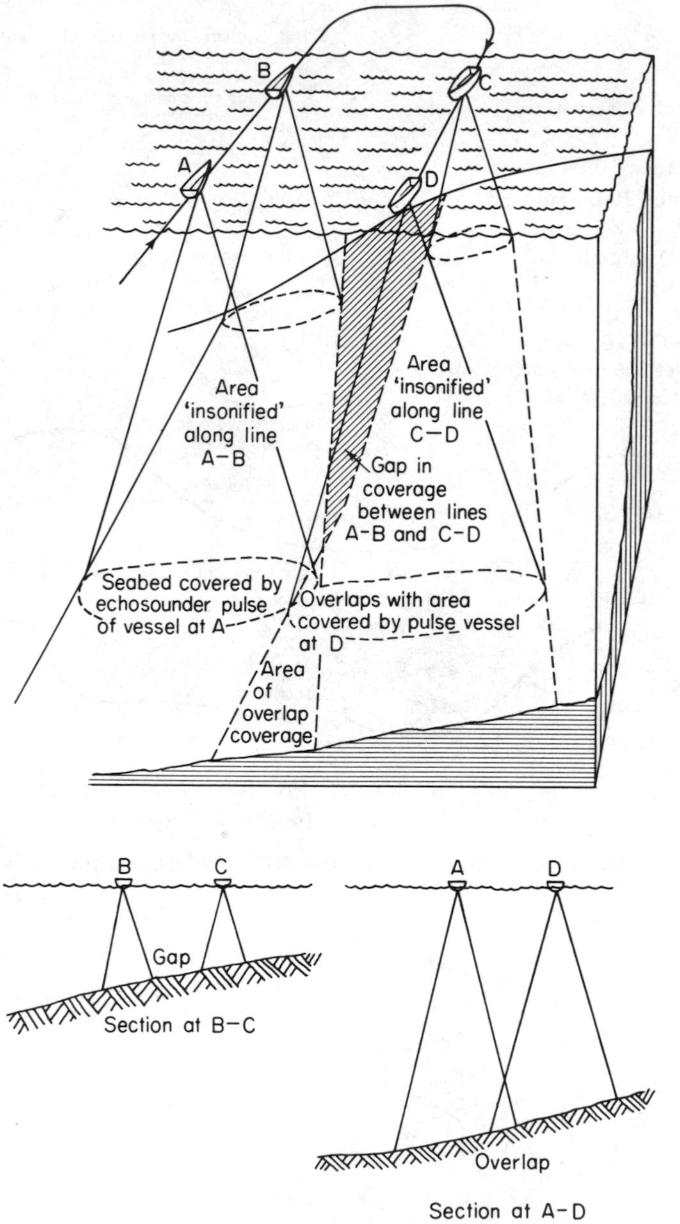

Figure 148. The dependence of seabed coverage on depth, beam width and line spacing.

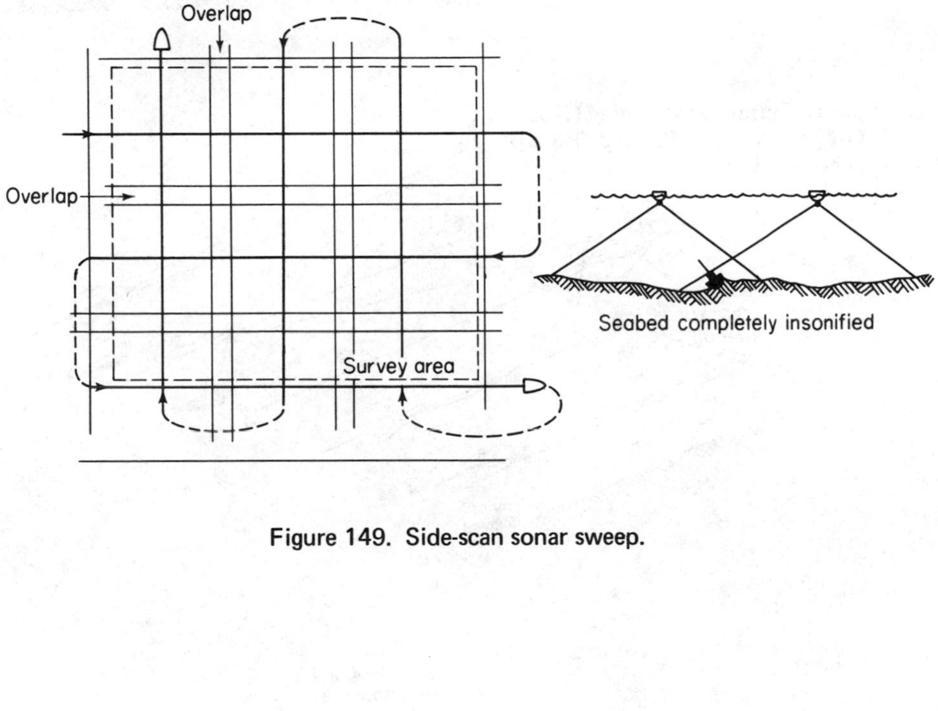
Overlap
Overlap
Survey area
Seabed completely insonified

Figure 149. Side-scan sonar sweep.

(a) Formation sounding (Sweden) suitable for rapid, high speed surveys of open sea areas. Small, fast, sounding boats in station on mother ship. Mother ship only fixes positions along sounding line.

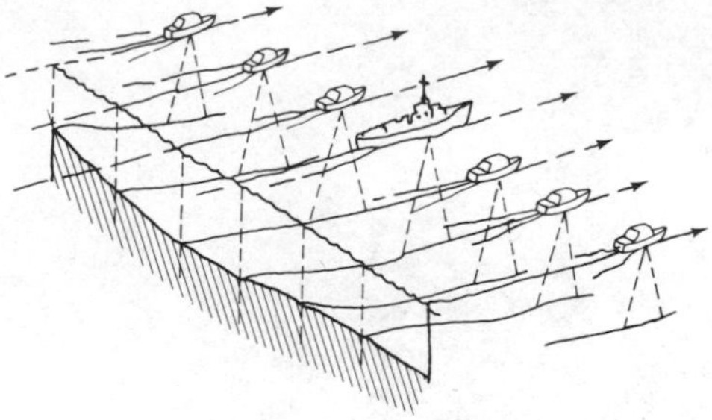

(b) Multi-transducer sounding (Germany) used in rivers and estuaries. Up to 41 transducers 2 m apart on outrigger booms, giving up to 80 m lane width.

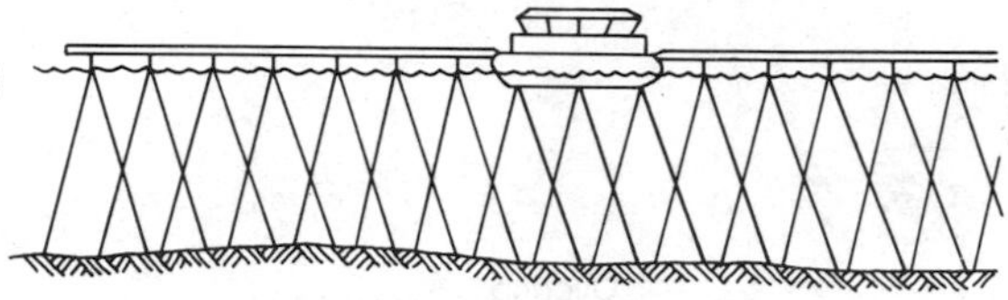

(c) Multi-transducer sounding (Holland); Oropesa sweeps of towed fish with echosounders.

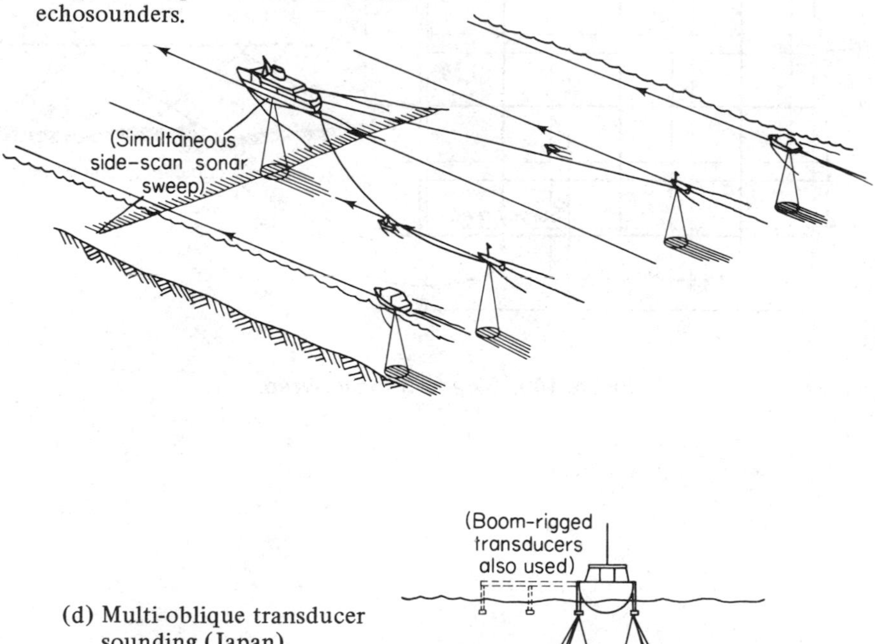

(d) Multi-oblique transducer sounding (Japan).

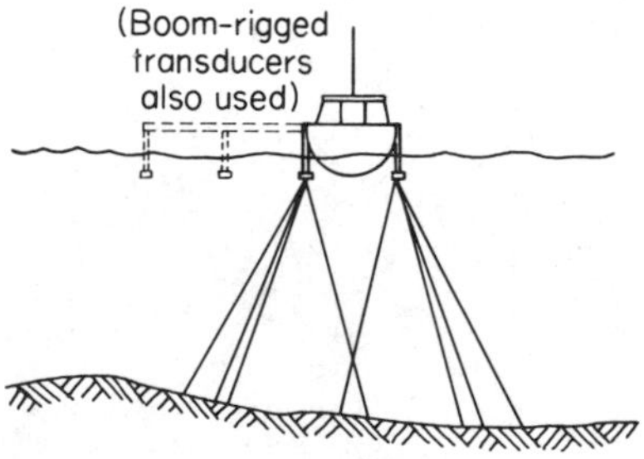

Figure 150. Examples of echosounder transducer configurations for economical wide line spacing, full coverage sounding.

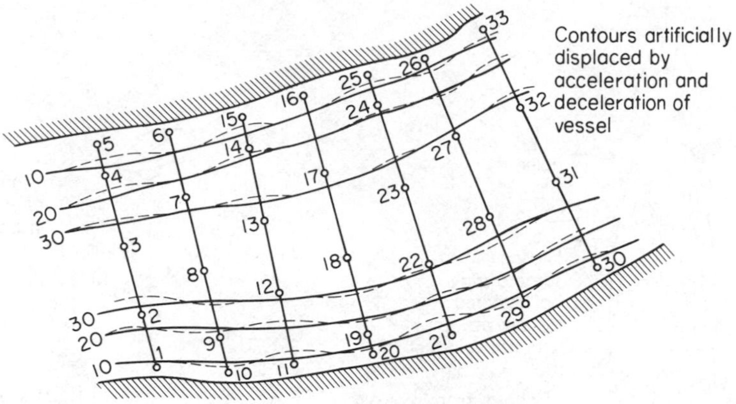

Figure 151. The dangers of transverse sounding lines in channels.

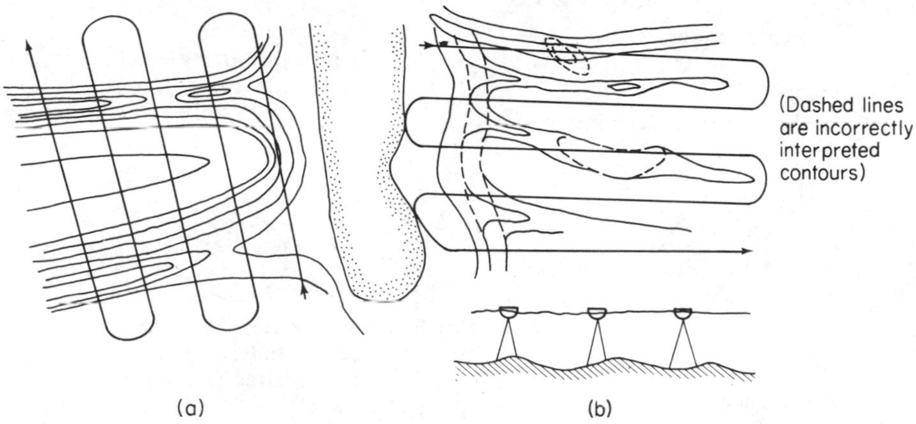

Figure 152. (a) Sand waves revealed by sounding lines.
(b) Sand waves not revealed by conventionally-run lines.

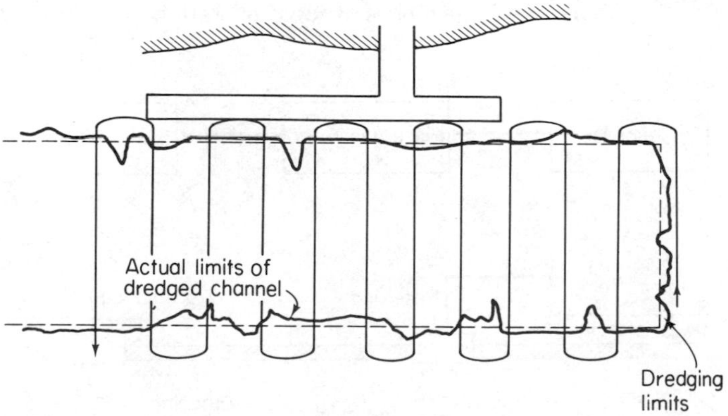

Figure 153. 'Teeth' protruding into a dredged area incompletely delineated by transverse sounding lines.

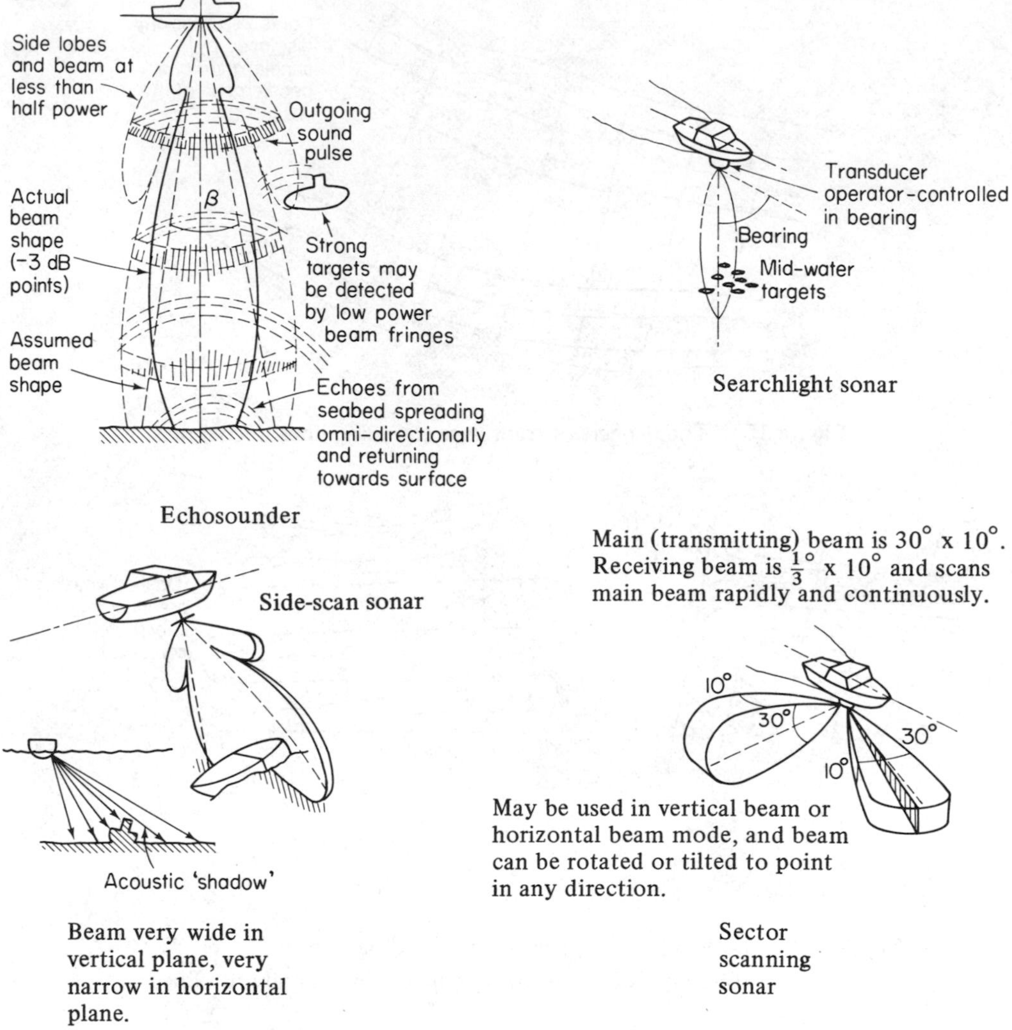
Side lobes
and beam at
less than
half power
Outgoing
sound
pulse
Actual
beam
shape
(−3 dB
points)
β
Strong
targets may
be detected
by low power
beam fringes
Assumed
beam
shape
Echoes from
seabed spreading
omni-directionally
and returning
towards surface
Echosounder
Transducer
operator-controlled
in bearing
Bearing
Mid-water
targets
Searchlight sonar
Main (transmitting) beam is 30° x 10°.
Receiving beam is $\frac{1}{3}$° x 10° and scans
main beam rapidly and continuously.
Side-scan sonar
10°
30°
30°
10°
Acoustic 'shadow'
May be used in vertical beam or
horizontal beam mode, and beam
can be rotated or tilted to point
in any direction.
Beam very wide in
vertical plane, very
narrow in horizontal
plane.
Sector
scanning
sonar

Figure 154. The modes of sonar operation.

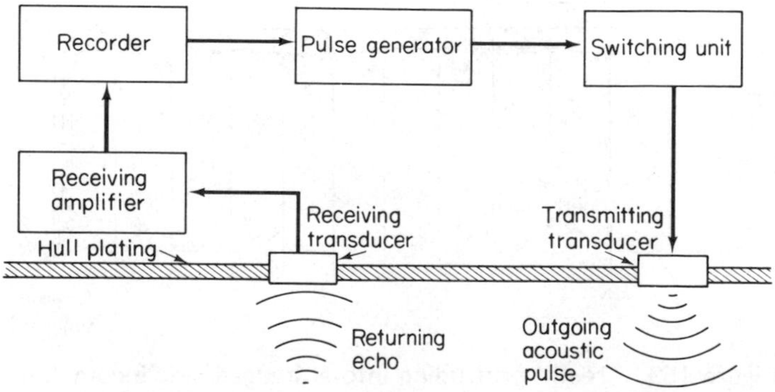
Recorder
Pulse generator
Switching unit
Receiving
amplifier
Receiving
transducer
Transmitting
transducer
Hull plating
Returning
echo
Outgoing
acoustic
pulse

Figure 155. The basic components of a sonar system.

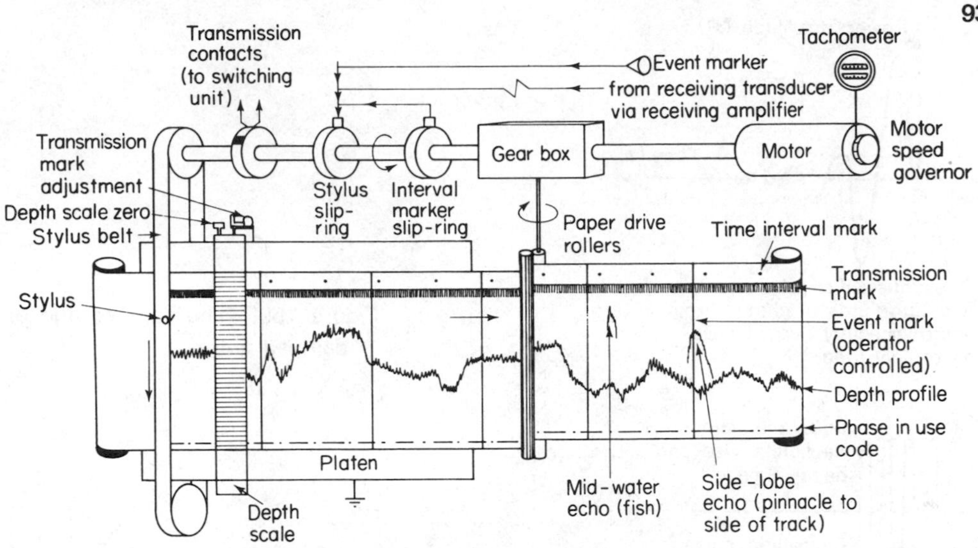

Figure 156. An exploded view of a simple graphic recorder.

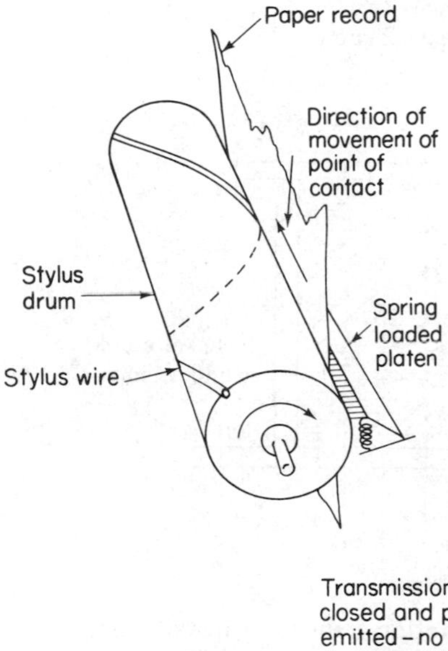

Figure 157. The helix stylus arrangement.

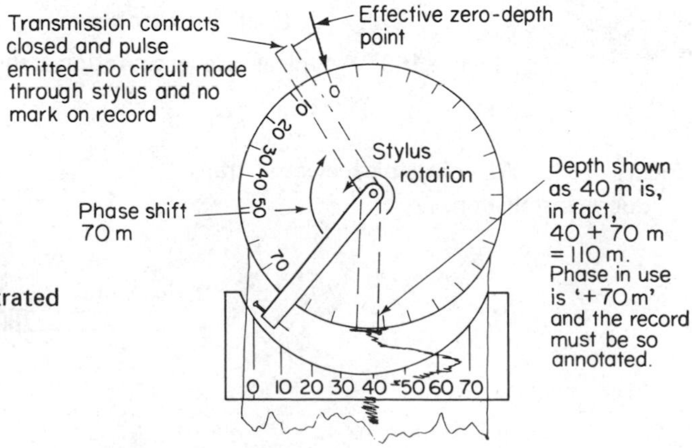

Figure 158. Phasing as demonstrated in the radial-arm recorder.

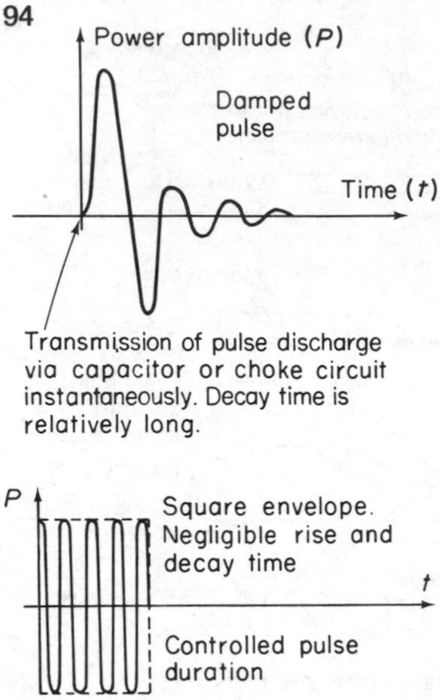

Figure 159. The damped and undamped pulses.

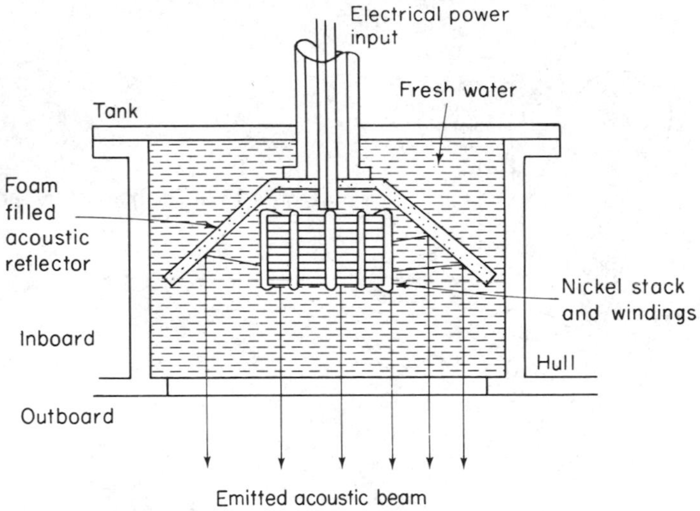

Figure 160. A typical magnetostriction transducer installation.

Figure 161. A typical piezoelectric transducer installation.

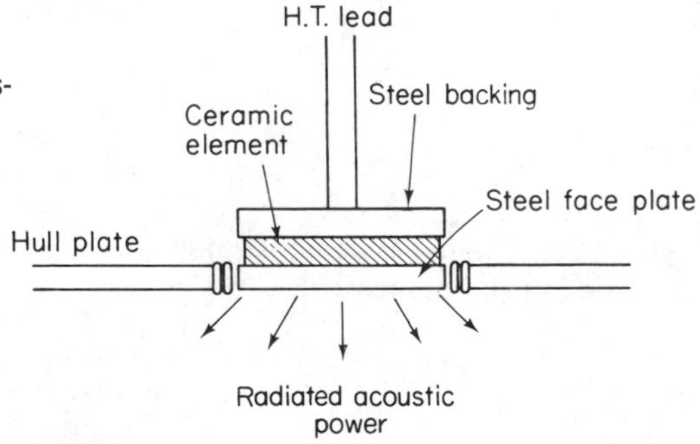

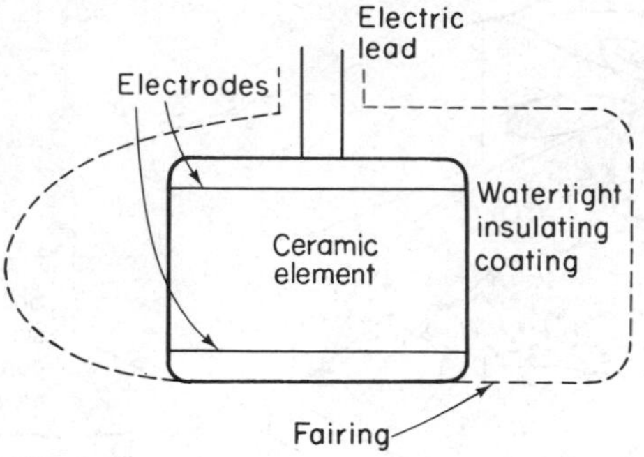

Figure 162. A typical electrostrictive transducer installation.

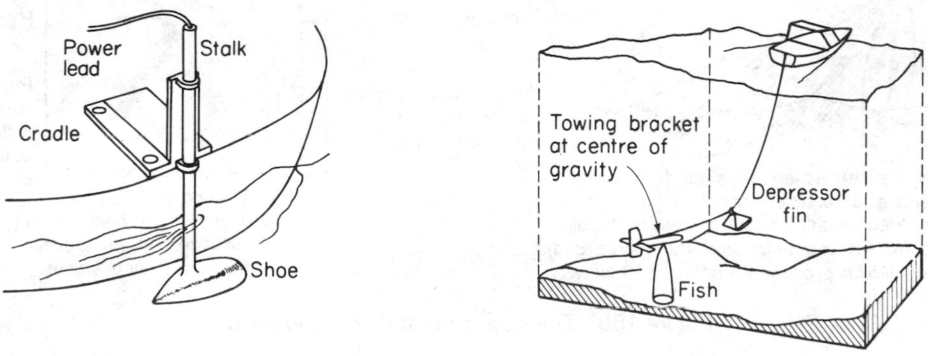

Figure 163. The shoe and fish installations.

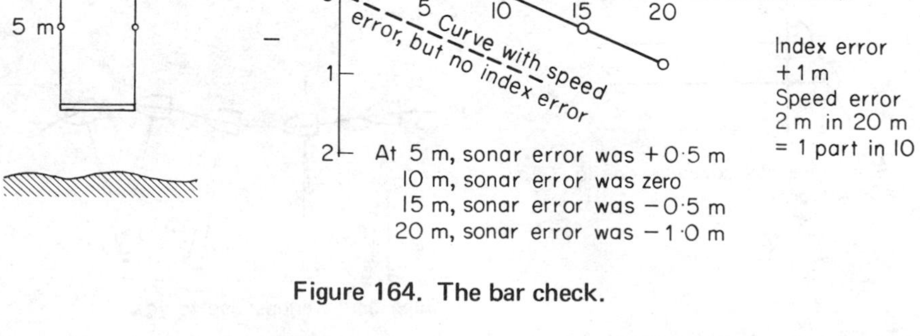

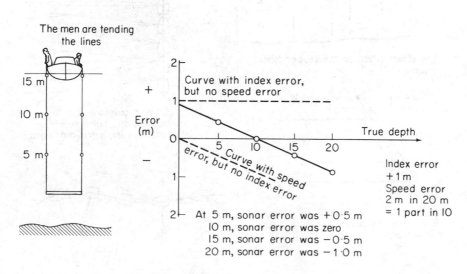

Figure 164. The bar check.

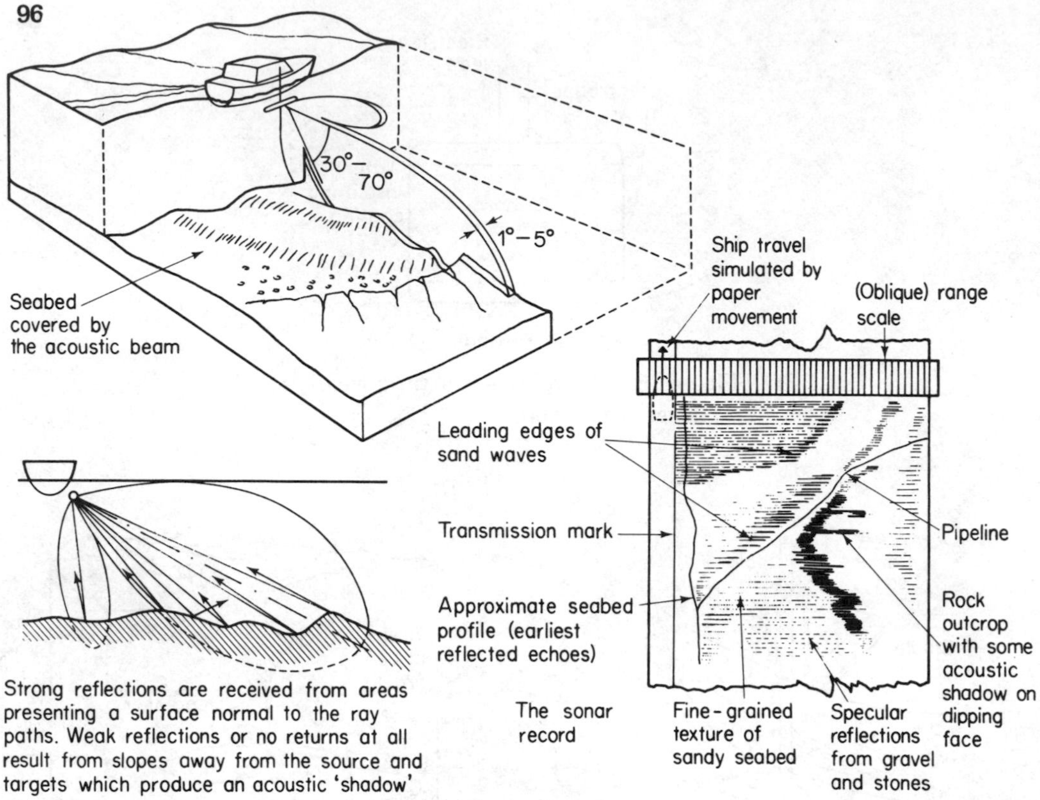

Figure 165. The side-scan mode of operation.

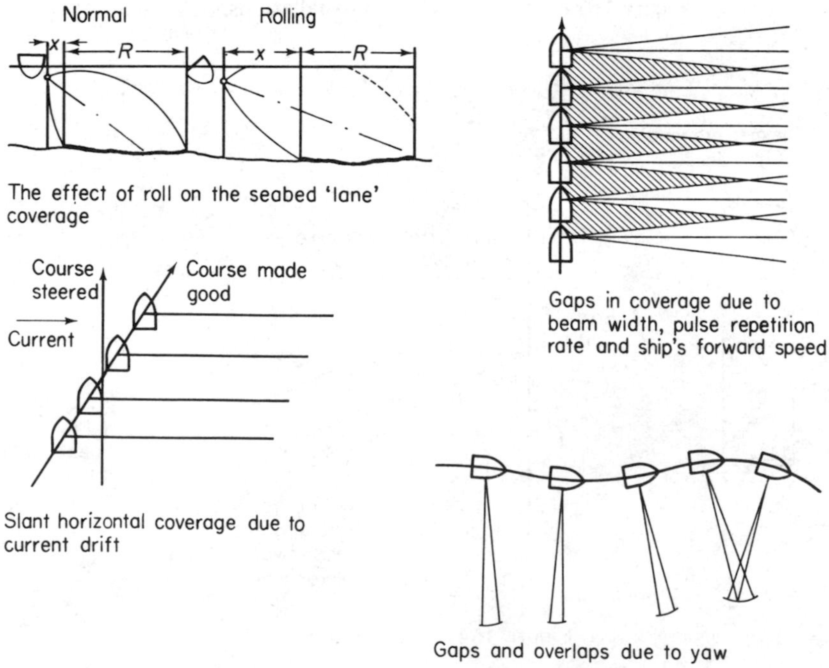

Figure 166. The factors affecting side-scan coverage and resolution.

Assumed geometry for calculation
of heights of obstructions

Exaggerated roll conditions which are
not allowed for in the record

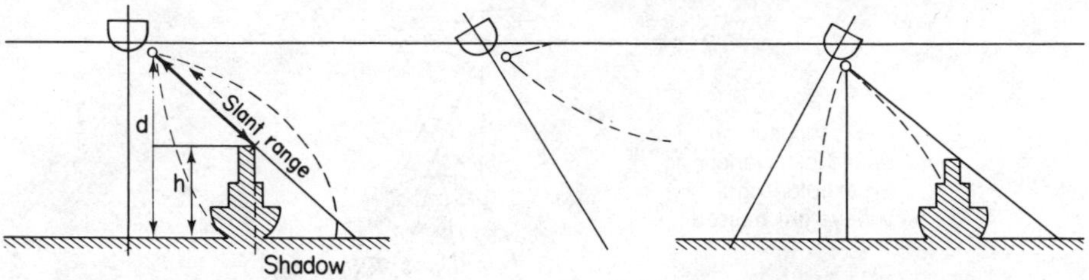

Figure 167. Quantitative interpretation and possibility of error.

$$R = h \tan \beta^\circ$$
$$(R + \Delta R) = (h + \Delta h) \tan \beta^\circ$$
$$\Delta R = (h + \Delta h) \tan \beta^\circ - R$$

Effective vertical beam width (β°) Transducer Pulse length

Altitude (h)

Extra altitude (Δh) Poor resolution Range (R) Good resolution ($\doteq$ Pulse length)

Extra range (ΔR)

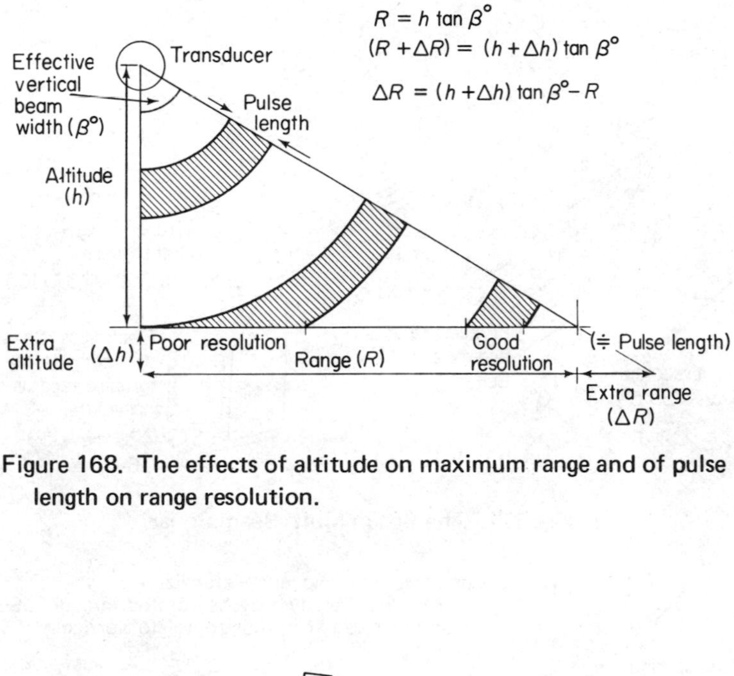

Figure 168. The effects of altitude on maximum range and of pulse length on range resolution.

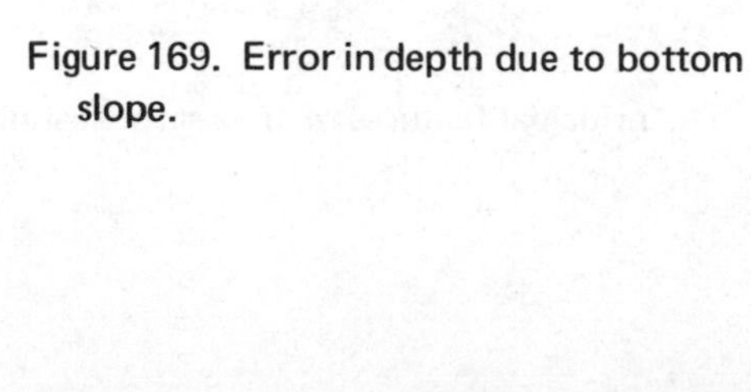

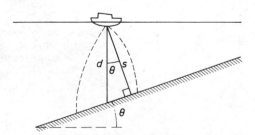

Figure 169. Error in depth due to bottom slope.

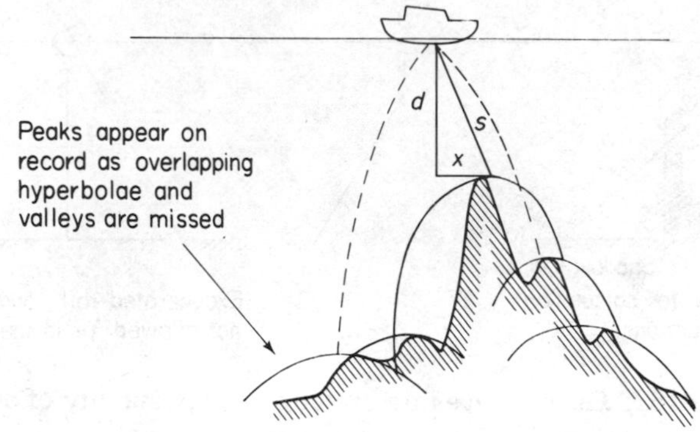

Figure 170. The hyperbolic echo problem: $s^2 = d^2 + x^2$, or $\frac{s^2}{d^2} - \frac{x^2}{d^2} = 1$, the latter being a classic formula of the hyperbola.

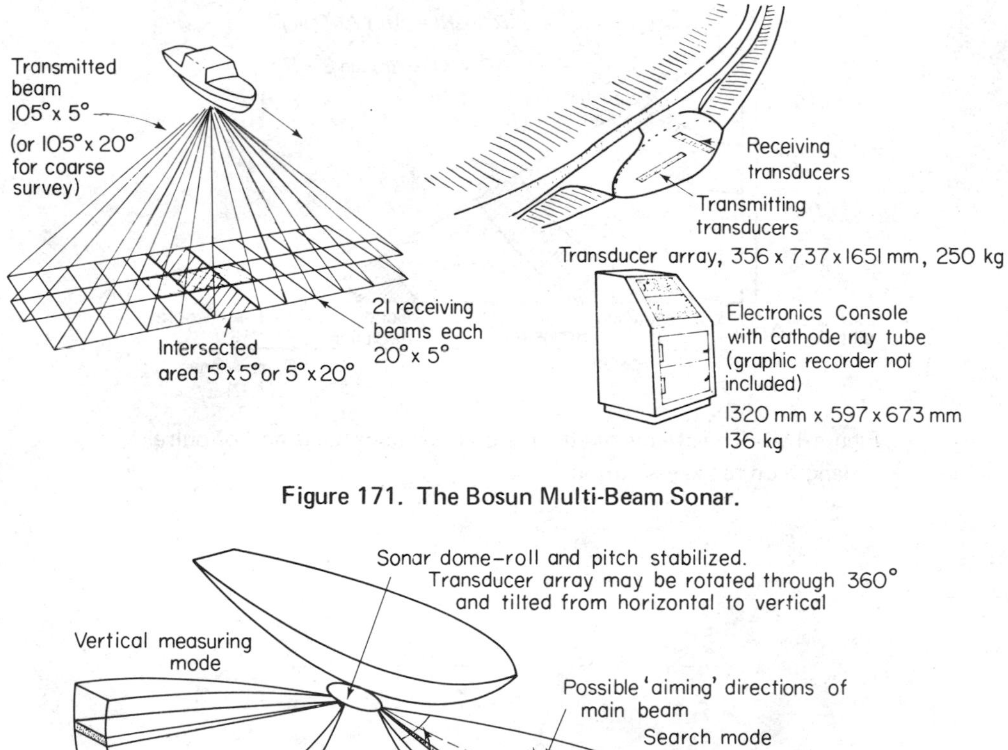

Figure 171. The Bosun Multi-Beam Sonar.

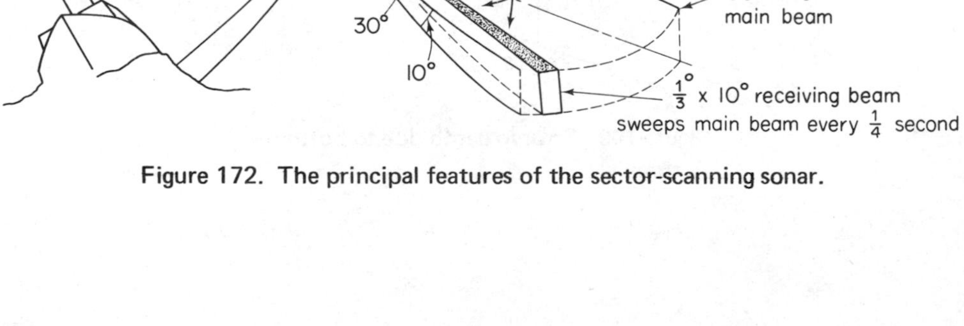

Figure 172. The principal features of the sector-scanning sonar.

SURVEYOR
(Electronic Laboratories Ltd.)

Recorder:
250 X 180 X 100 mm, 4.5 kg.

Transducers (2):
38 X 22 mm (fairing not supplied)

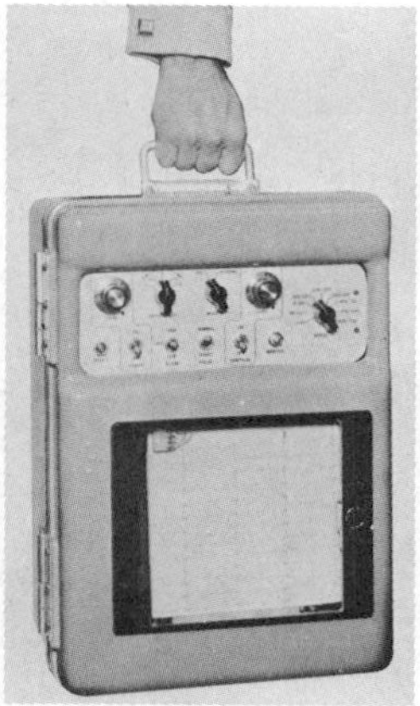

EDO 4034
(Messrs. Edo Western Corpn.)

Recorder:
483 X 356 X 254 mm, 23.4 kg

Transducer:
222 X 117 mm, 15.7 kg

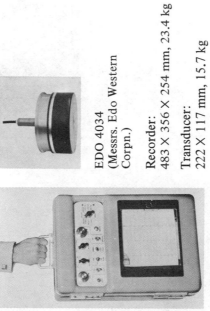

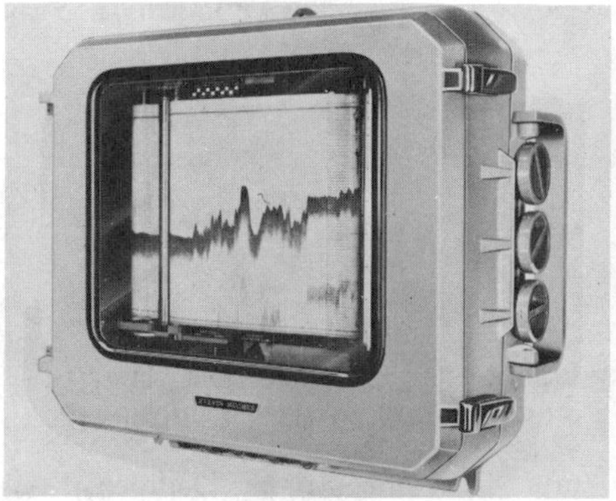

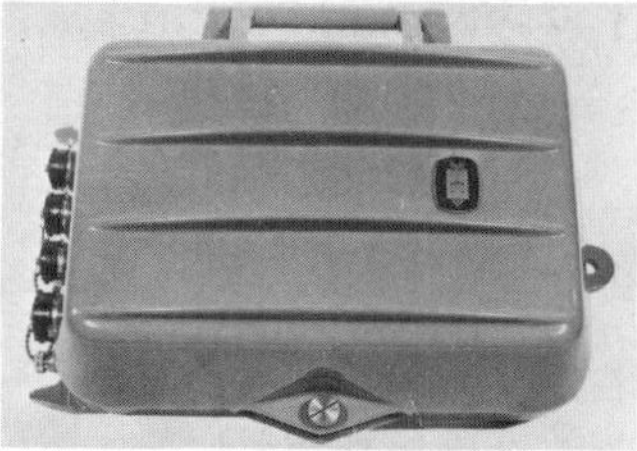

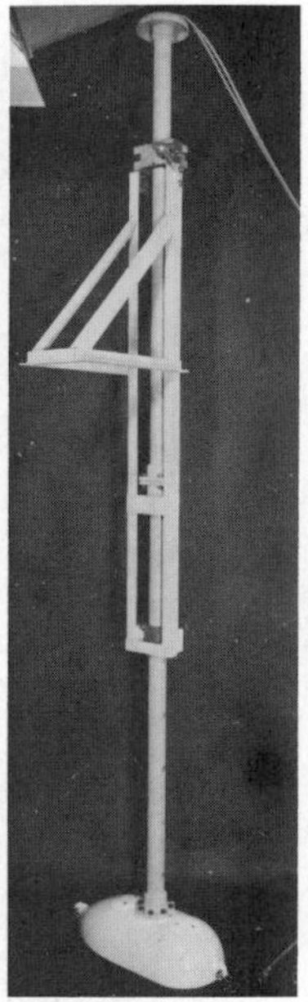

MS 36 M
(Kelvin Hughes Divn. of Smiths Industries Ltd.)

Recorder:
59 X 546 X 176 mm, 35.4 kg

Power unit:
392 X 303 X 150 mm, 8.8 kg

Transducers (2) in fairing for overside rig:
514 X 121 X 241 mm, 32.2 kg

Figure 173. Examples of echosounders (continued on page 100).

ATLAS-DESO 10
(Fried. Krupp GmbH)

Recorder:
440 X 440 X 195 mm, 21 kg

Control unit:
440 X 220 X 390 mm, 18 kg
(choice of transducers,
two frequency operation)

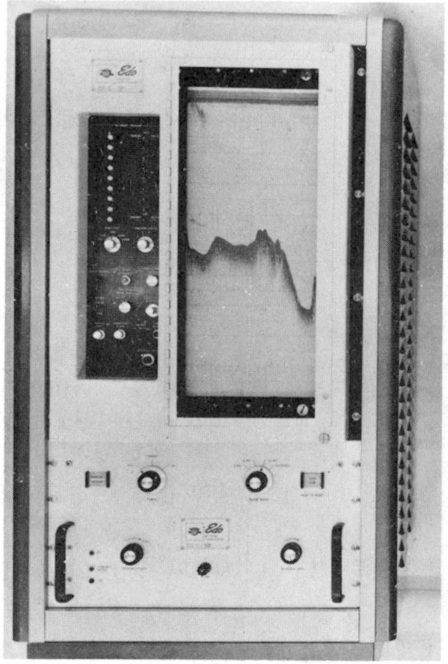

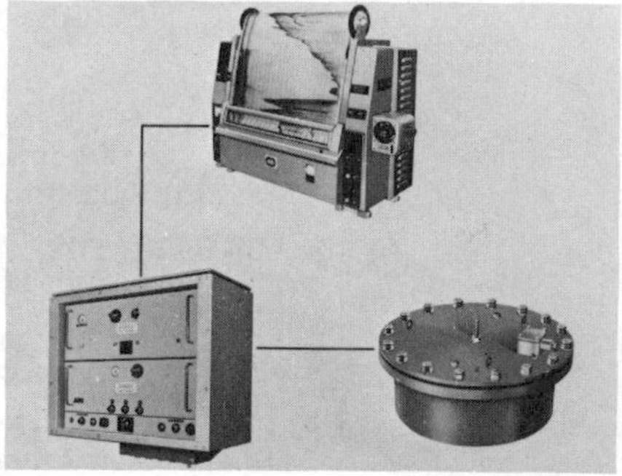

MS 38 (Kelvin Hughes Divn. of Smiths Industries Ltd.)

Recorder: 724 X 584 X 356 mm, 73 kg
Power unit: 698 X 737 X 356 mm, 70 kg
Transducer (in housing) 406 X 724 X 356 mm, 177 kg
(without housing, 26 kg)

EDO 349 PBS (Messrs. Edo Western Corp.)

Recorder: 670 X 500 X 327 mm, 84.4 kg
Transducer: 220 X 260 mm, 68 kg
Transceiver: 483 X 222 X 380 mm 29.5 kg

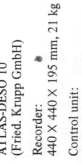

Figure 173 (continued). Examples of echosounders.

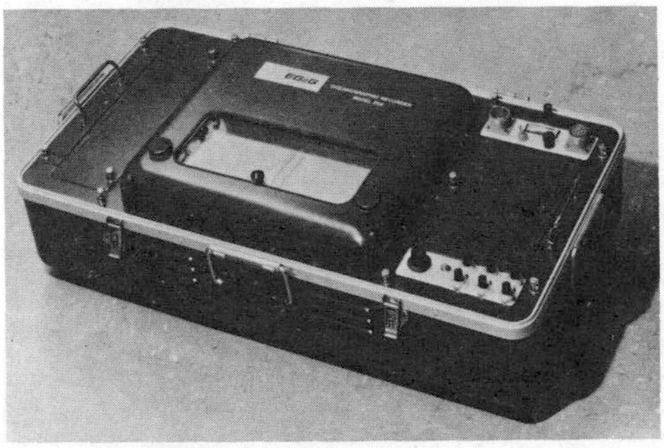

EG and G Mk 1A
(EG and G Geophysical Ltd.)
Recorder: 838 X 444 X 280 mm, 39 kg
Transducer (in towed body) 1270 X 102 mm, 17.7 kg

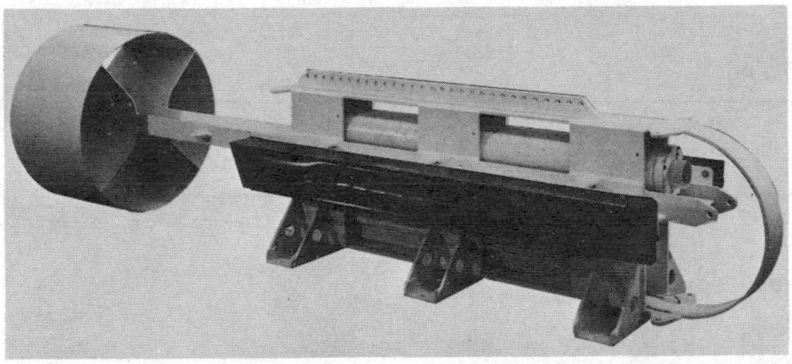

WESTINGHOUSE
(Westinghouse Underseas Divn.)
(Towed transducer body only, 1730 mm)

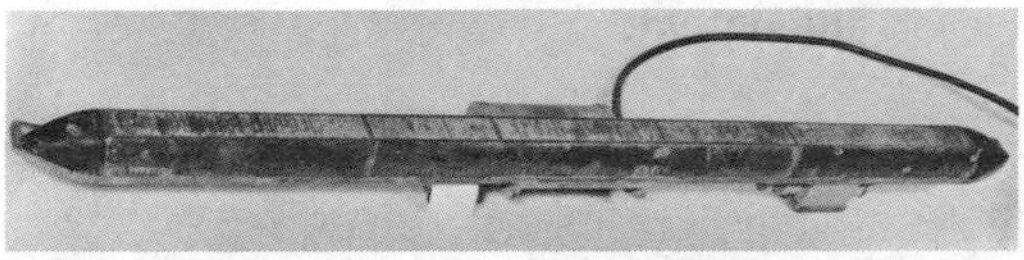

MS 47
(Kelvin Hughes Divn. of Smiths Industries Ltd.)
Recorder: 409 X 305 X 178 mm, 18.1 kg

Transducer:
1117 X 100
X 152 mm
24 kg

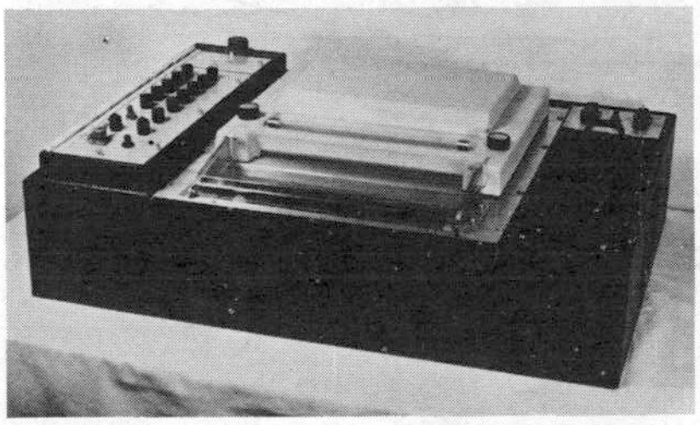

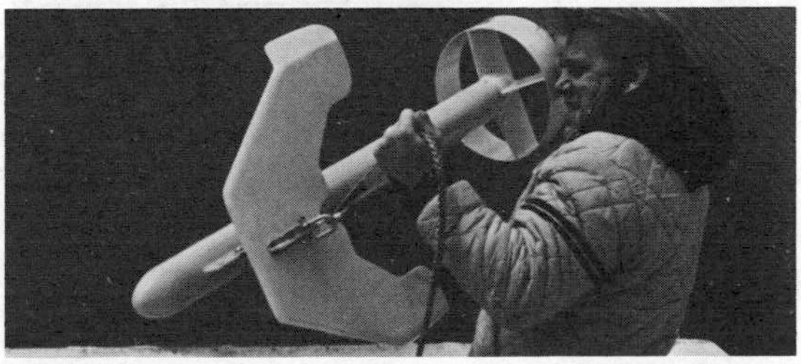

HYDROSCAN 400 (Klein Associates Inc.)
Recorder: 844 X 597 X 254 mm, 45.4 kg
Transducer: 1219 X 89 mm (305 mm over tail), 14 kg

Figure 174. Examples of side-scan sonar systems.

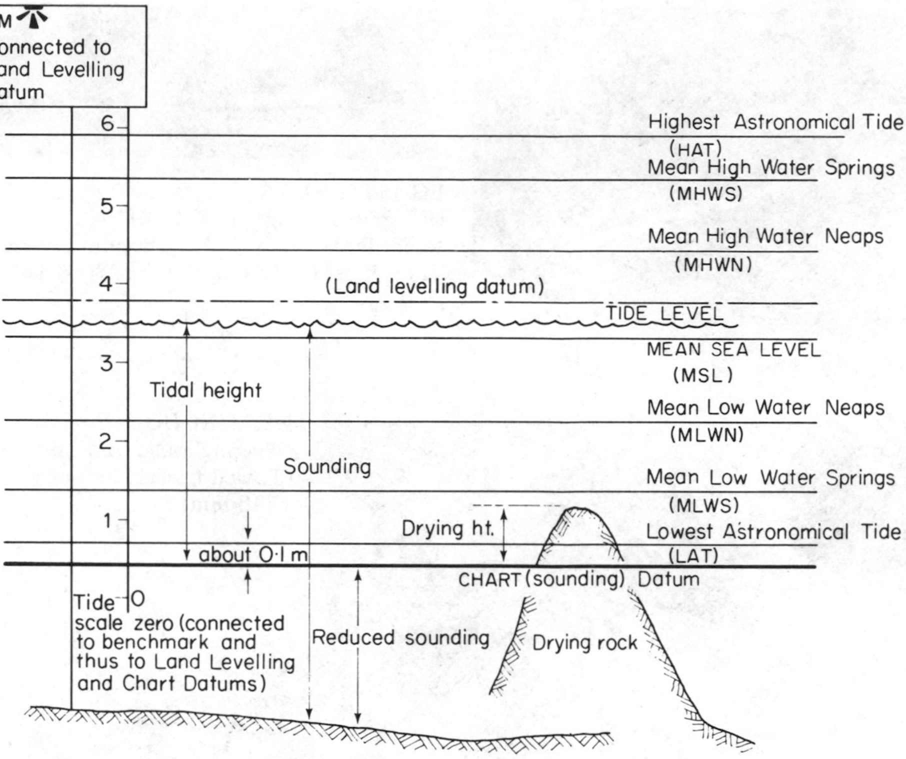

Figure 175. Tidal levels and datum.

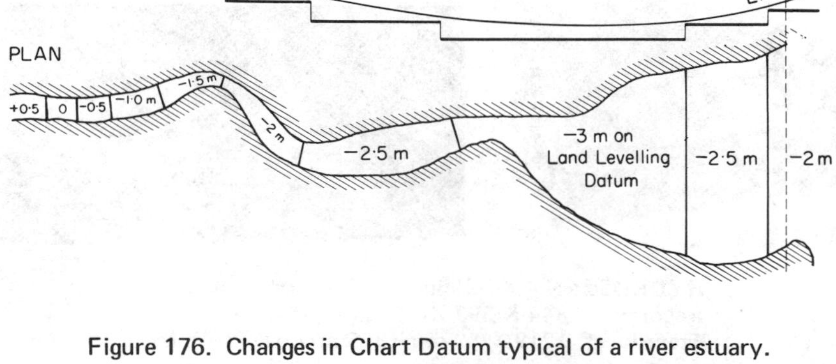

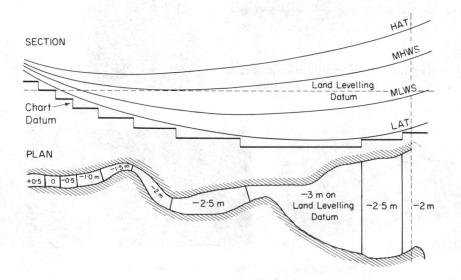

Figure 176. Changes in Chart Datum typical of a river estuary.

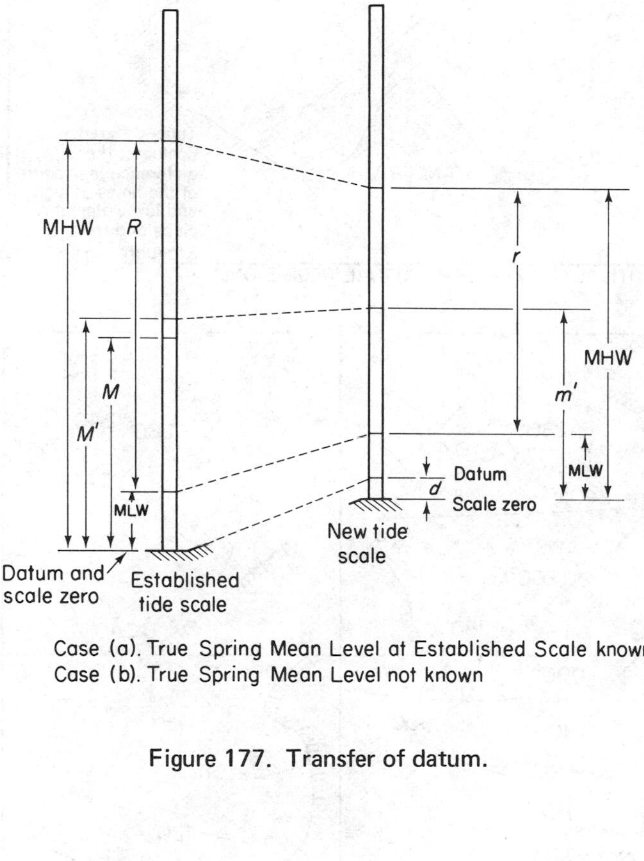

Case (a). True Spring Mean Level at Established Scale known
Case (b). True Spring Mean Level not known

Figure 177. Transfer of datum.

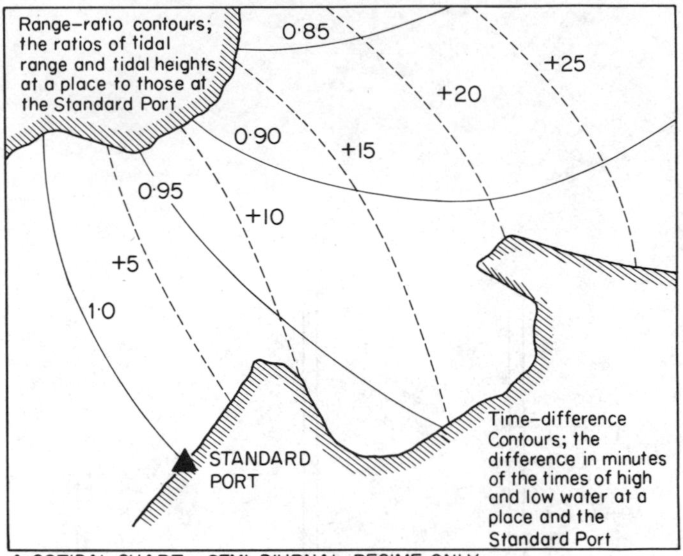

A COTIDAL CHART – SEMI-DIURNAL REGIME ONLY

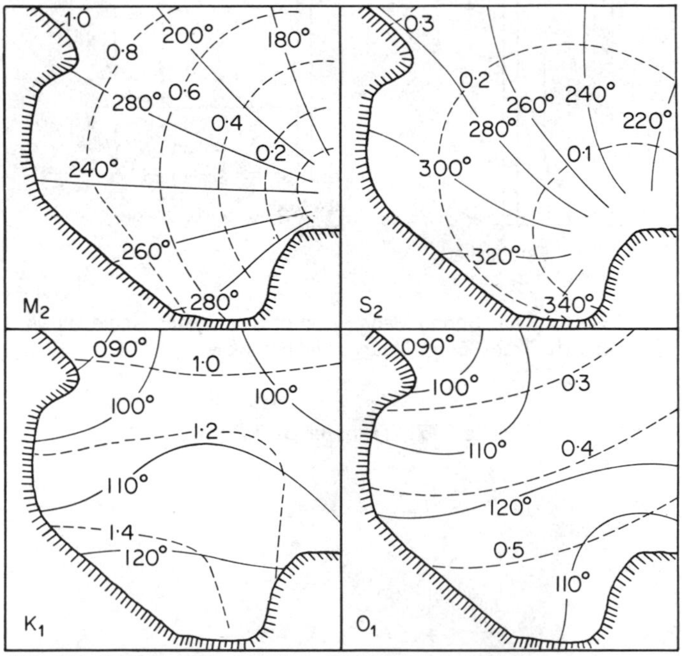

COTIDAL CHART – SEMI-DIURNAL OR DIURNAL REGIMES

Figure 178. Examples of cotidal charts (after *Admiralty Tidal Handbook No. 2,* 'Datums for Hydrographic Surveys').

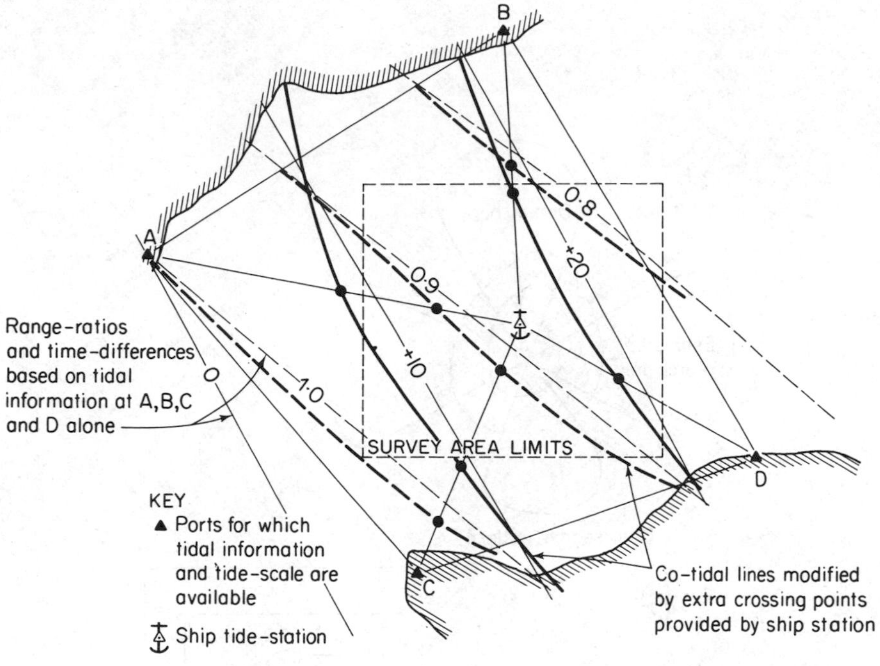

Figure 179. Using shore and ship tidal stations to construct a cotidal chart for the area to be surveyed.

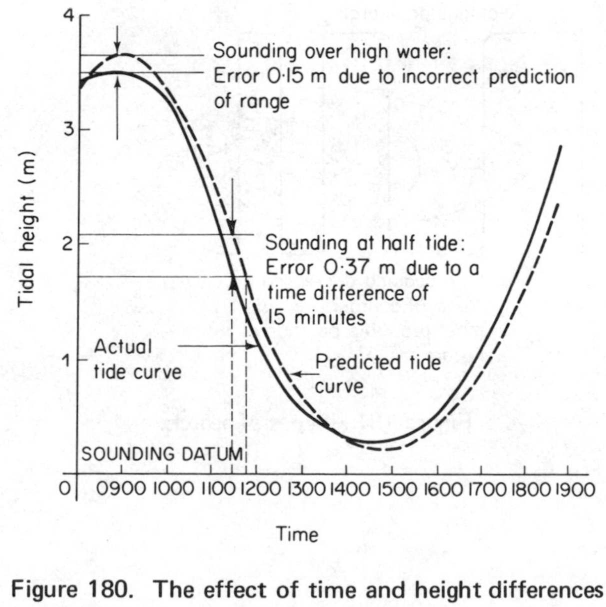

Figure 180. The effect of time and height differences on the tidal reductions.

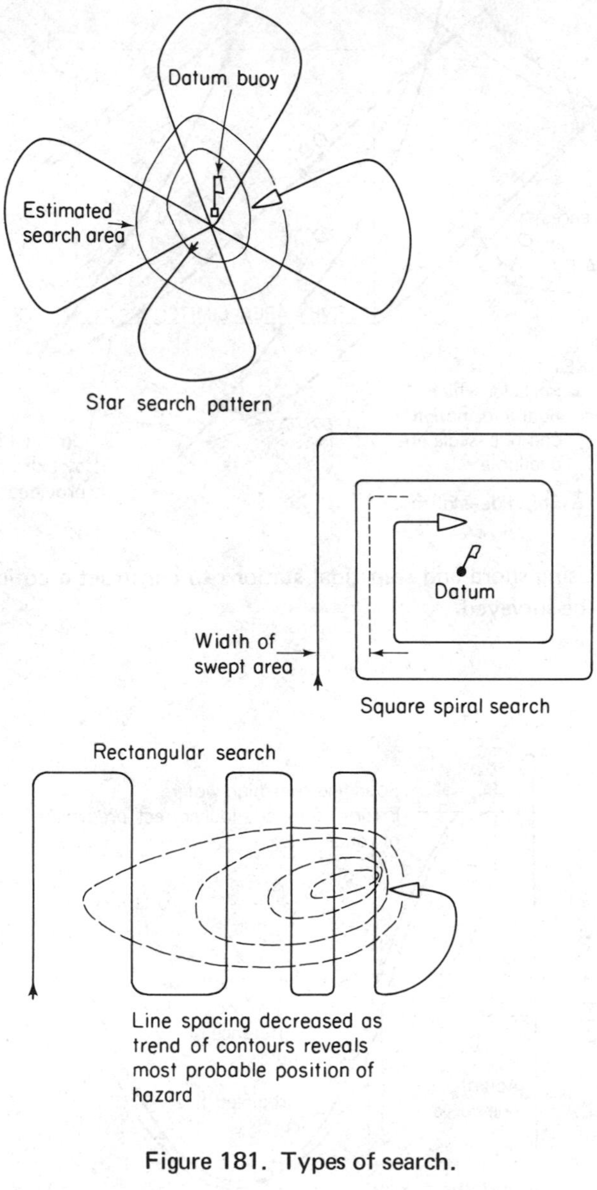

Figure 181. Types of search.

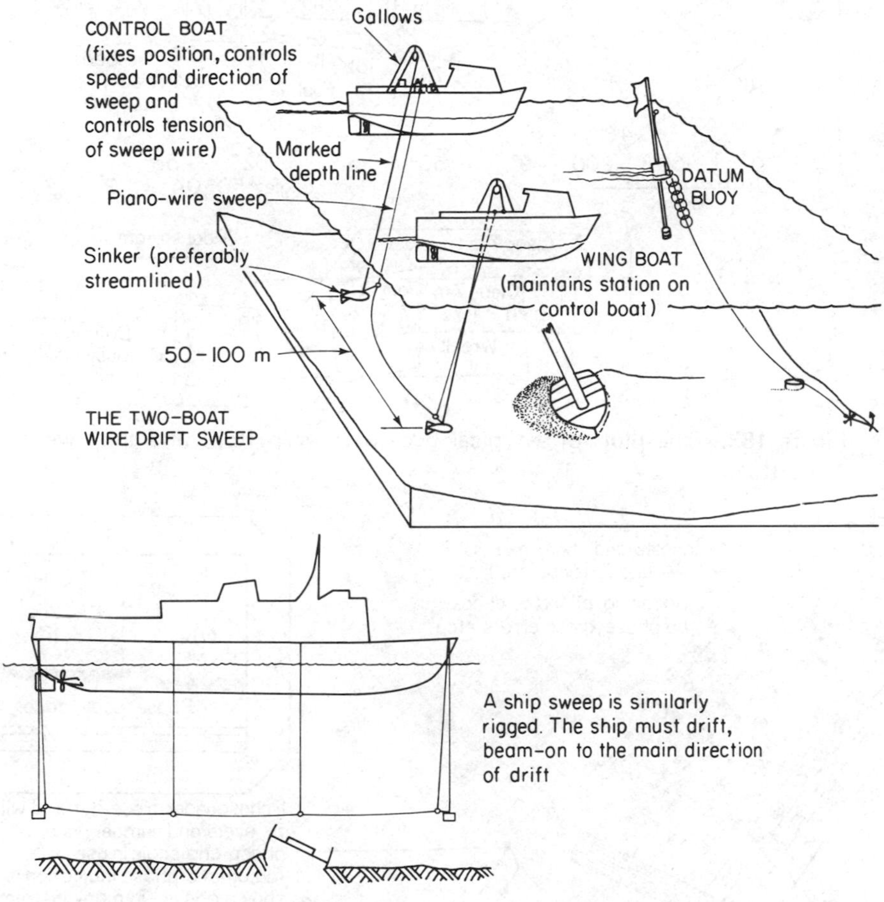
CONTROL BOAT
(fixes position, controls speed and direction of sweep and controls tension of sweep wire)
Gallows
Marked depth line
Piano-wire sweep
Sinker (preferably streamlined)
DATUM BUOY
WING BOAT
(maintains station on control boat)
50-100 m
THE TWO-BOAT WIRE DRIFT SWEEP
A ship sweep is similarly rigged. The ship must drift, beam-on to the main direction of drift

Figure 182. The wire drift sweep.

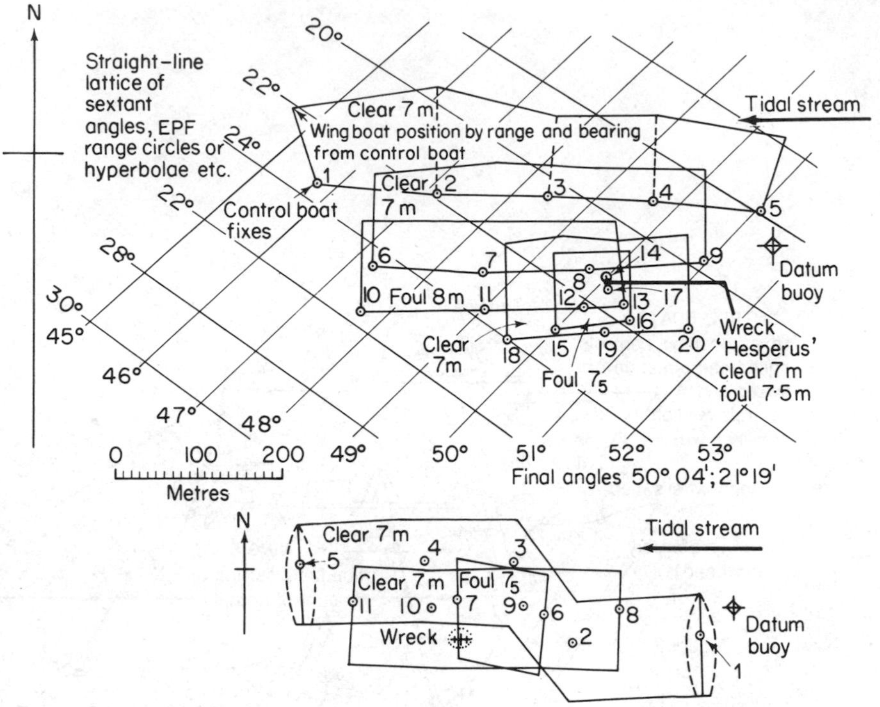

Figure 183. The plots of a typical two-boat sweep (top) and ship sweep (bottom).

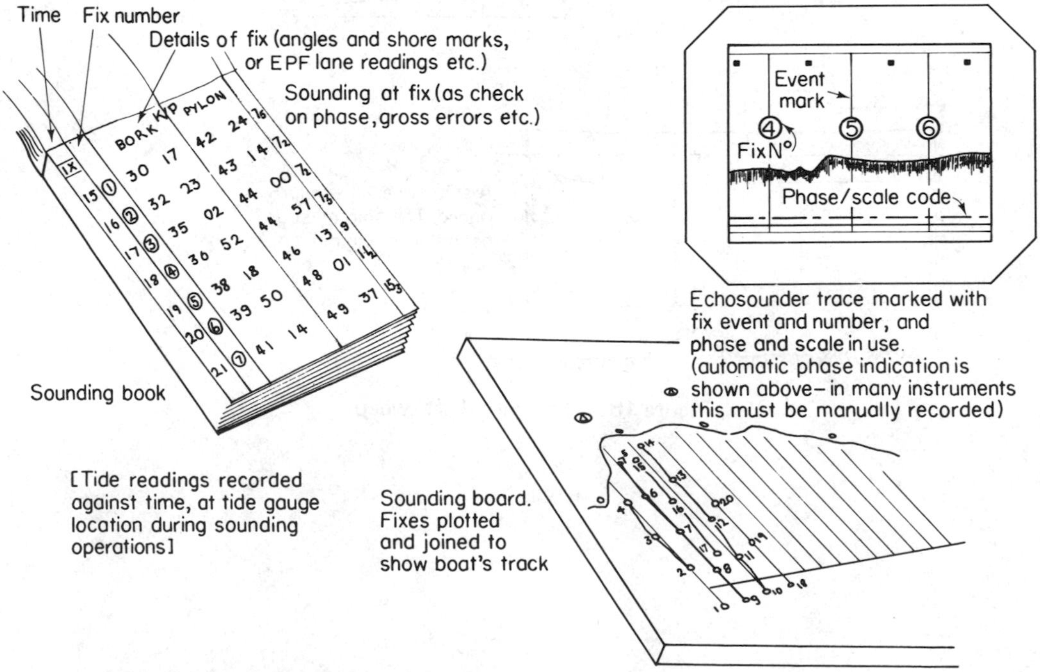

Figure 184. The field records kept during a survey using double sextant angle fixing.

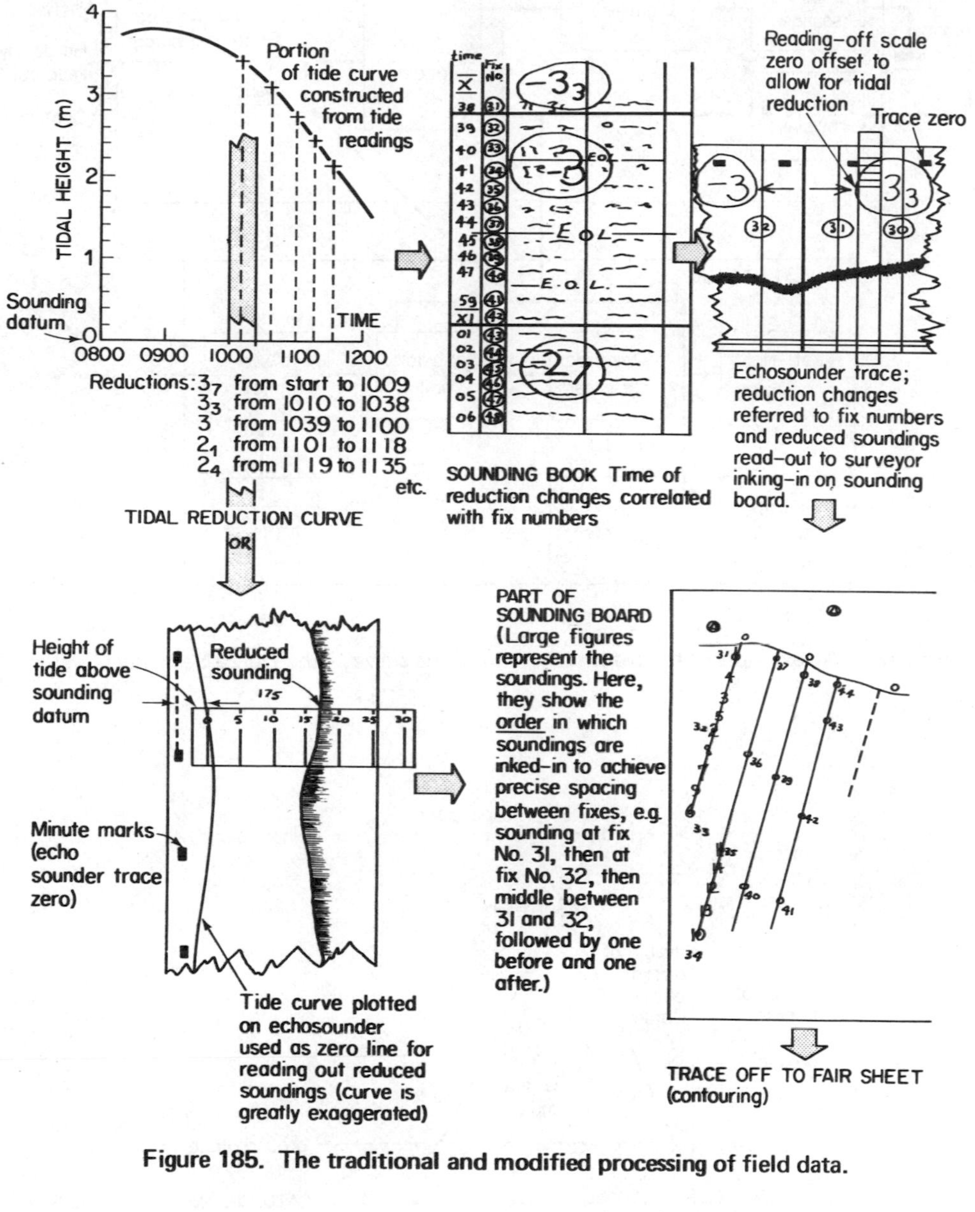

Figure 185. The traditional and modified processing of field data.

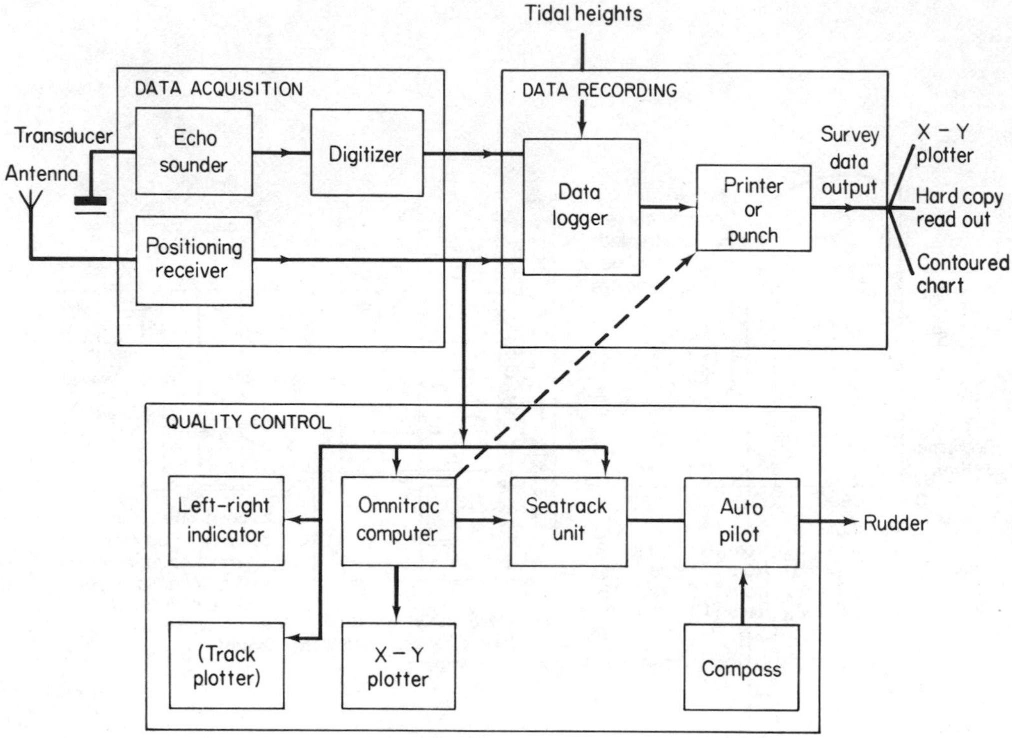

Figure 186. An example of the automation of the sea survey—the Decca Sea Chart System.

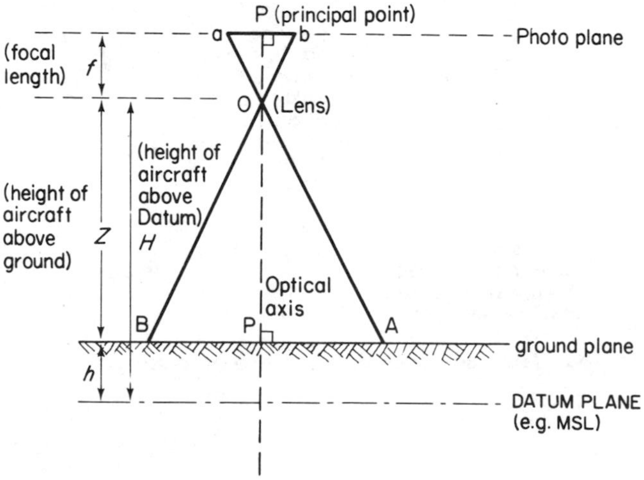

Figure 187. The geometry of the vertical air photograph.

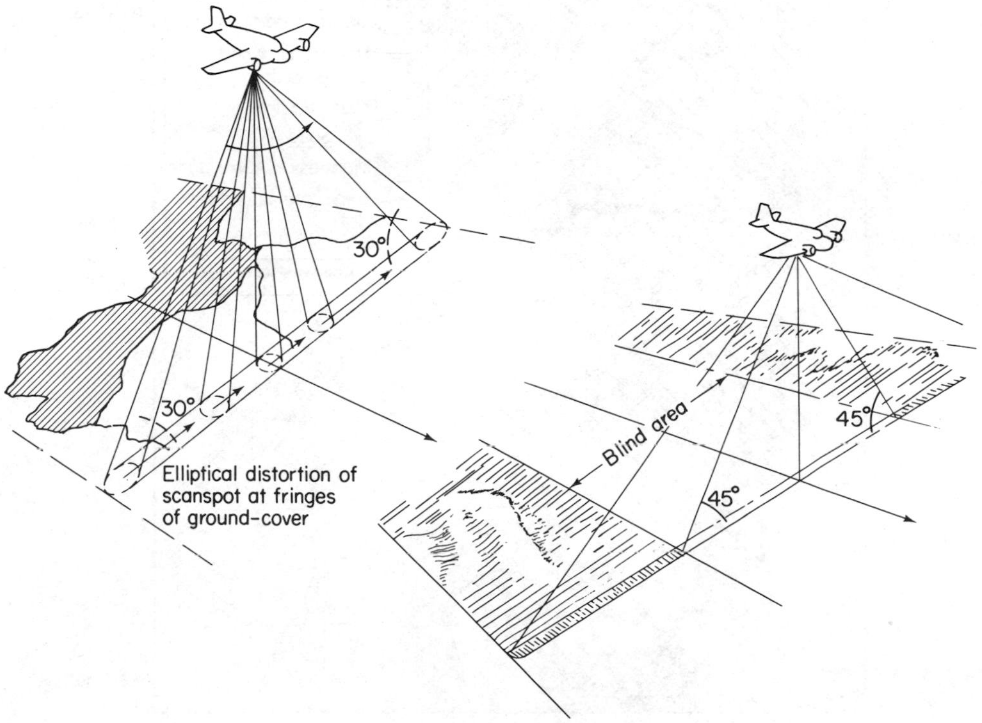

Figure 188. (a) Infrared linescan imagery.
(b) Sideways-looking airborne radar.

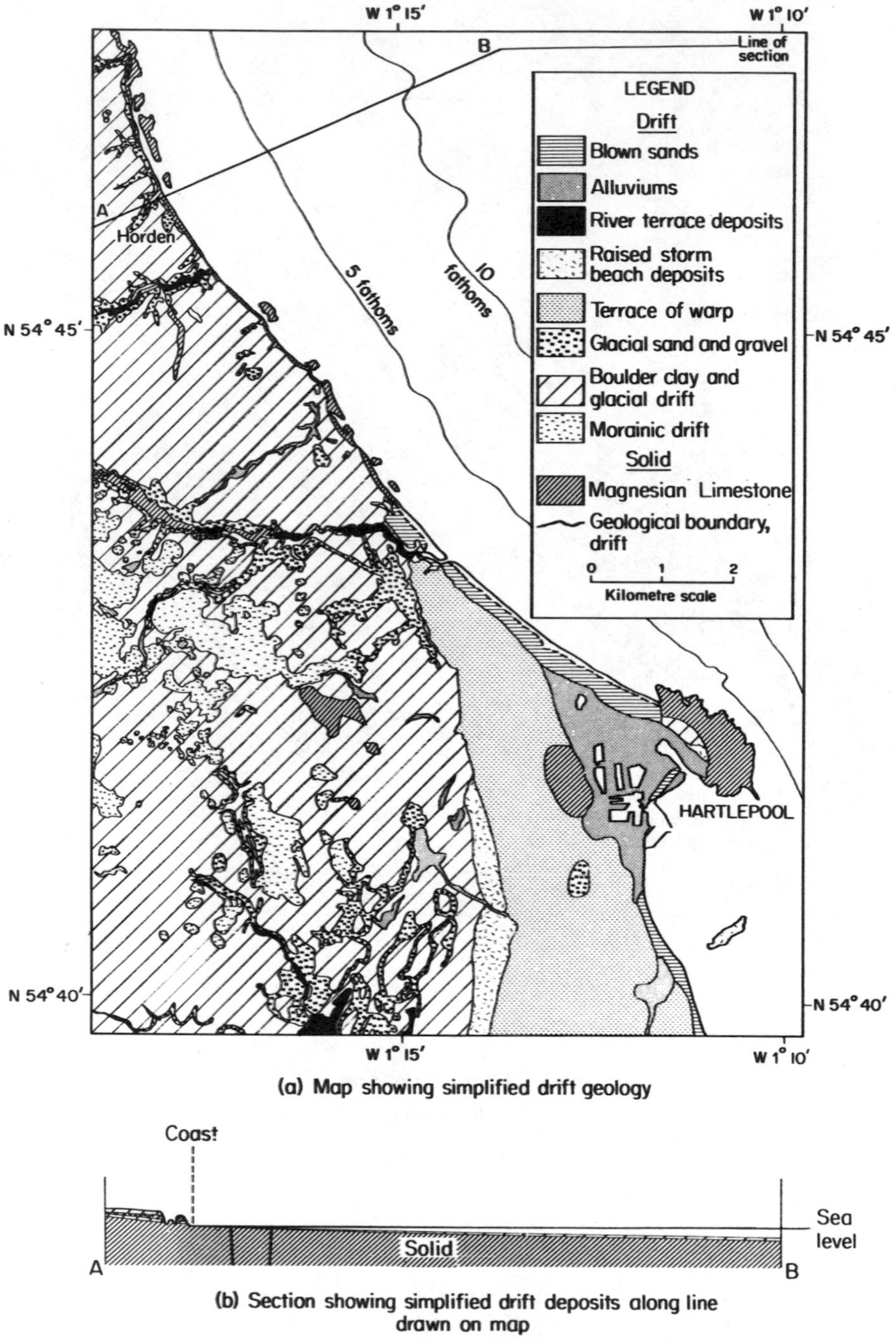

Figure 189. Map and section showing simplified drift geology. Based on Geological Survey Durham (27 drift) Sheet, reproduced by permission of The Director, Institute of Geological Sciences.

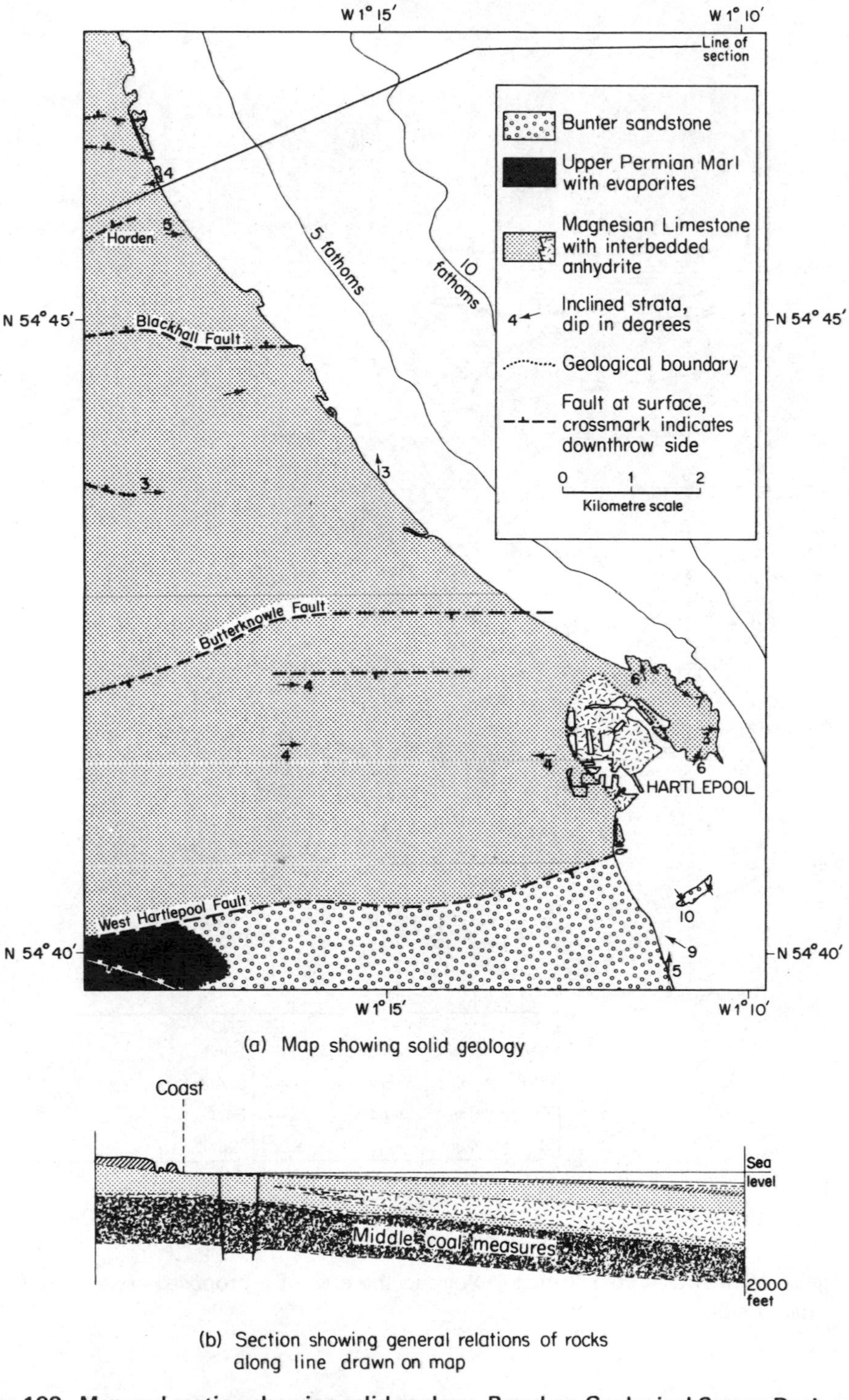

(a) Map showing solid geology

(b) Section showing general relations of rocks along line drawn on map

Figure 190. Map and section showing solid geology. Based on Geological Survey Durham (27 solid) Sheet, reproduced by permission of The Director, Institute of Geological Sciences.

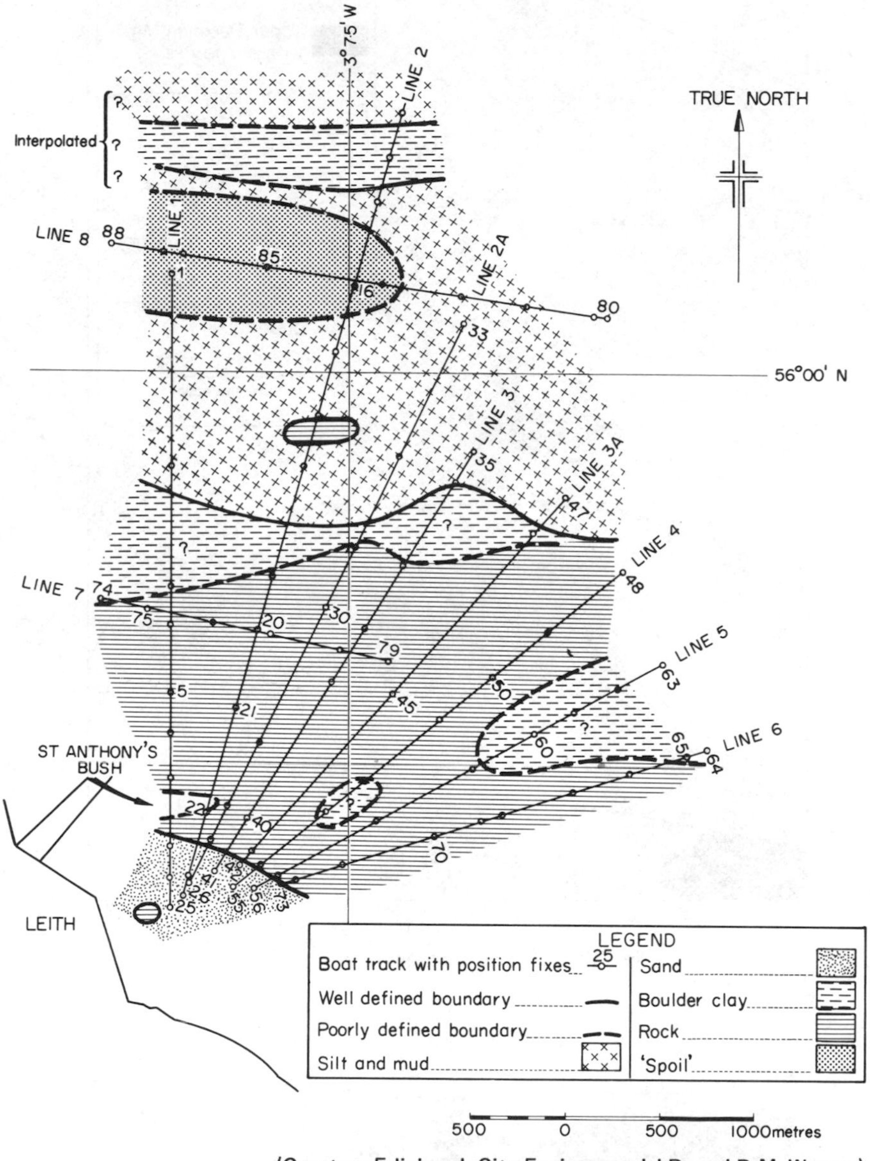

(Courtesy Edinburgh City Engineer and J.D. and D.M. Watson)

Figure 191. Interpreted seabed geology in the area of a proposed sewer outfall near Leith.

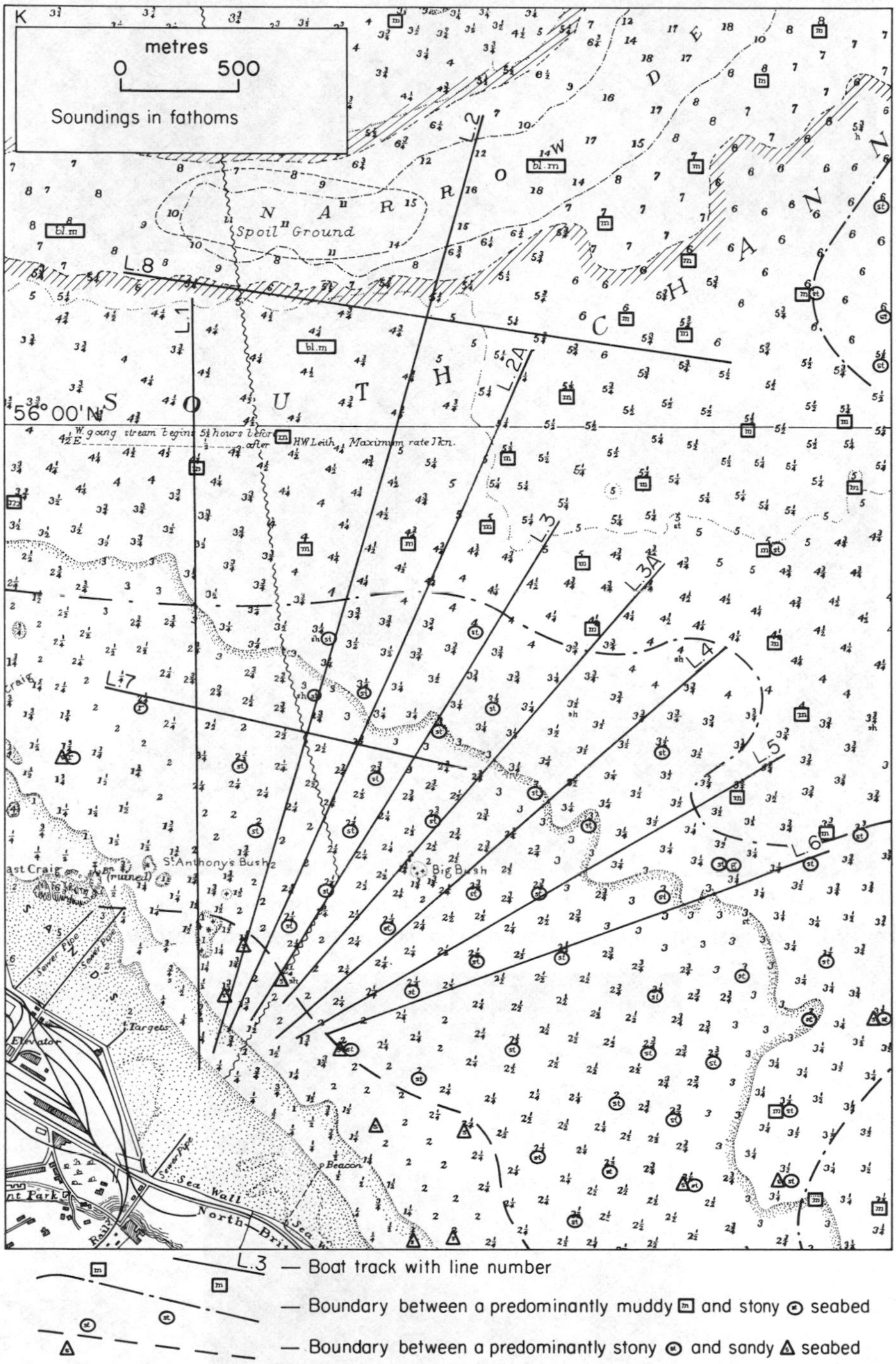

Figure 192. Seabed sampling information on an Admiralty chart. (Reproduced from British Admiralty Chart No 114B with the sanction of the Controller, H.M. Stationery Office and of the Hydrographer of the Navy.)

(Courtesy Hunting Surveys and Consultants Ltd.)

Figure 193. Vertical photograph showing changes in the seabed geology off Abu Dhabi.

(Courtesy La-Coste and Romberg, Inc.)

Figure 194. La-Coste and Romberg seabed gravimeter.

(Courtesy Siemens A.G.)

Figure 195. Graf type Askania ship-borne gravimeter mounted on an Anschutz gyro-stabilized platform.

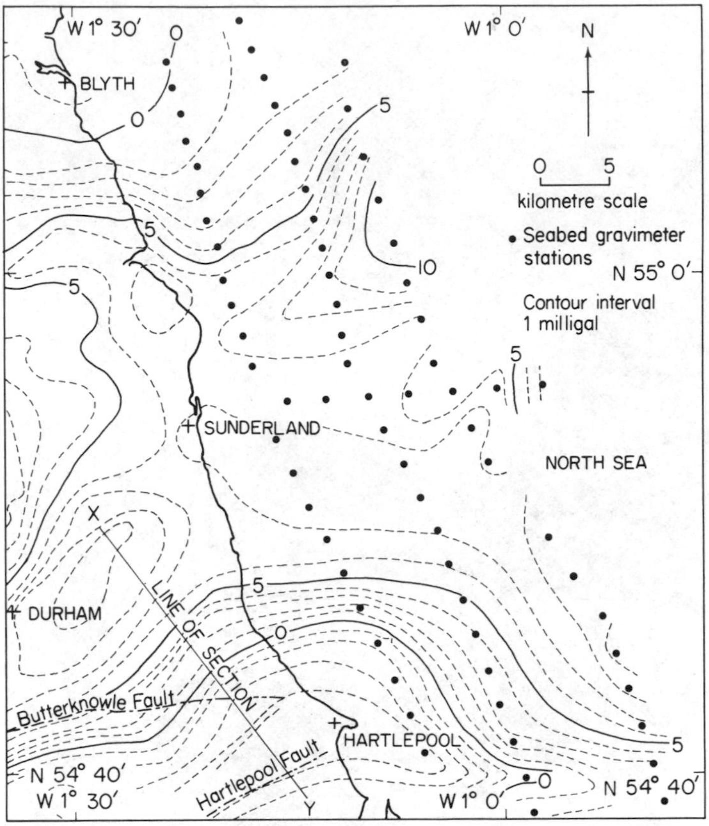

(a) Bouguer Anomalies Map of the Durham Coastal Region (see figure 190 for solid geology of area)

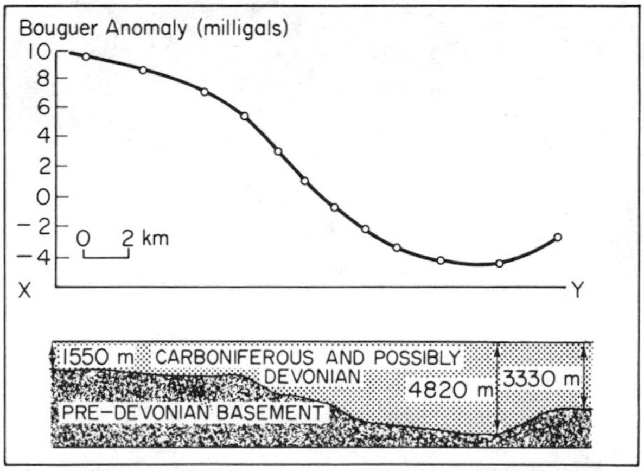

(b) Gravity profile along section XY and a possible geological interpretation based on an assumed uniform density contrast of − 0·15 g/cm^3

(Courtesy M.H.P. Bott)

Figure 196. Gravity information based on land and seabed gravimeter values. (Taken from 'A Gravity Survey off the coast of NE England' published in *the Proc. of the Yorkshire Geol. Soc.* Vol. 33, Pt 1, No. 1.)

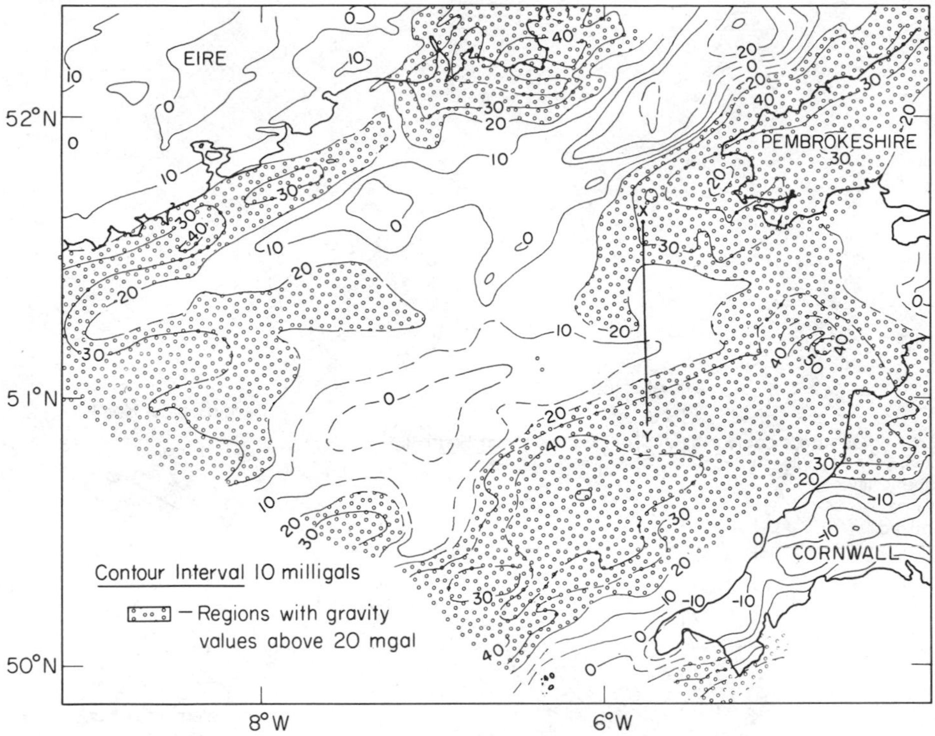

(a) Bouguer Anomaly Map of North Celtic Sea

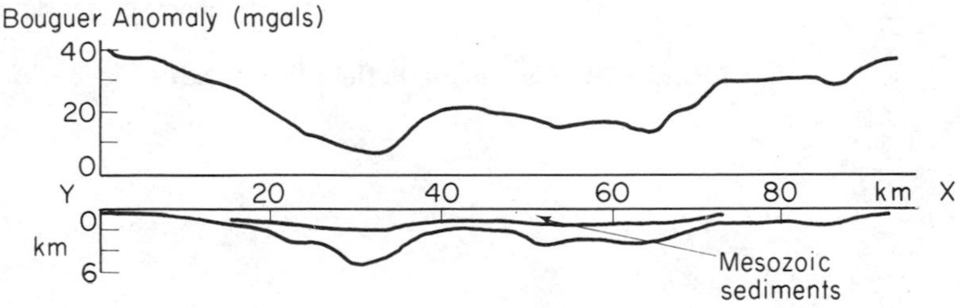

(b) Gravity profile along section XY and a two dimensional model assuming density contrasts of 0·4 and 0·2 g/cm^3

(Courtesy F.J. Davey)

Figure 197. Gravity information based on land and ship-borne gravimeter values. (Taken from 'Bouguer Map of the North Celtic Sea and Entrance to the Bristol Channel,' published in *Geophysics J. Roy. Astro. Soc.* (1970), **22**, 277–282.)

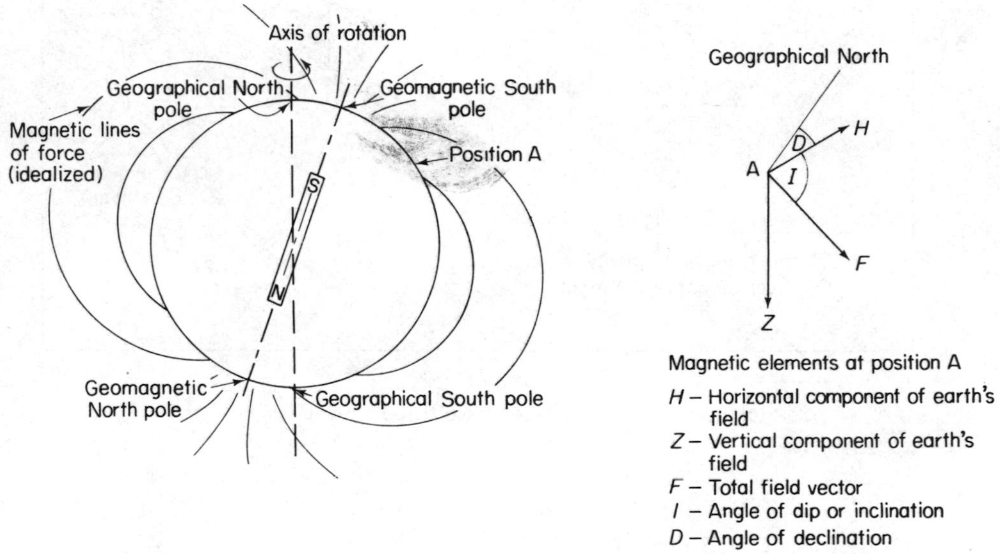
Axis of rotation
Geographical North pole
Geomagnetic South pole
Position A
Magnetic lines of force (idealized)
S
N
Geomagnetic North pole
Geographical South pole
Geographical North
H
D
A
I
F
Z
Magnetic elements at position A
H – Horizontal component of earth's field
Z – Vertical component of earth's field
F – Total field vector
I – Angle of dip or inclination
D – Angle of declination

Figure 198. Geomagnetic field (idealized)

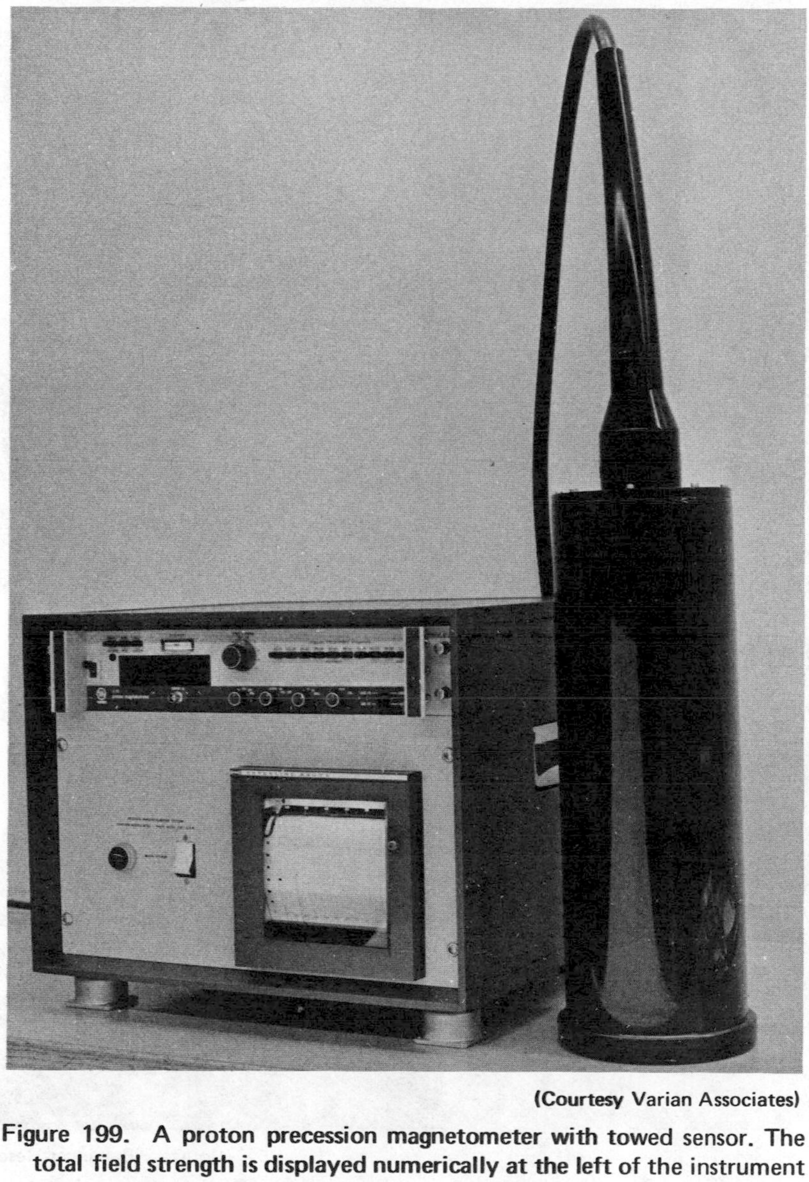

(Courtesy Varian Associates)

Figure 199. A proton precession magnetometer with towed sensor. The total field strength is displayed numerically at the left of the instrument panel and is also recorded on a paper recorder.

(Courtesy Barringer Research)

Figure 200. Launching a proton precession magnetometer towed sensor.

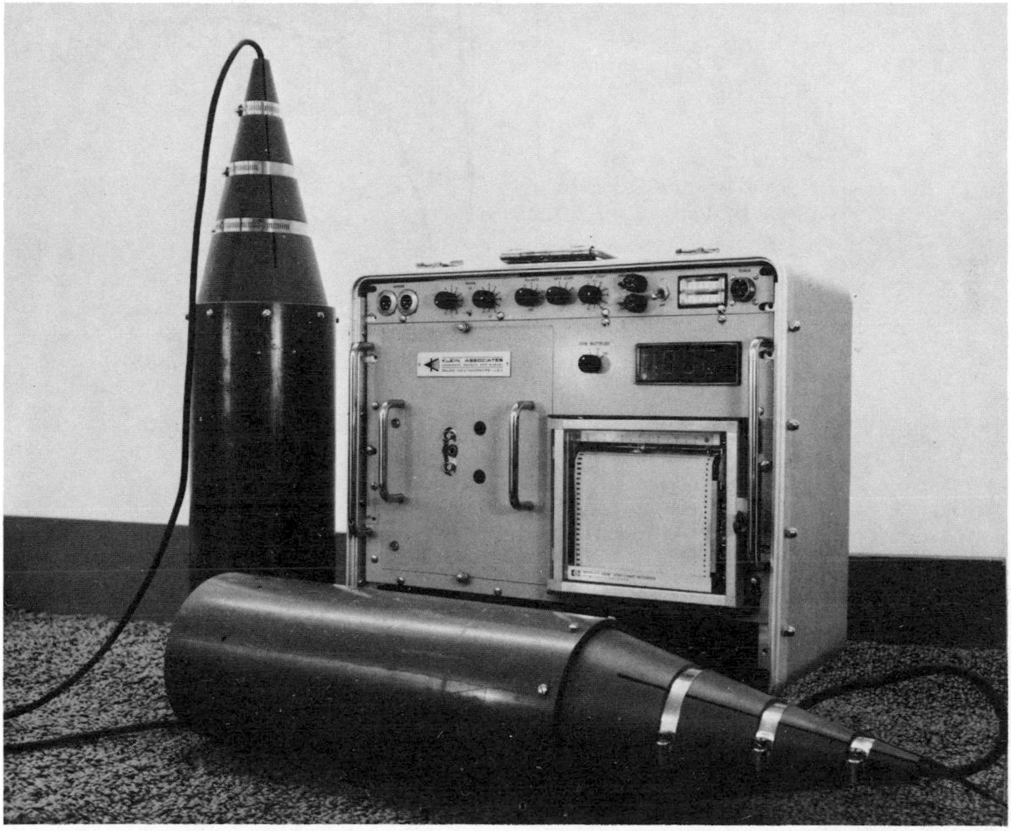

(Courtesy Klein Associates, Inc.)

Figure 201. Differential search proton precession magnetometer and sensors.

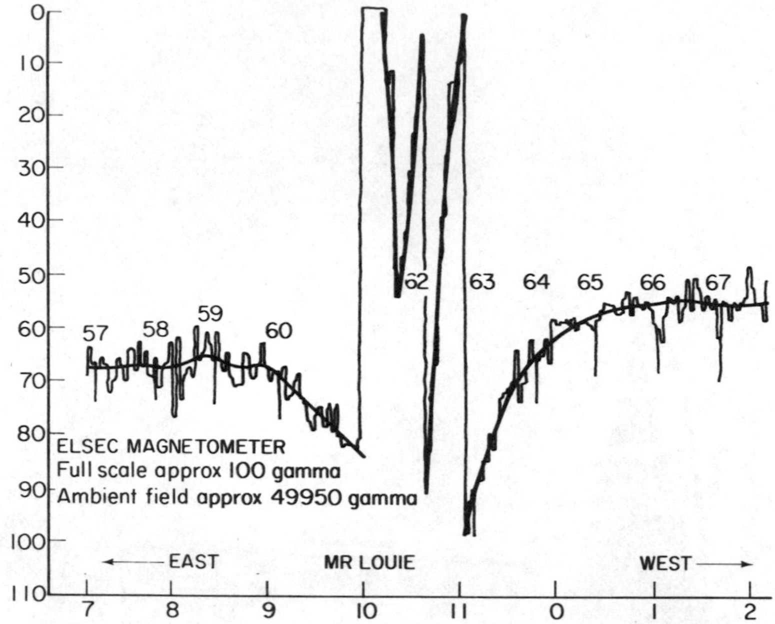

(Courtesy Hunting Surveys and Consultants Ltd.)

Figure 202. Analogue record of a magnetic anomaly associated with a drilling rig (Mr. Louie).

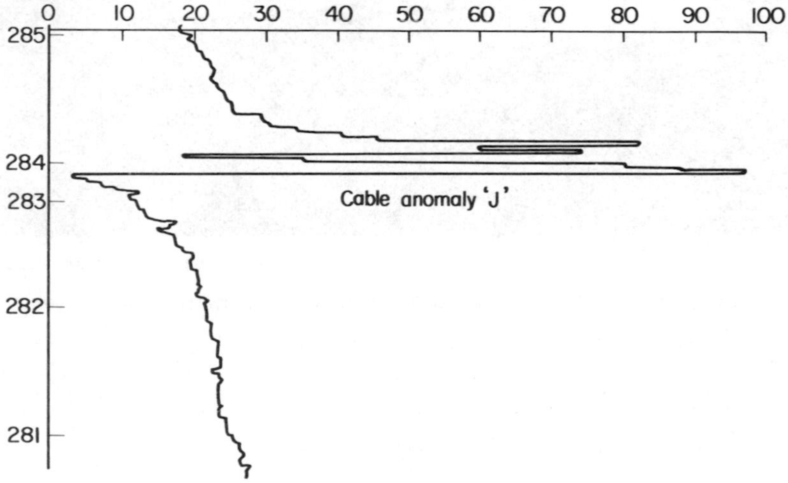

(Courtesy Gardline Shipping Ltd.)

Figure 203. Analogue record of a magnetic anomaly associated with an exposed submarine telephone cable in the North Sea.

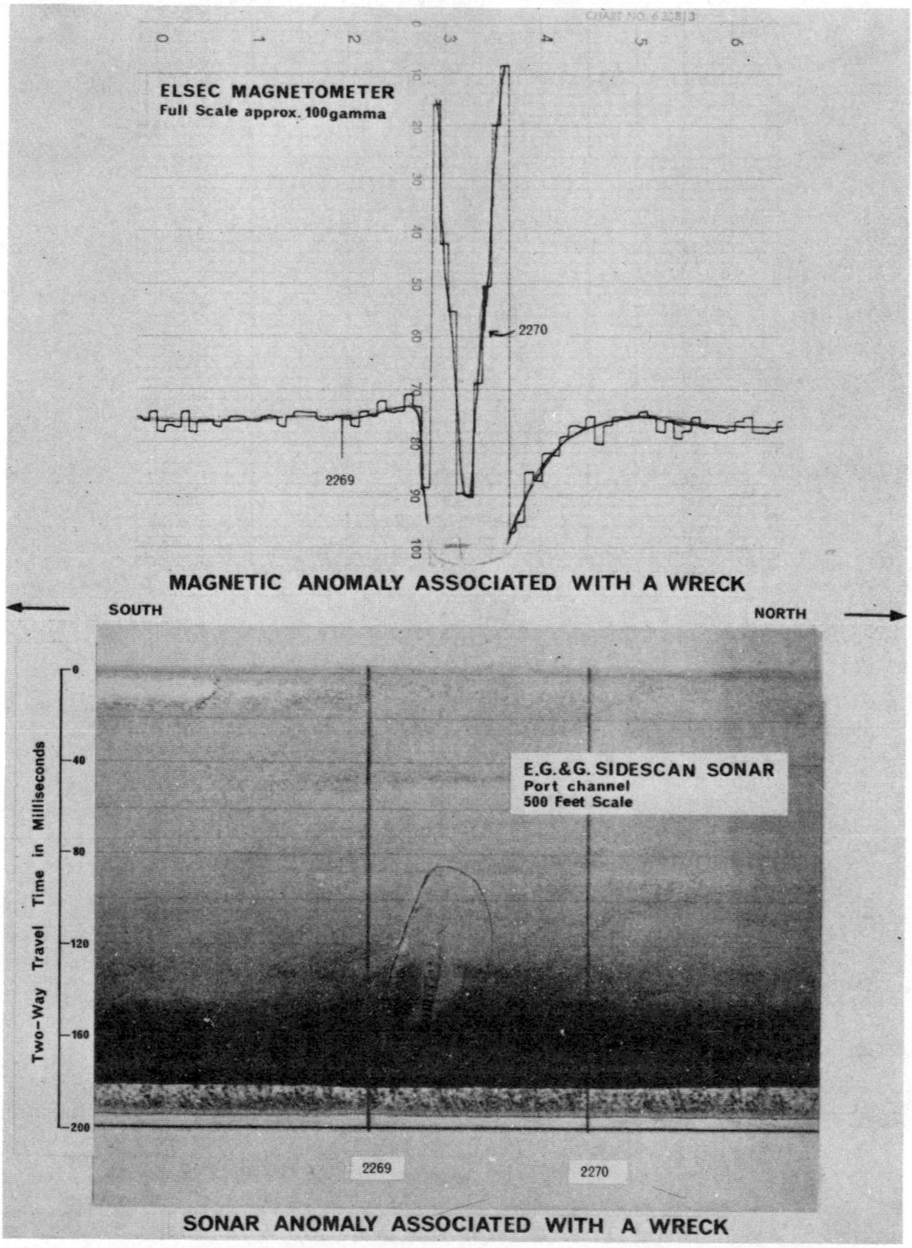

Figure 204. Magnetometer and side-scan sonar records obtained simultaneously near a wreck in the North Sea. (Taken from 'The Diver as a Geophysical Tool', published in *Hydrospace* (now *Offshore Services*) December 1970.)

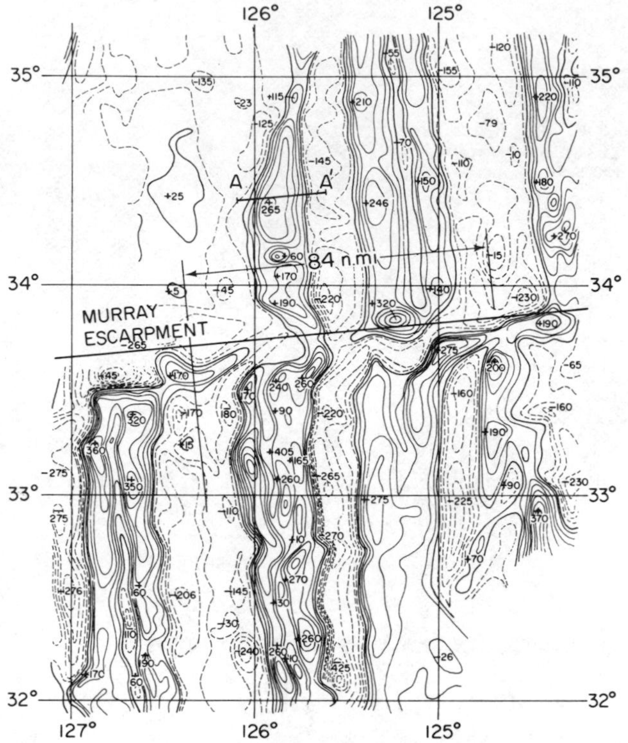

(a) Part of the magnetic anomaly map of the North-East Pacific showing the lateral displacement across the Murray Fracture Zone. (Contour interval 50 gammas.)

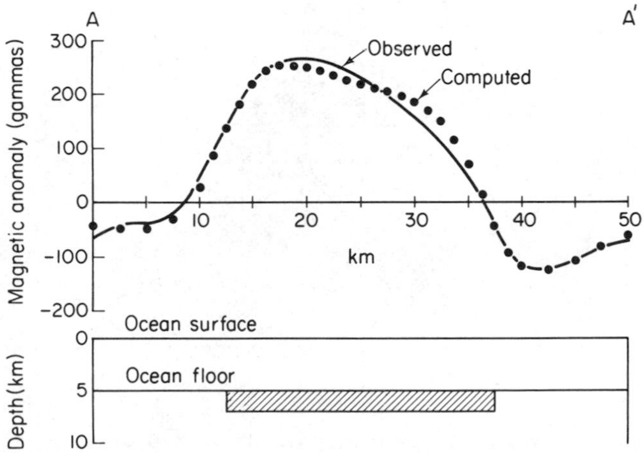

(b) Magnetic profile along line A–A′ and its interpretation in terms of an infinite slab of rectangular cross-section. (It is assumed that the magnetic susceptibility is 0.01 and that the magnetization is in the present direction of the earth's field.)

(Courtesy R.G. Mason)

Figure 205. Magnetic information based on marine proton precession magnetometer values. (Taken from 'Investigations of the Sea Floor', published in the *Liverpool and Manchester Geological Journal,* Vol. 2, part 3 (1960) pp. 389-406.)

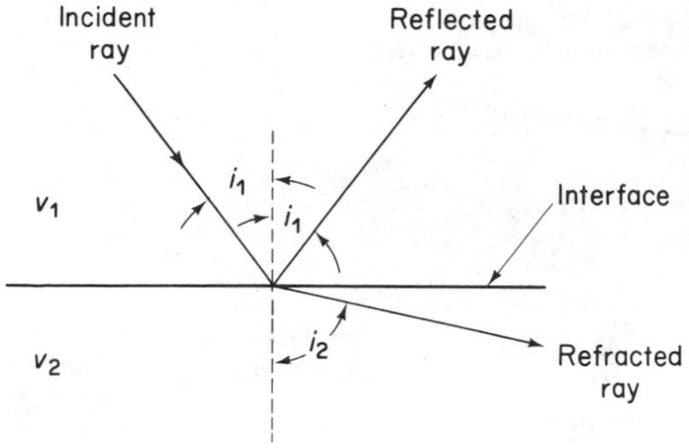

Figure 206. Paths of the incidence ray of an elastic wave and of rays which are reflected and refracted at a velocity interface.

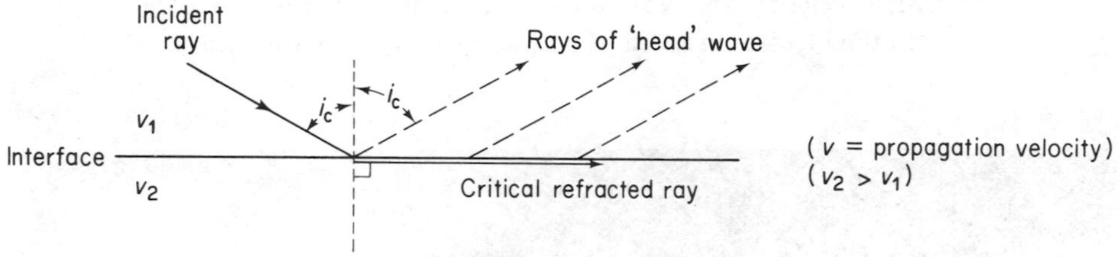

Figure 207. Paths of the incidence ray which intersects a velocity interface at the critical angle, the critically refracted ray and rays of the associated 'head' wave.

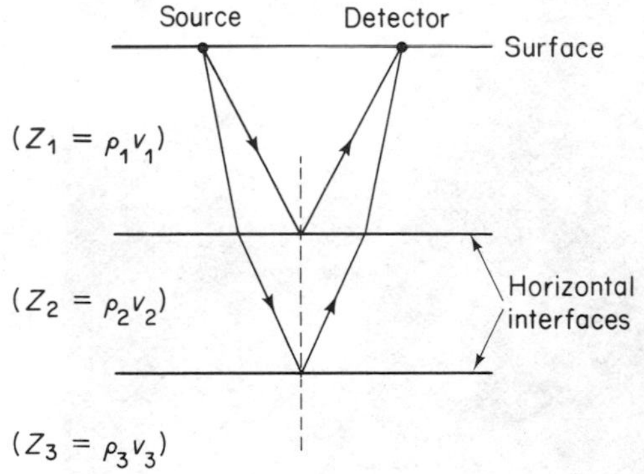

Figure 208. Paths of the incident and reflected rays associated with two horizontal interfaces.

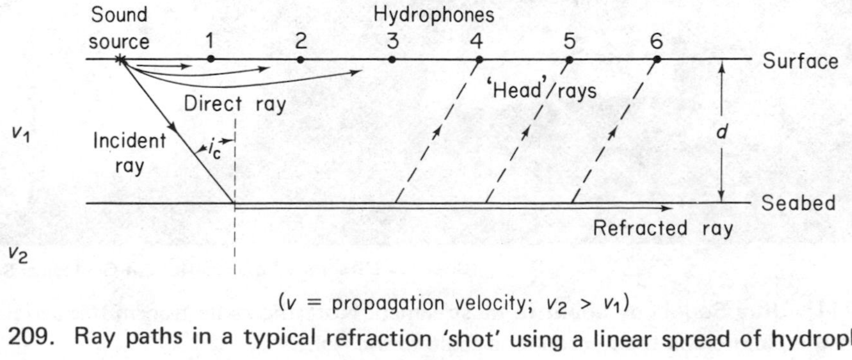

Figure 209. Ray paths in a typical refraction 'shot' using a linear spread of hydrophones.

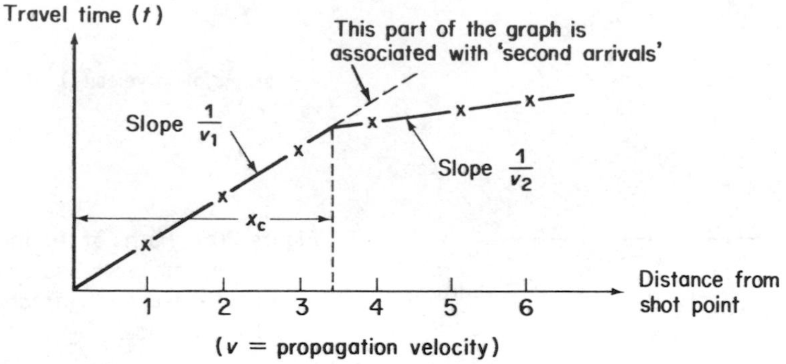

Figure 210. Travel time graph associated with the refraction 'shot' shown in Figure 209. Note the change in the slope of the graph due to the increase in speed of propagation of the seismic wave.

(Courtesy Director of the Institute of Geological Sciences)

Figure 211. Ultra Sonobuoy about to be streamed. Note the radio transmitter aerial to the left of the picture and the hydrophone cable to the right.

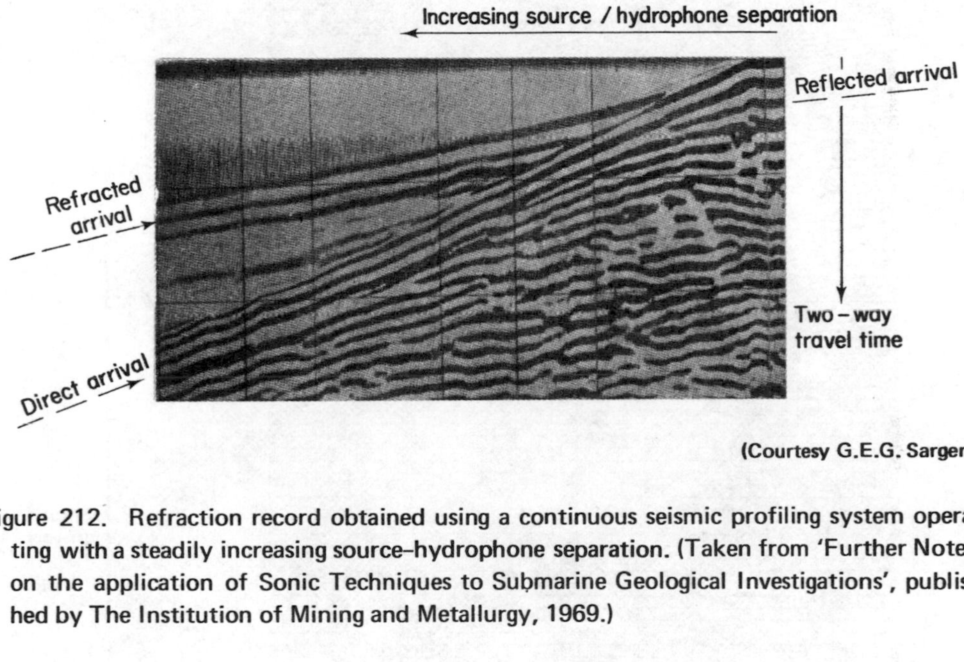

(Courtesy G.E.G. Sargent)

Figure 212. Refraction record obtained using a continuous seismic profiling system operating with a steadily increasing source–hydrophone separation. (Taken from 'Further Notes on the application of Sonic Techniques to Submarine Geological Investigations', published by The Institution of Mining and Metallurgy, 1969.)

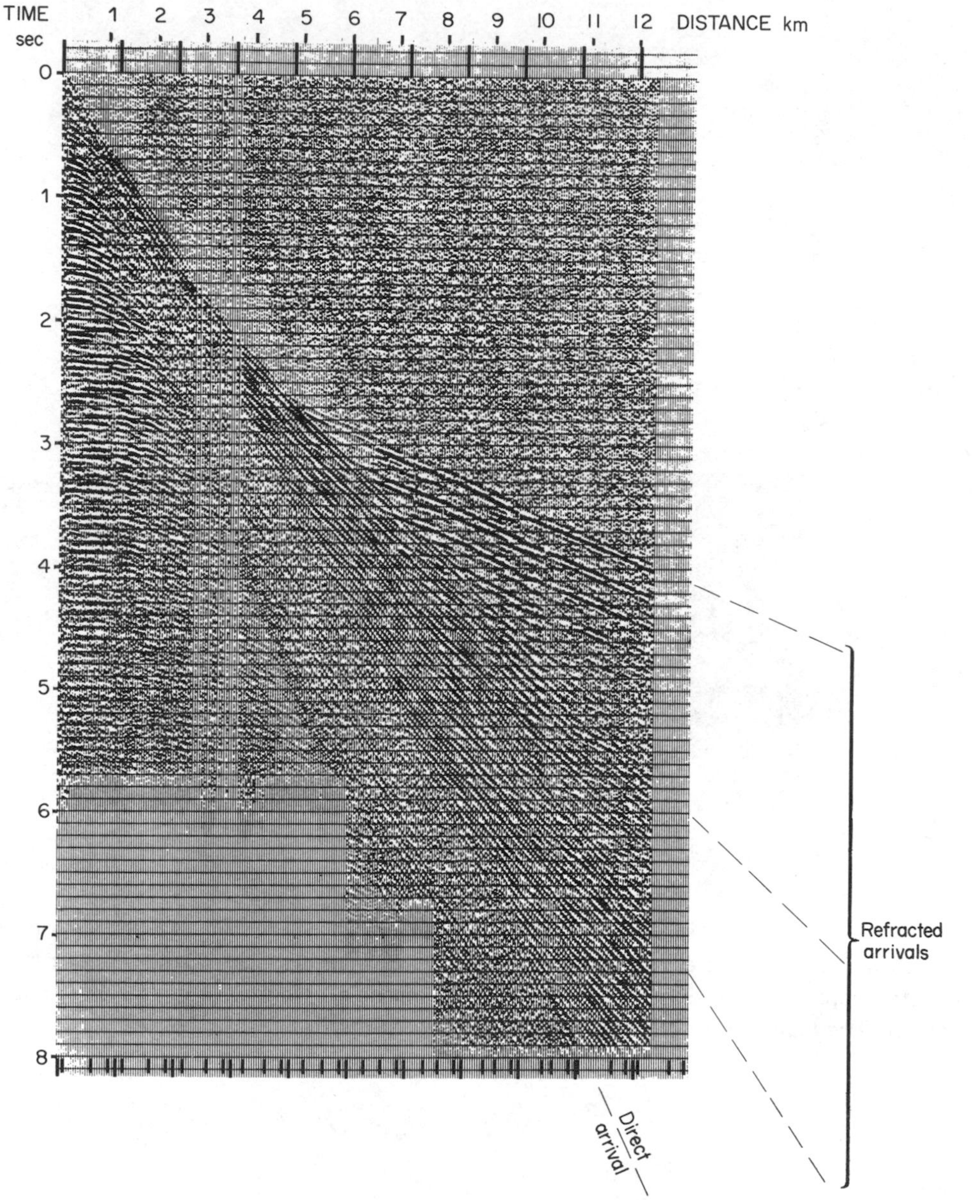

(Courtesy Compagnie Générale de Géophysique)

Figure 213. Processed refraction record obtained using a Vaporchoc® seismic source and sonobuoy. This information is presented in variable area format. Each vertical line running across the record represents the refraction trace corresponding to an individual shot position. Note the change between the slopes of the direct arrival and different refracted arrivals. Also observe the noise in the top right of the record. (® CGG Registered Trade Mark).

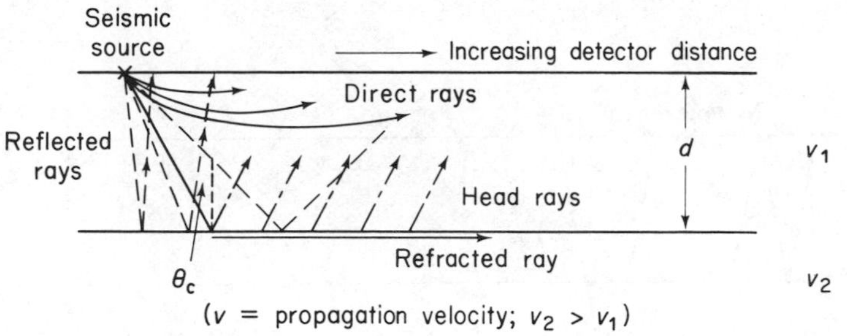

Figure 214. Ray paths associated with a single horizontal interface.

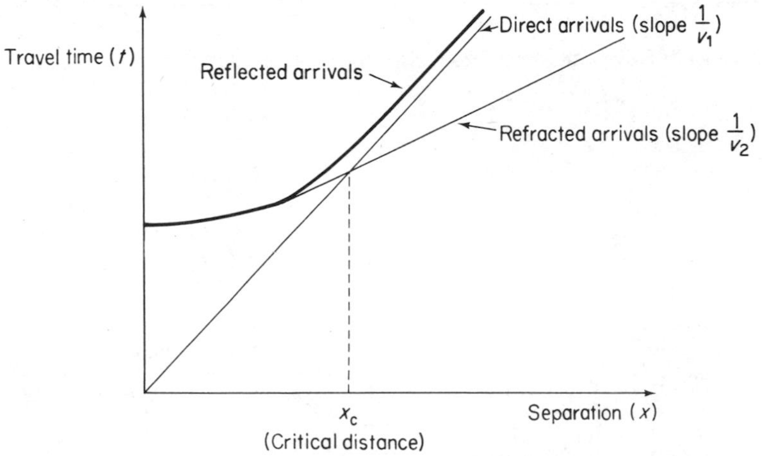

Figure 215. Travel time curves for the direct wave, reflected wave and head waves associated with a single horizontal interface separating layers with propagation velocities v_1 and v_2 ($v_1 < v_2$).

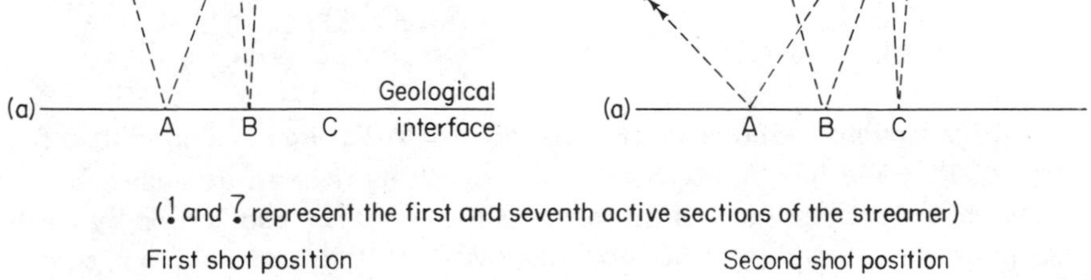

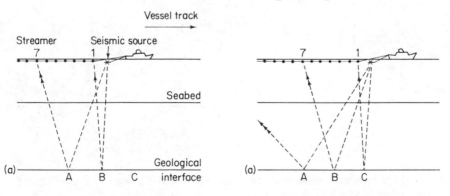

Figure 216. Ray paths of the incident and reflected waves associated with particular depth points (A, B and C) and hydrophone sections (Nos. 1 and 7) in adjacent shot positions.

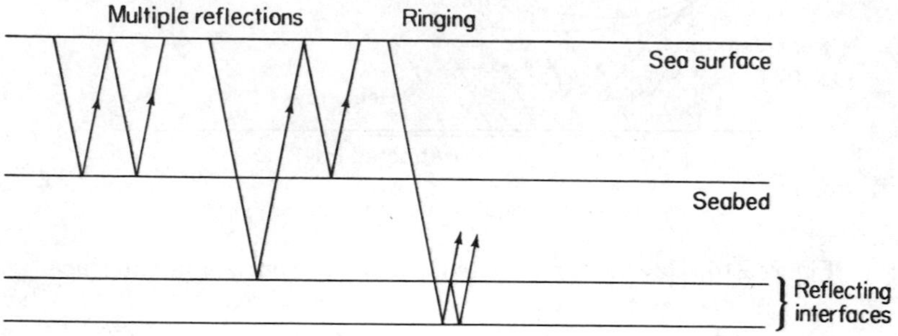

Figure 217. Ray paths associated with simple examples of reverberation.

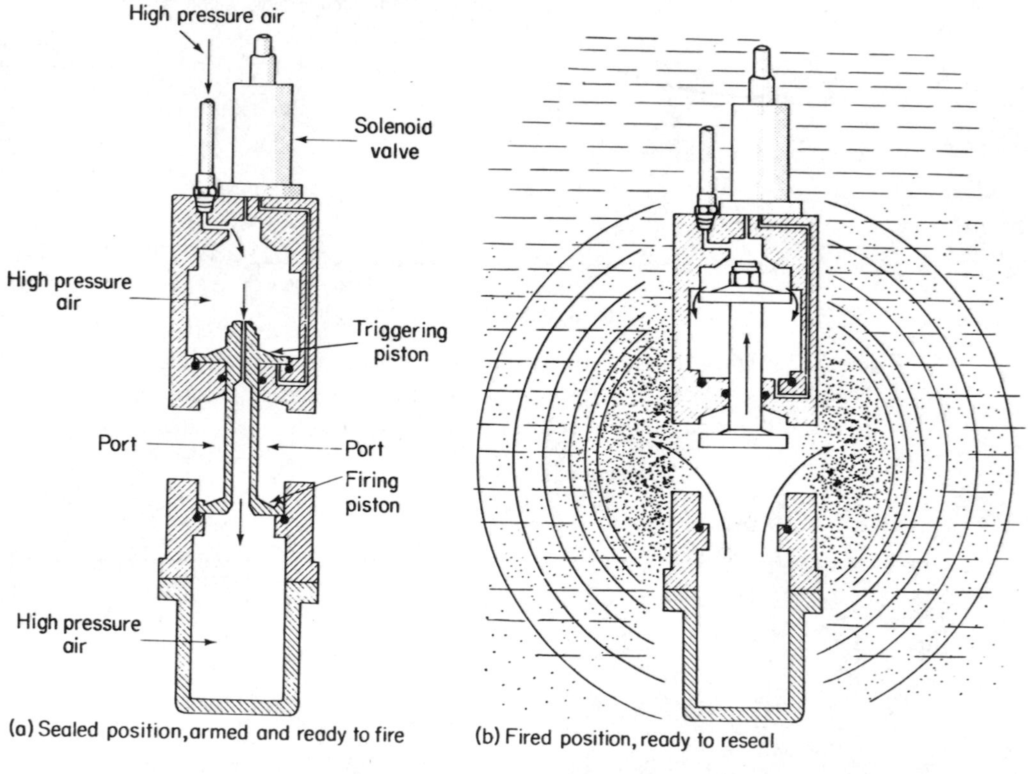

Figure 218. Cross-sections and principle of operation of a PAR® airgun sound source. Basically, high pressure air stored in a firing chamber is explosively released through four portholes through the action of a shuttle with pistons at either end. The gun is fired by electrically triggering the solenoid valve. (® Registered Trade Mark of Bolt Associates, Inc; courtesy Bolt Associates, Inc.)

(Courtesy Seismograph Service Ltd.)

Figure 219. Airgun array about to be streamed.

(Courtesy Compagnie Générale de Géophysique)

Figure 220. Triggered ejection valve of the Vaporchoc® steam sound source. (® Registered Trade Mark of Compagnie Générale de Géophysique.)

(Courtesy Seismograph Service Ltd.)

Figure 221. Drum carrying 2400 metres 48 trace seismic streamer.

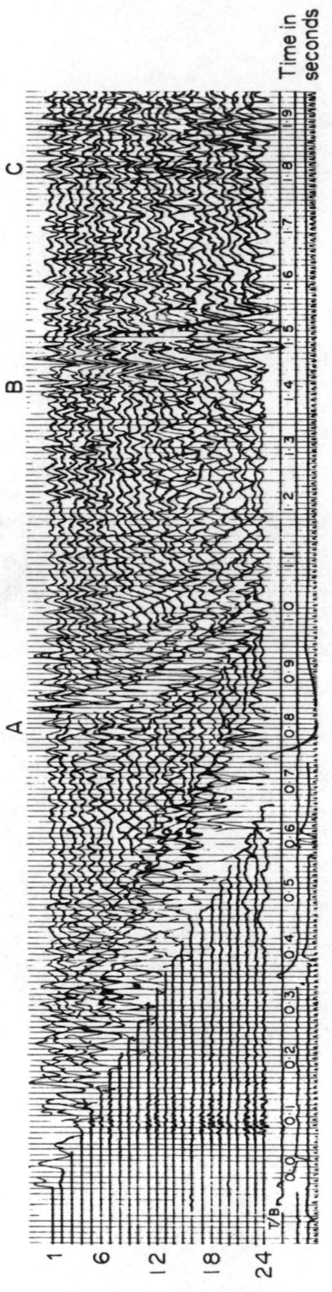

A. Reflections from the base of the Tertiary.
B. Reflections from the base of the Cretaceous.
C. Reflections from the top of the Permian Salt.

(Courtesy J.T. Hornabrook and British Petroleum)

Figure 222. Typical 'wiggly trace' monitor record obtained during a seismic reflection survey in the North Sea using a 24 trace 1200 metre cable. Note the effect of normal move-out on the positions of adjacent reflections across the record and that the effect becomes less with depth (Taken from 'Seismic Interpretation Problems in the North Sea', published in the Proceedings of the 7th World Petroleum Congress by Elsevier.)

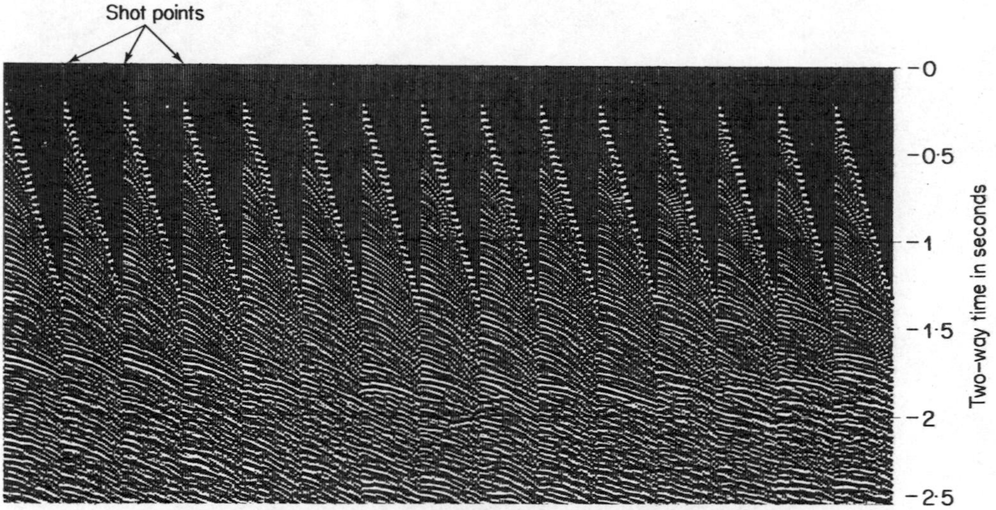

(a) Playout of raw field data after gain recovery and correction for spherical divergence. (Note the effects of normal move-out on the positions of adjacent reflections across the reflection section for each shot point. Also observe that the effect becomes less with depth.)

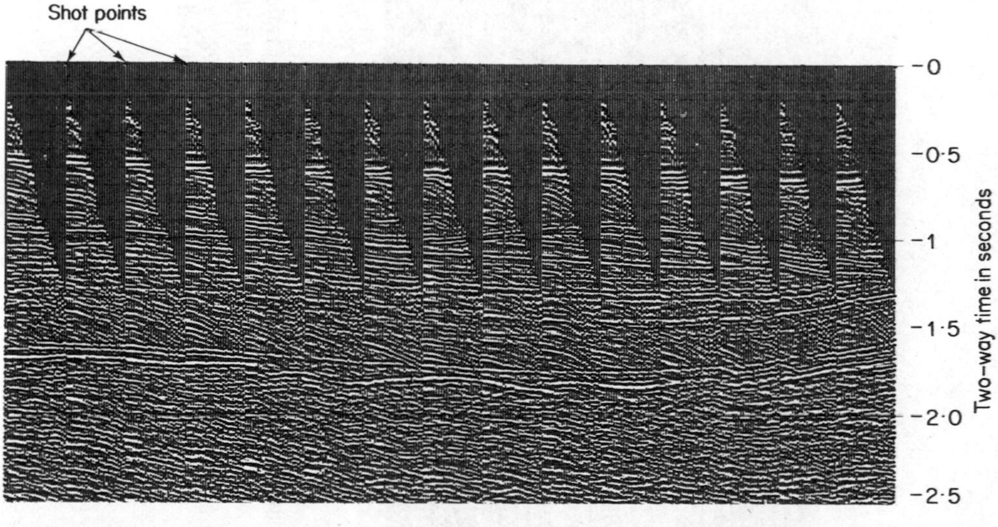

(Courtesy Seismograph Service Ltd.)

(b) Playout of single cover deconvoluted section corrected for normal move-out effects immediately prior to stacking. (Adjacent reflections can now be followed across the section with comparative ease. Note that some of the reflections are associated with reverberation effects.)

Figure 223. Effect of applying normal move-out corrections and deconvolution to seismic reflection data.

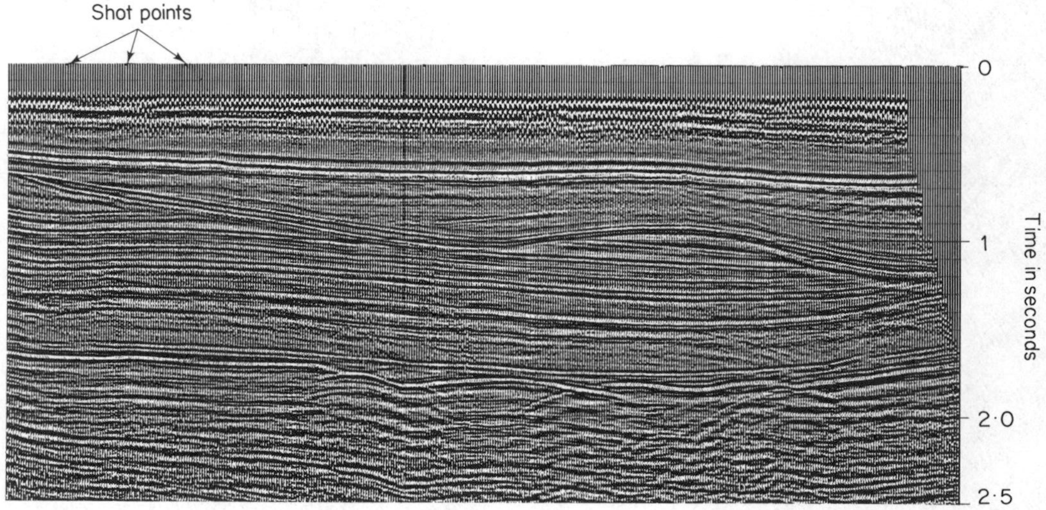

(Courtesy Seismograph Service Ltd.)

Figure 224. Final processed time section of reflection data shown in Figure 223. Stacking, deconvolution, time variant filtering and time variant amplitude equalization have been applied during processing. The presentation mode is wiggly trace on variable area. Reflections are now traced clearly across the section and are not confused by reverberation effects.

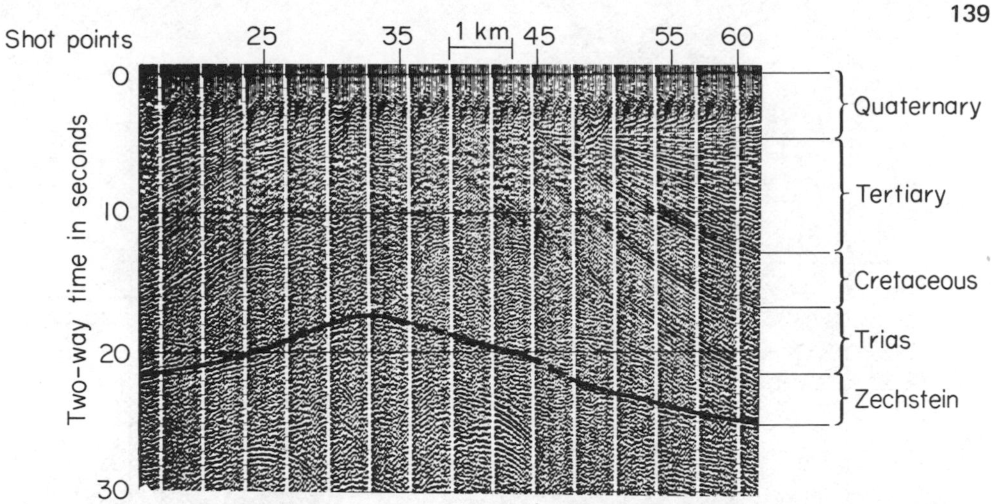

(Courtesy J.T. Hornabrook and British Petroleum)

(a) Interpreted final processed time section. (The heavy line marks the position of the reflection from the Base of the Zechstein salt deposits. The 'time high' of the reflection is a false structure arising from the effects of overburden velocity on the time profile. Observe that it is difficult to identify clearly any reflections over the top of this structure.)

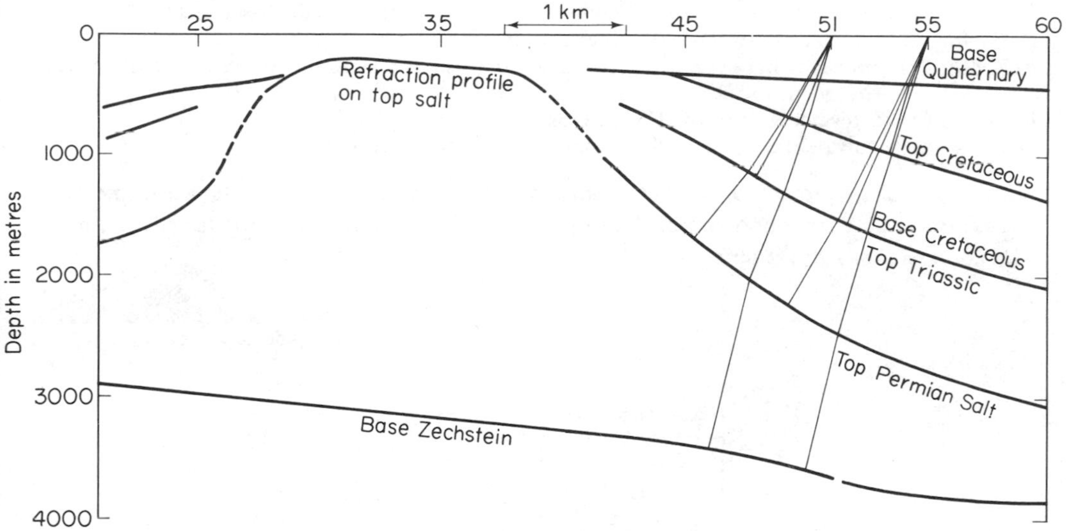

(b) Interpreted depth section. (The section has been corrected for migration effects—observe the migrated ray paths drawn for shot points 51 and 55. Note the disappearance of the 'time structure' in the Base Zechstein interface. The position of the top of the Permian salt over the 'salt dome' was derived from an analysis of refraction profiles constructed using the first arrival refraction breaks on the reflection records.) (Taken from 'Seismic Interpretation Problems in the North Sea', published in the Proceedings of the 7th World Petroleum Congress by Elsevier.)

Figure 225. Comparisons of interpreted time and depth sections from the North Sea. Note that the Zechstein salt deposits form the cap rock to many of the North Sea gas reservoirs.

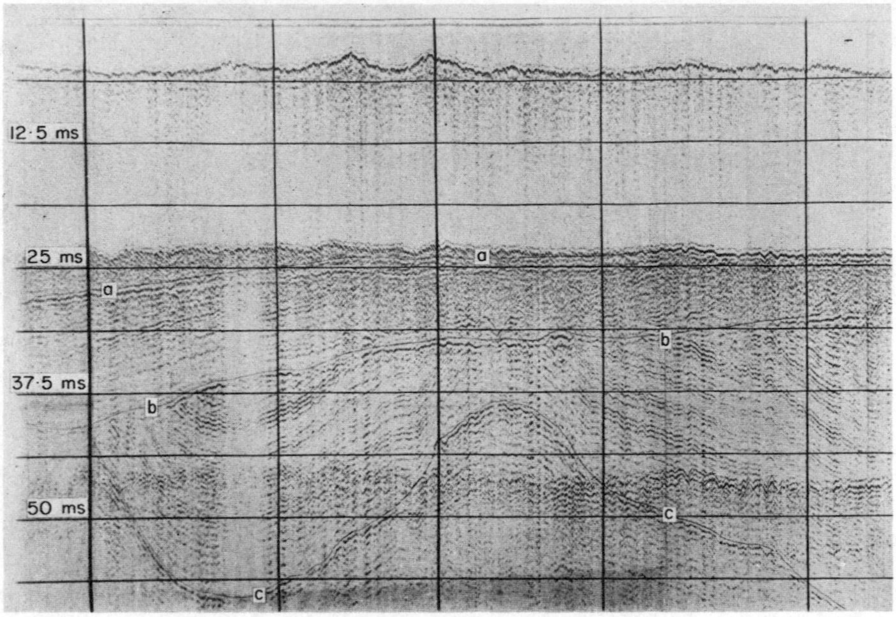

(Courtesy Director of the Institute of Oceanographic Sciences)

Reflector (a)–Interface between overlying sand and gravel deposits.
Reflector (b)–Interface between overlying gravel deposit and a truncated sequence of folded Cretaceous rocks.
Reflector (c)–Interface within the Cretaceous rocks.
(Timing lines equivalent to 6.25 milliseconds two-way travel time.)

Figure 226. Interpreted continuous seismic profiling record obtained using a Huntec high resolution boomer sound source. Note that the effective pulse length is less than 1 millisecond two-way travel time.

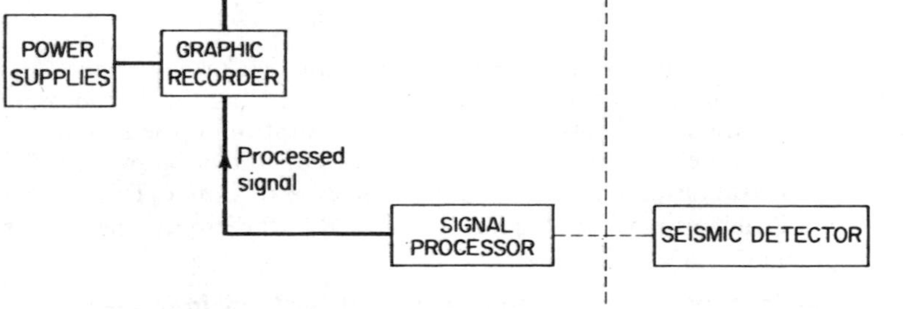

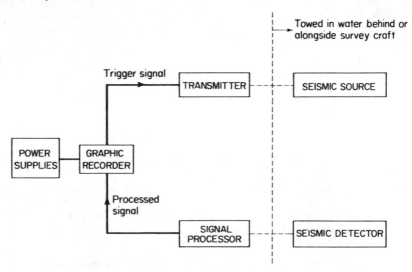

Figure 227. Block diagram of a typical continuous seismic profiling system.

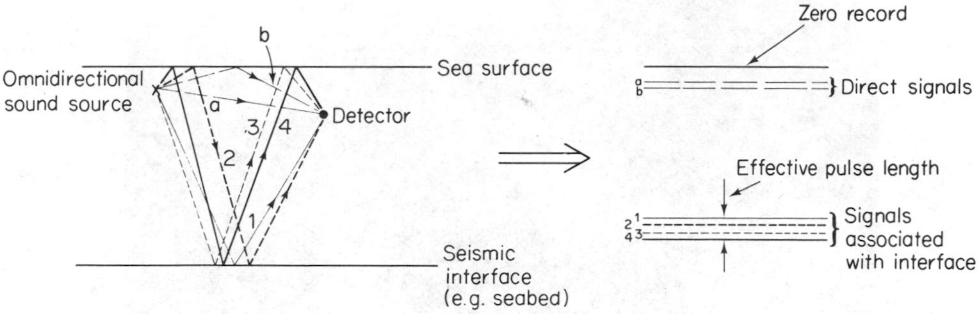

(a) Ray path diagram (b) Equivalent record section

Figure 228. Ray paths and equivalent record section illustrating the effect of ghost reflections from the sea surface when operating with an omnidirectional sound source.

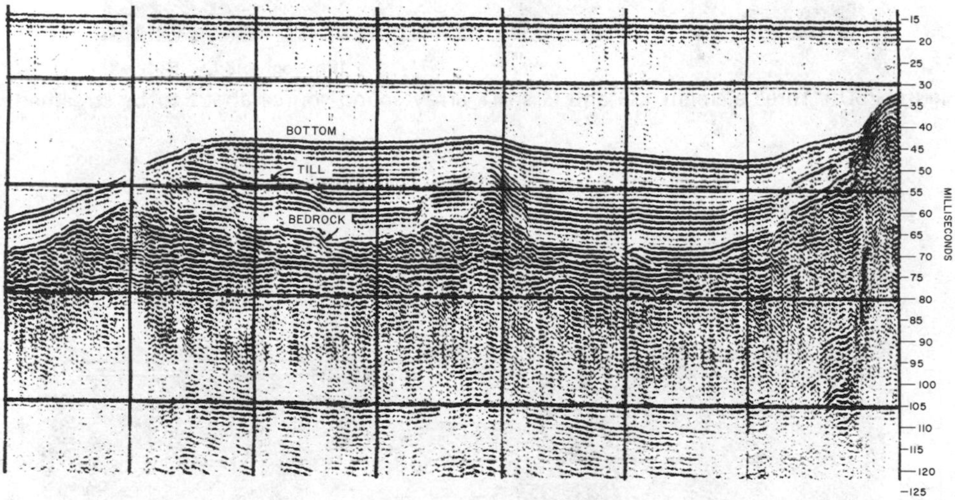

(Courtesy Huntec Ltd.)

Figure 229. Continuous seismic profiling record obtained using a 5 cubic inch PAR® sound source operating at 400 p.s.i. of recent deposits overlying a layered limestone. Note that layering can be identified in both the overburden and bedrock. (® Registered Trade Mark of Bolt Associates Ltd.)

(Photograph J.R. Bidwell)

Figure 230. Three element EG and G spark-array sound source about to be streamed.

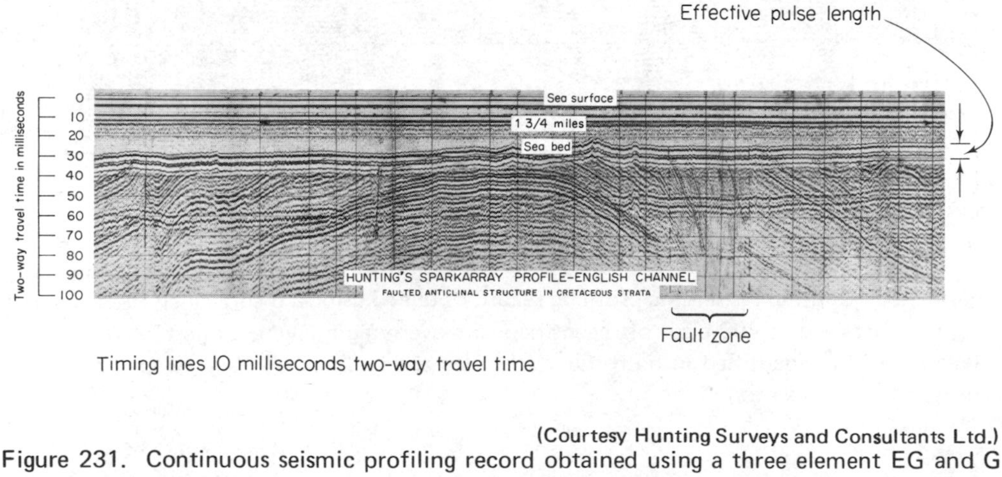

(Courtesy Hunting Surveys and Consultants Ltd.)

Figure 231. Continuous seismic profiling record obtained using a three element EG and G spark-array of folded and faulted Cretaceous strata in the English Channel. Note that the effective pulse length is approximately 9 milliseconds two-way travel time.

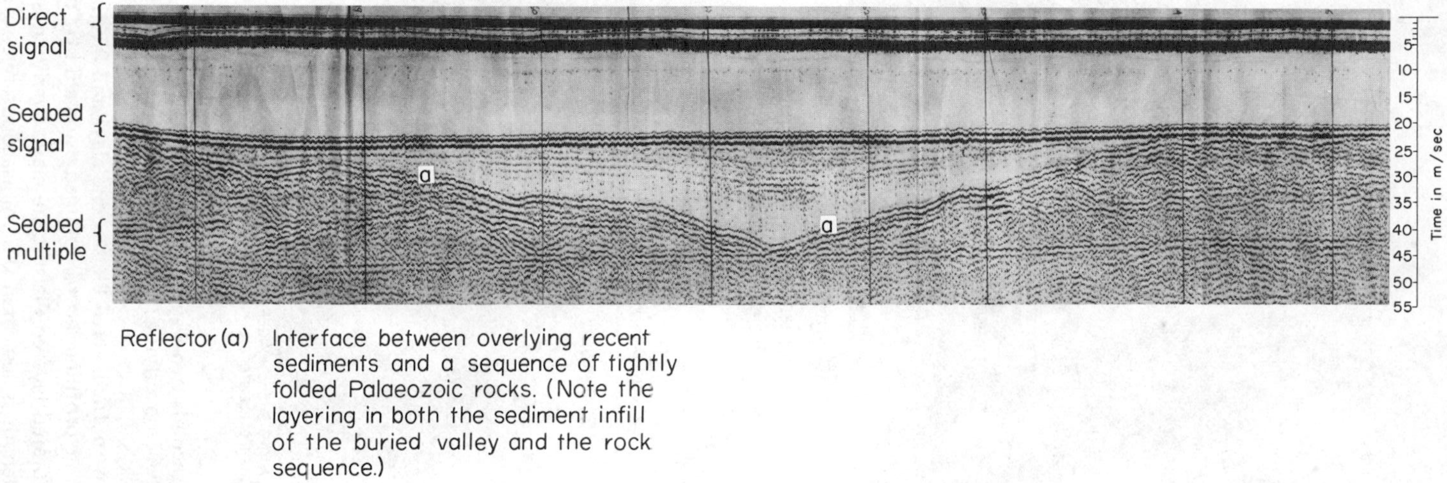

(Courtesy G.E.G. Sargent)

Figure 232. Continuous seismic profiling record obtained using a Huntec sparker over an infilled buried channel cut into Palaeozoic rocks. Note that the effective pulse length is approximately 3 milliseconds two-way travel time. (Taken from 'Review of acoustic equipments for studying submarine sediments', published in *Transactions/Section B of the Institution of Mining and Metallurgy,* Vol. 75, 1966.)

(Courtesy Director of the Institute of Oceanographic Sciences)

Figure 233. Huntec high resolution boomer sound source about to be streamed. Note the towing catamaran and protective outriggers. (Photograph: Author)

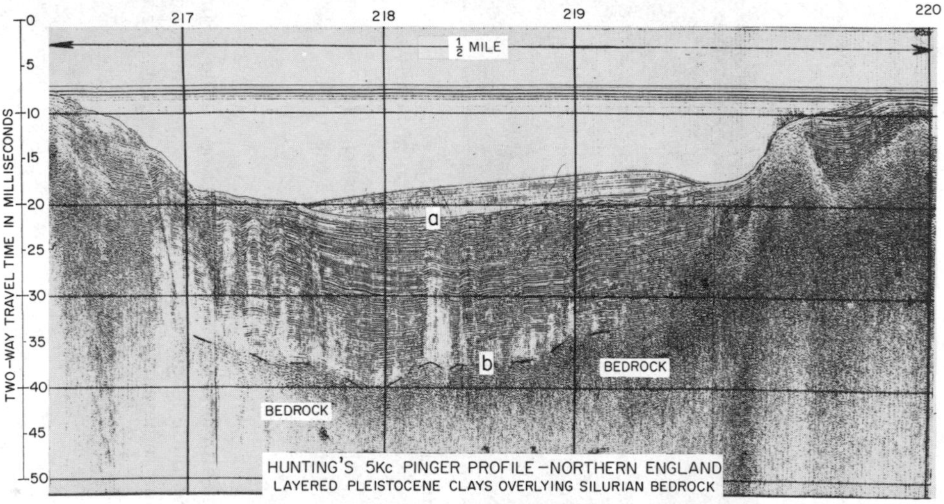

(Courtesy Hunting Surveys and Consultants Ltd.)

Timing line spacing 10 milliseconds two-way travel time.

Reflector (a)—Interface between overlying fine mud and a sequence of saturated layered Pleistocene clays.

Reflector (b)—Interface between the layered Pleistocene clays and Silurian bedrock.

Figure 234. Continuous seismic profiling record obtained using a 5 kHz pinger sound source showing fine mud and saturated layered Pleistocene clays overlying Silurian rocks. Note that the effective pulse length is less than 0·5 milliseconds two-way travel time. Also observe that very little energy has penetrated to the rockhead.

(Courtesy Director of the Institute of Oceanographic Sciences)

Figure 235. ORE pinger sound source about to be streamed. Note that the source is mounted in a hydrodynamically designed 'fish' (Photograph: Author).

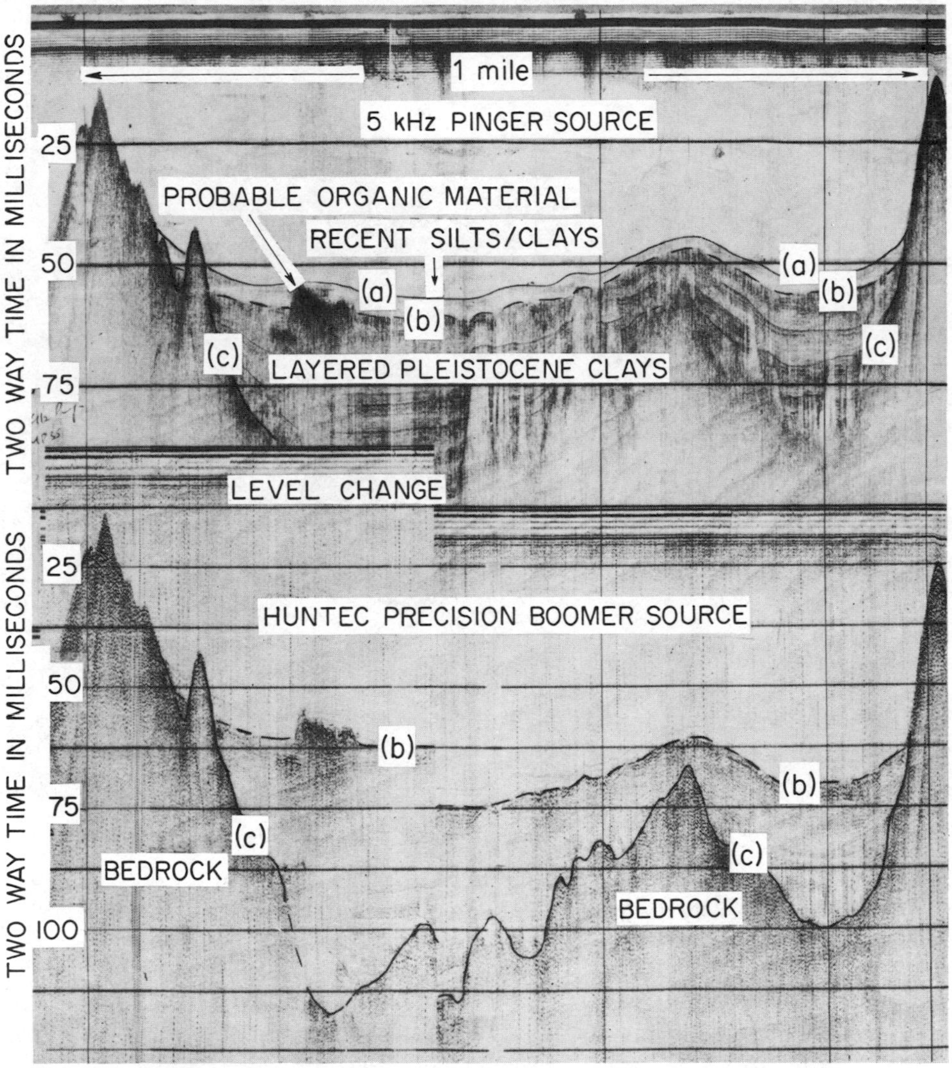

Reflector (a)–Interface between the water and soft recent silts and clays.
Reflector (b)–Interface between overlying recent silts and clays, and layered Pleistocene clays.
Reflector (c)–Interface between the Pleistocene clays and Silurian rocks.

Figure 236. Dual channel continuous seismic profiling record obtained using 5 kHz pinger and high resolution boomer sound sources operated simultaneously. Note the more detailed layering apparent on the pinger profile but the improved penetration achieved with the boomer. (Taken from 'The Diver as a Geophysical Tool', published in *Hydrospace* (now *Offshore Services*) December 1970.)

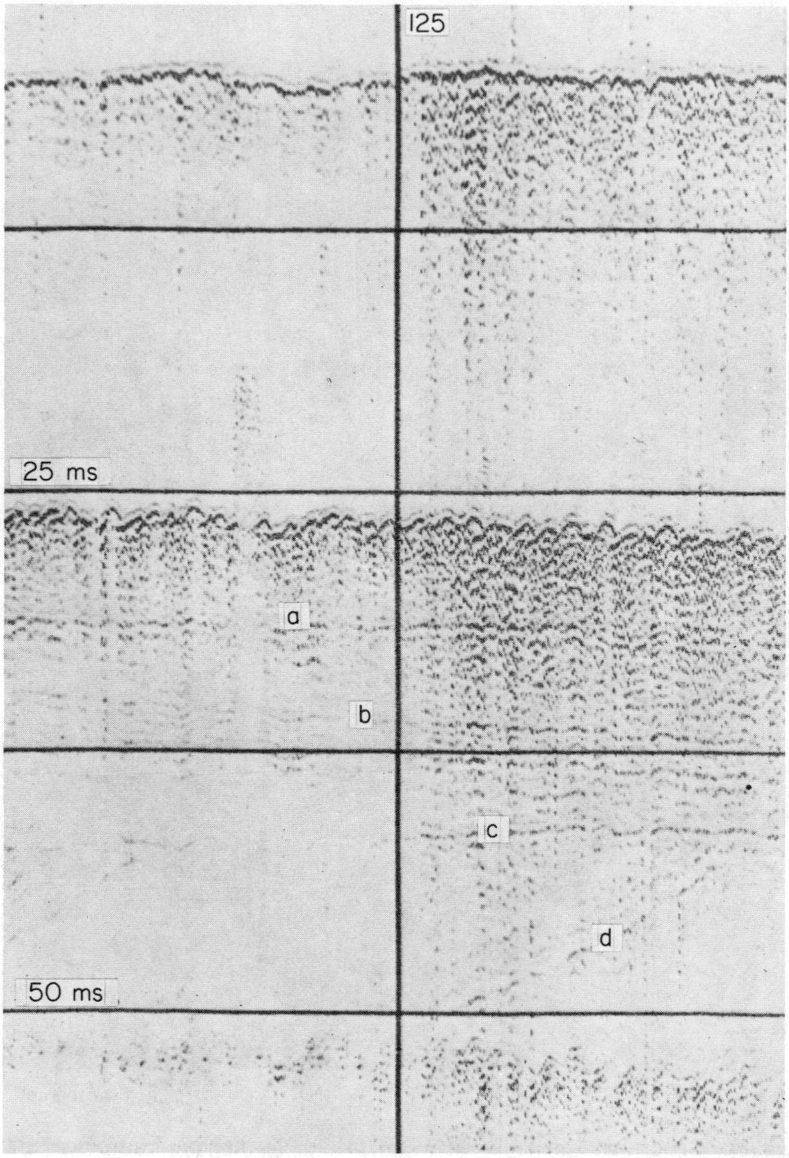

(Courtesy Director of the Institute of Oceanographic Sciences)

Reflector (a)–Interface between overlying sand and gravel deposits.
Reflector (b)–Interface within gravel deposit.
Reflector (c)–Interface between gravel deposit and a truncated sequence of folded Cretaceous strata.
Reflector (d)–Interface within Cretaceous strata.

Figure 237. Interpreted continuous seismic profiling record obtained at different gain settings using a Huntec high resolution boomer sound source. Note the increased information content in the right-hand part of the profile where the gain was increased.

(Photograph: Author)

Figure 238. Geophysical installation in an inflatable dinghy. Equipment includes a Huntec continuous seismic profiling system, Kelvin Hughes Transit Sonar and MS 36 echosounder, 5 kVA ONAN Generator and two Evinrude outboard motors. (Taken from 'The Diver as a Geophysical Tool', published in *Hydrospace* (now *Offshore Services*) December 1970.)

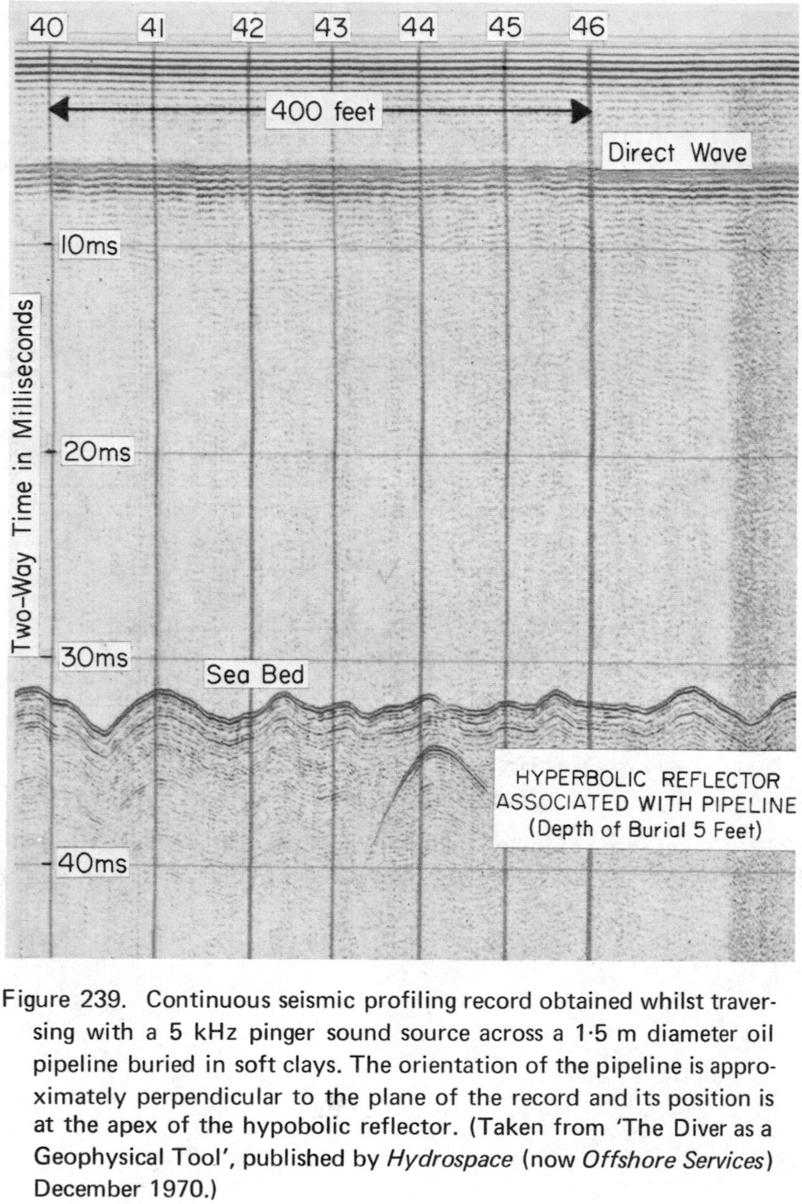

Figure 239. Continuous seismic profiling record obtained whilst traversing with a 5 kHz pinger sound source across a 1·5 m diameter oil pipeline buried in soft clays. The orientation of the pipeline is approximately perpendicular to the plane of the record and its position is at the apex of the hypobolic reflector. (Taken from 'The Diver as a Geophysical Tool', published by *Hydrospace* (now *Offshore Services*) December 1970.)

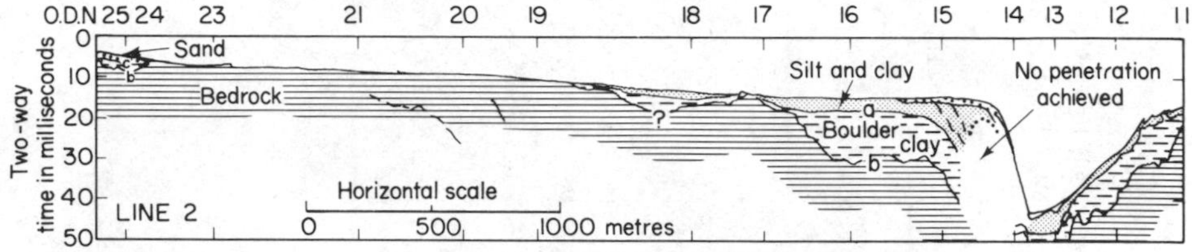

(a) Interpreted continuous seismic profile plotted out as a time section.

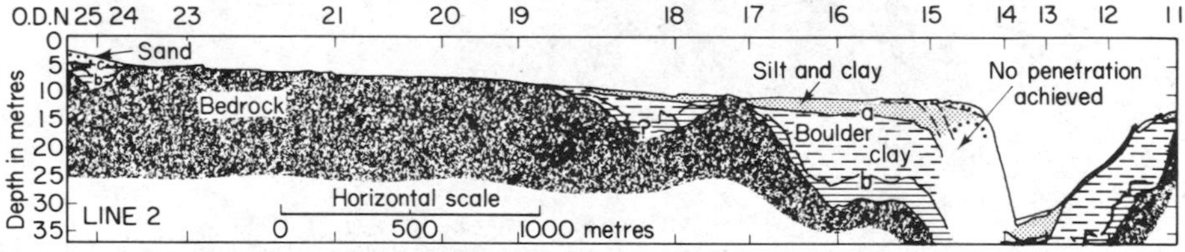

(b) Interpreted continuous seismic profile plotted out as a corrected depth section using velocity information.

(Courtesy Edinburgh City Engineer and J.D. and D.M. Watson)

Figure 240. Comparison of interpreted time and depth sections. Note the difference in the positions of the interface between the boulder clay and bedrock on each profile.

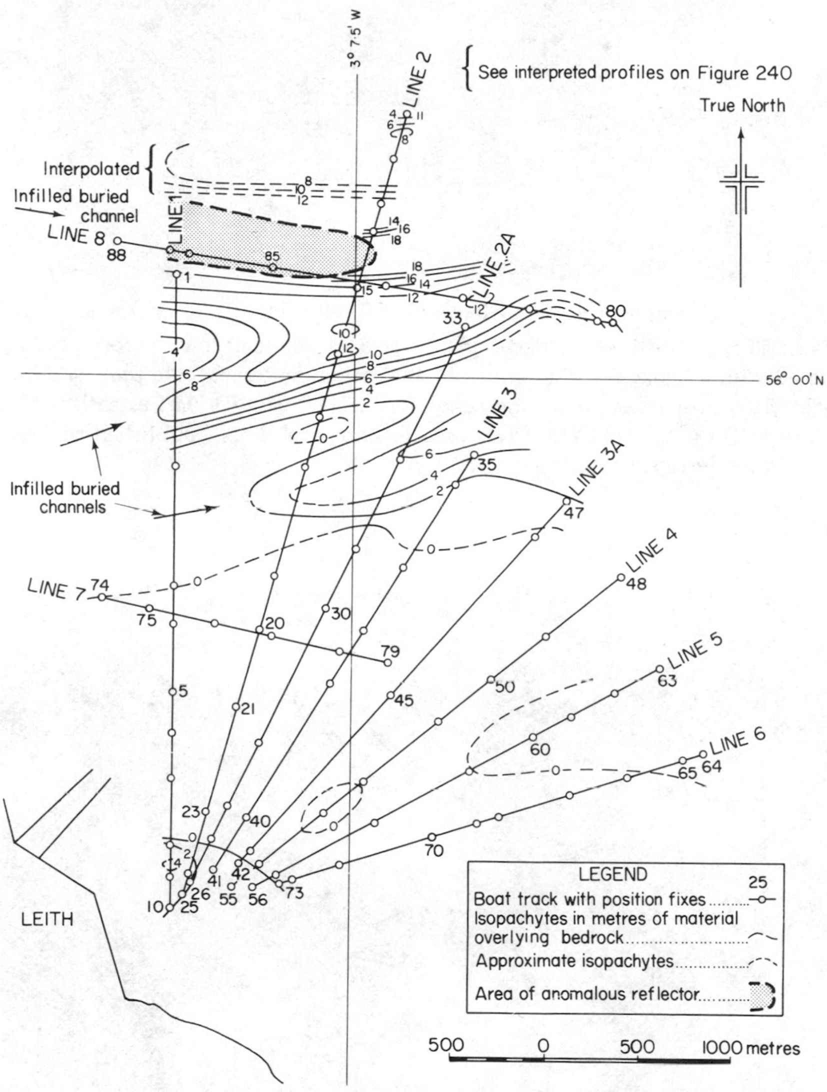

(Courtesy Edinburgh City Engineer and J.D. and D.M. Watson)

Figure 241. Isopachyte map showing variation in the thickness of material overlying bedrock in the area of a proposed sewer outfall near Leith. Note the position of infilled buried channels which have been cut into the bedrock during a period of lower sea level.

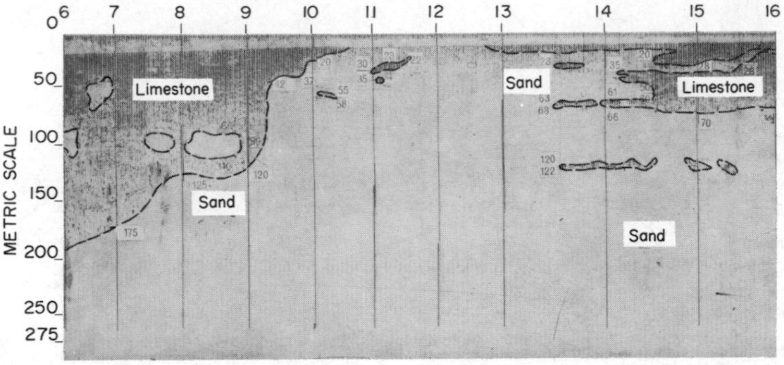

(Record obtained with a Kelvin Hughes Transit Sonar.)

Figure 242. Changes in the recorded signal level on a side-scan sonar record predominantly associated with a change in the nature of the seabed material. Note that sand has weaker backscattering properties than limestone. The seabed geology was established by diving observation. (Taken from 'The Diver as a Geophysical Tool', published by *Hydrospace* (now *Offshore Services*) December 1970.)

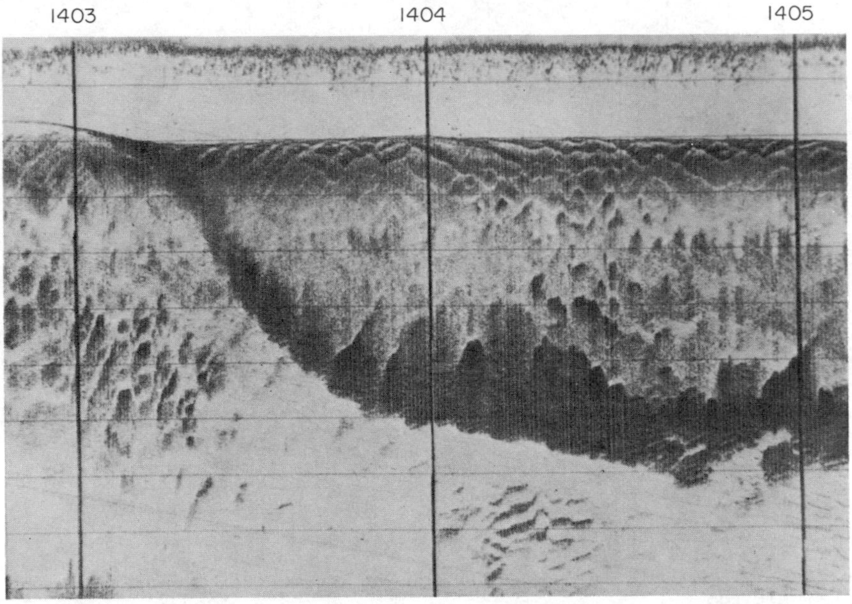

(Courtesy Director of the Institute of Oceanographic Sciences)

(Timing line spacing 20 millisecond two-way travel time. Record obtained with an E.G. and G. equipment).

Figure 243. Changes in the recorded signal level on a side-scan sonar record predominantly associated with changes in the shape of the seabed. Note that there is a stronger back-scattered signal from the facing edge (lee slope) of the sand wave running across the record. Also observe the presence of an 'acoustic shadow zone' behind the crest of the sand wave.

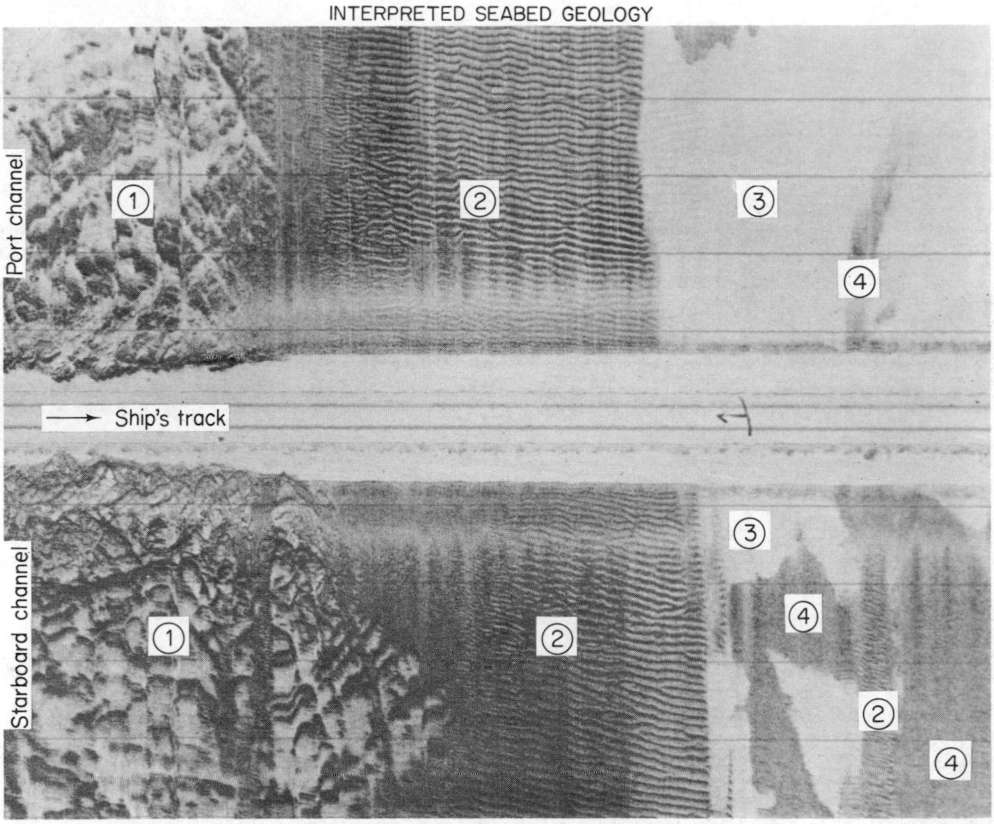

(Courtesy Director of the Institute of Oceanographic Sciences)

(1) Rock outcropping at seabed

(2) Seabed predominantly covered by sand ripples

(3) Seabed predominantly composed of fine silty sand

(4) Seabed predominantly composed of coarse sands and shells

(Timing line spacing 20 milliseconds two-way travel time)
(Record obtained with an E.G. and G. equipment.)

Figure 244. Side-scan sonar record obtained over a seabed exhibiting variations in both morphology and lithology.

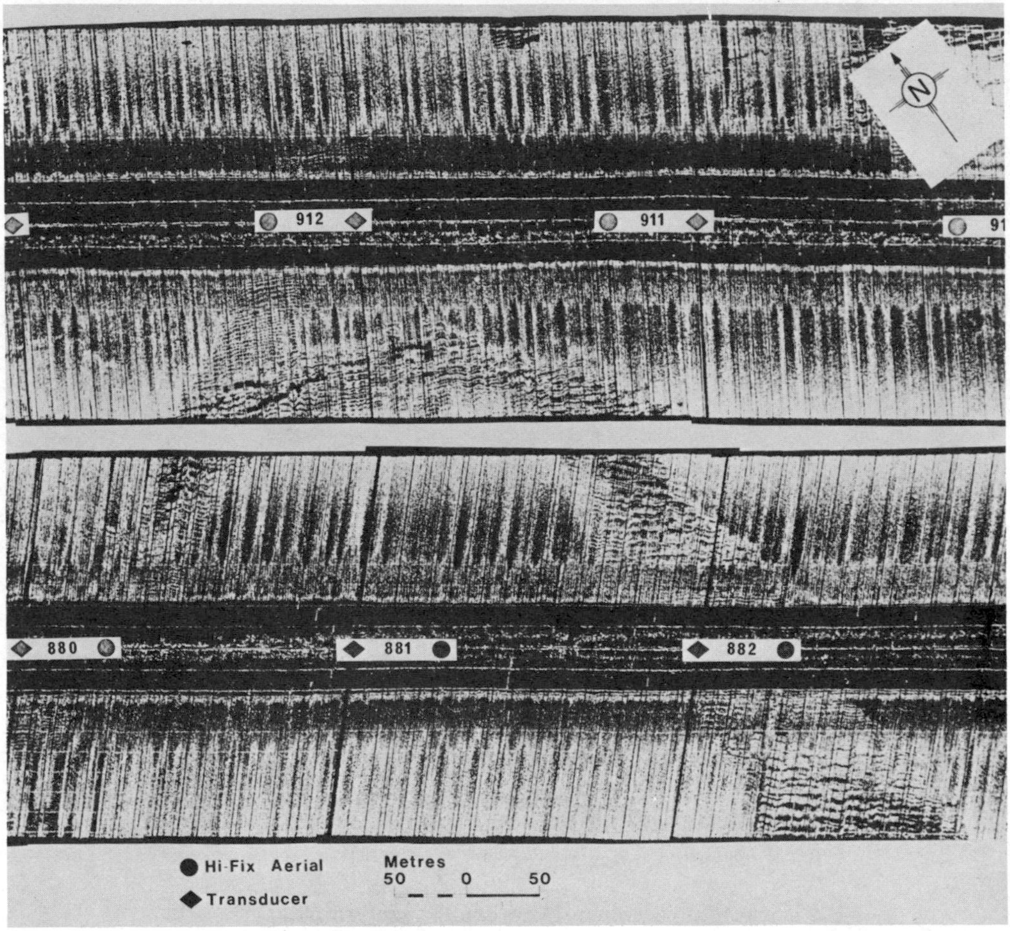

(Courtesy Director of the Institute of Oceanographic Sciences)

Figure 245. High resolution side-scan sonar mosaic of an area in the southern North Sea. Prepared using Bath University Playback System. Note that each sonar channel has been corrected for scale distortions and heading variations, but not for slant range effects. (Taken from 'Mosaicing High Resolution Sonar Data', published in *Offshore Services*, October 1972.)

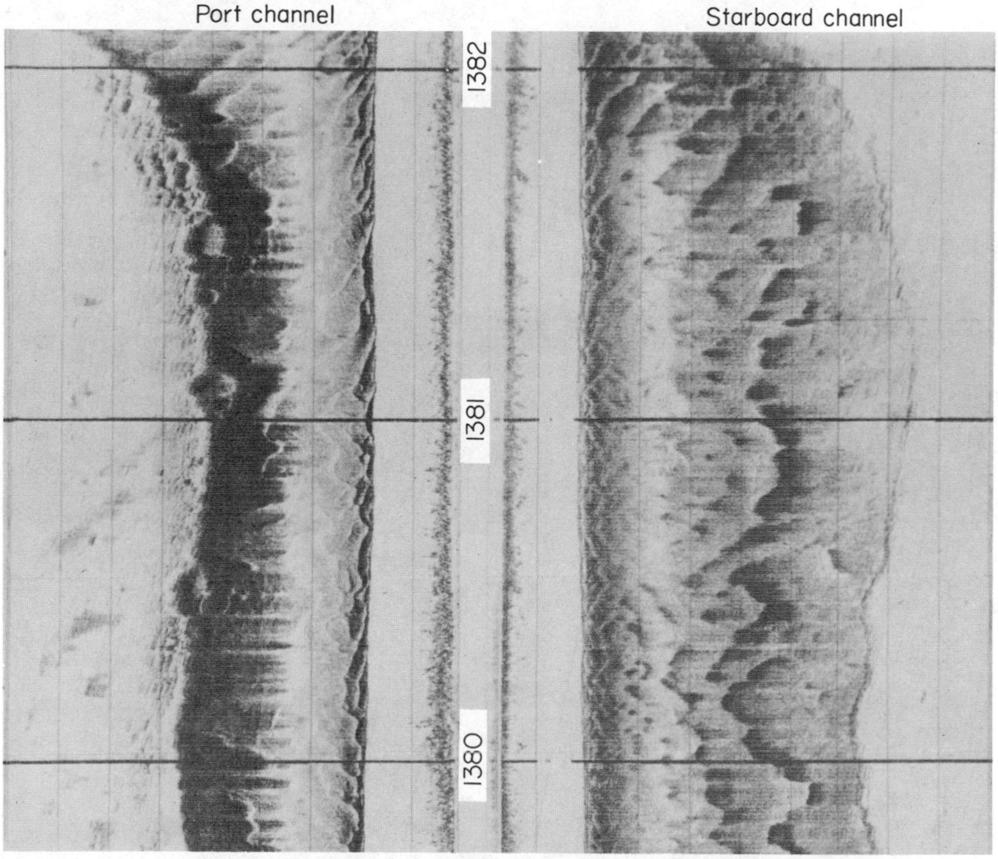

(Courtesy Director of the Institute of Oceanographic Sciences)

(a) Dual channel side-scan sonar record. (Timing line spacing 20 milliseconds two-way travel time. Record obtained with an E.G.&G. equipment.)

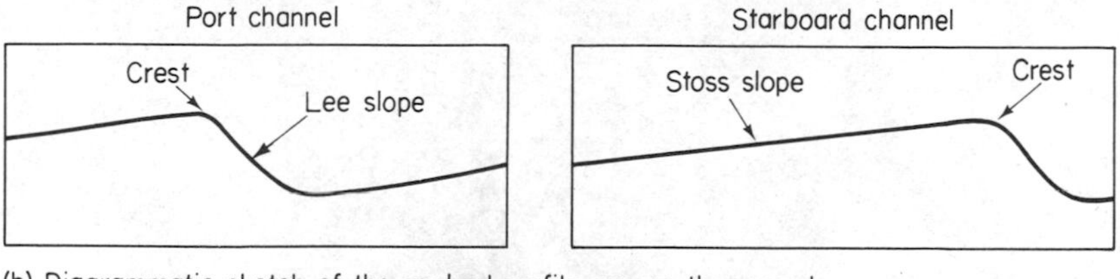

(b) Diagrammatic sketch of the seabed profile across the record

Figure 246. Side-scan sonar record obtained whilst running approximately parallel to the crests of asymmetrical sand waves. Note the stronger back-scattered signal from the lee slope of the sand wave recorded on the port channel, and the 'acoustic shadow zone' behind the sand wave crest recorded on the starboard channel.

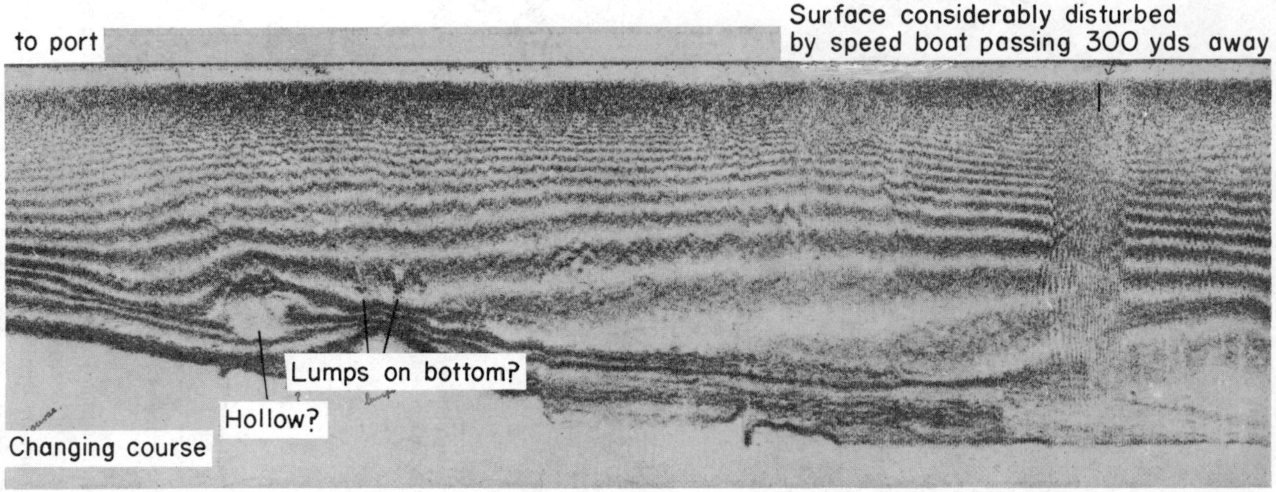

(Courtesy Kelvin Hughes)

Figure 247. Side-scan sonar record illustrating the effect of Lloyd's mirror interference.

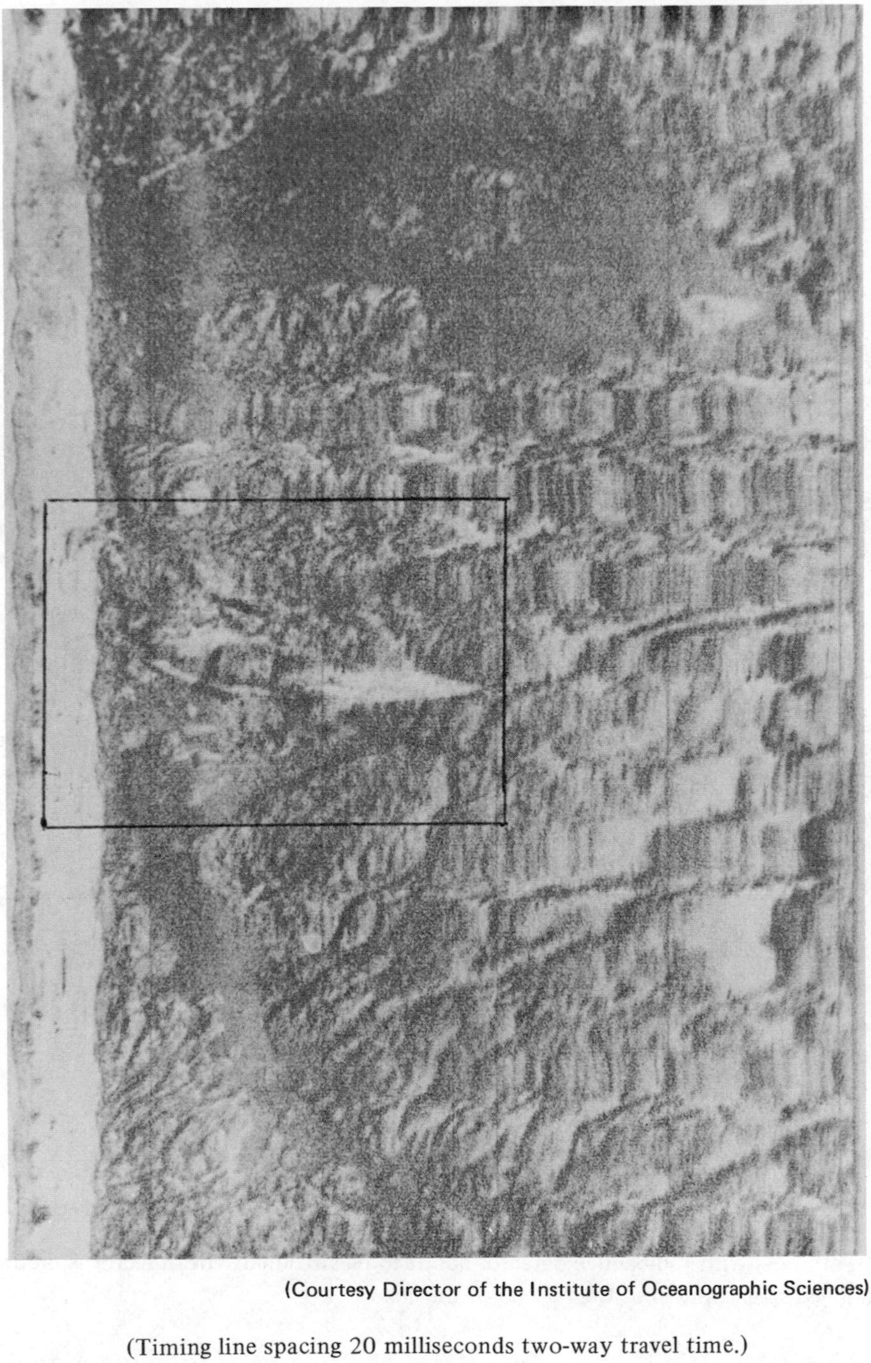

(Courtesy Director of the Institute of Oceanographic Sciences)

(Timing line spacing 20 milliseconds two-way travel time.)

Figure 248. Side-scan sonar record of a shipwreck lying on a rocky seabed. The shipwreck lies in the centre of the marked rectangle. Note the similarity between the shapes and recorded signal levels from the rocks and wreck. The wreck in this instance is characterized by the typical shape of its acoustic shadow.

(Courtesy Director of the Institute of Geological Sciences)

Figure 249. Eel with radioactive detector about to be streamed. The detector is mounted in a stainless steel cylinder at the far end of the eel.

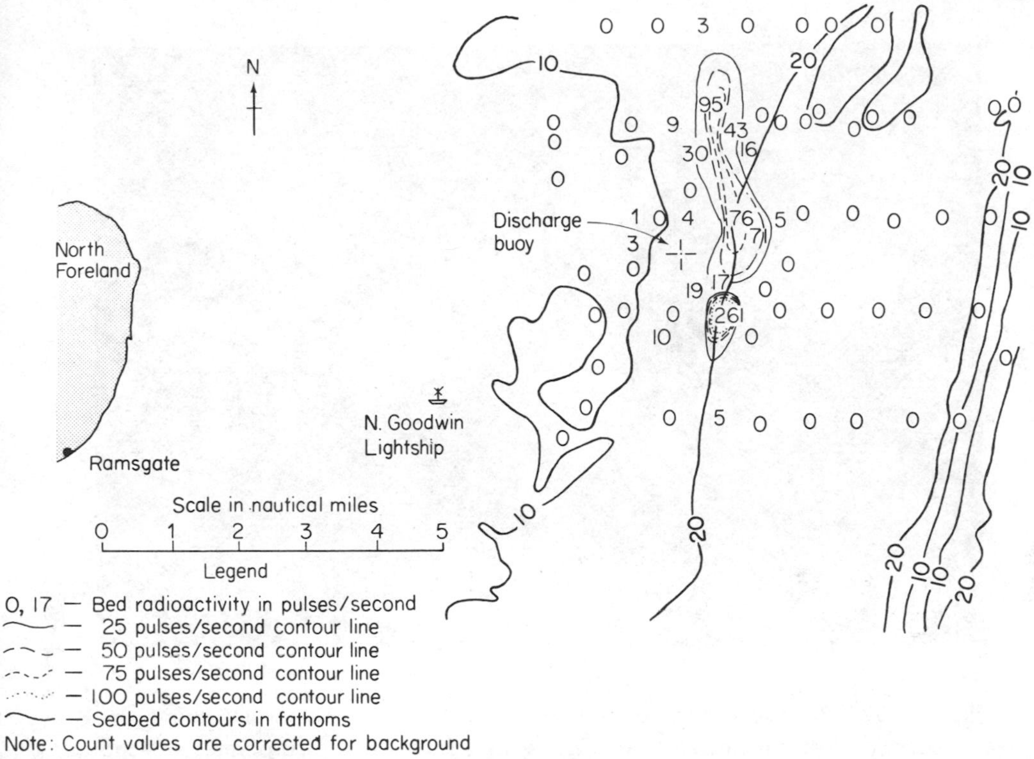

Figure 250. Trace distribution measured 2, 3 and 4 days after the start of a tracer injection study during the survey of a proposed GLC sewer outfall line. Note that the imprint of the tracer on the seabed is streamed in the tidal direction without any tendency to expand towards the coast or join in the Goodwins circulation. (Reproduced by permission HMSO, courtesy the Hydraulic Research Station, Wallingford, the Greater London Council and Sir Frederick Snow and partners.)

(Courtesy Director of the Institute of Oceanographic Sciences)

Figure 251. Radioactive backscatter and transmission gauges used in experiments to determine geotechnical properties of seabed materials. The twin probes are the transmission gauge and single probe the backscattered gauge. (Photograph: R. Parker).

(Courtesy Mobell Marine Development Co. Ltd.)

Figure 252. Pipelok system comprising the towed seabed sensor and ship-borne recording instruments. Note that the seabed sensor also incorporates an echo-sounder transducer.

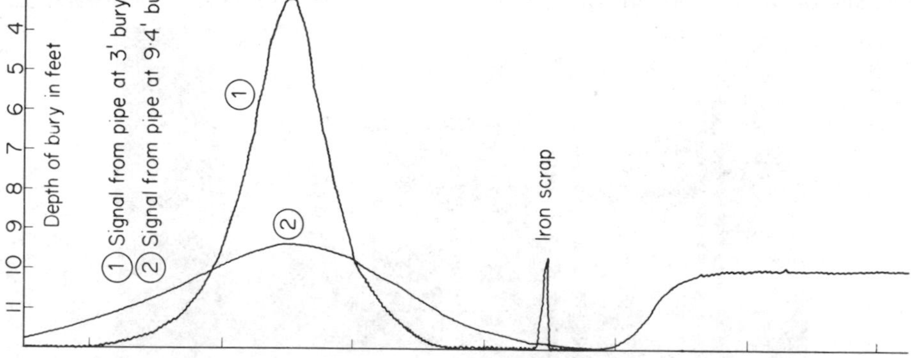

(Courtesy Mobell Marine Development Co. Ltd.)

Figure 253. Record obtained with the Pipelok system when passing over pipelines buried at different depths. Note the change in signal base level on the right of the record. This occurred when the seabed sensor 'flew' above the seabed through being towed too fast.

(Courtesy Geo Electronics)

Figure 254. Diver using a metal detector to search for metal artifacts buried beneath the seabed. (Photograph taken on S. Wignal's archaeological expedition to the Azores, 1972.)

(Courtesy Director of the Institute of Geological Sciences)

Figure 255. Recovered seabed rock dredge with sample. Note the strengthened metal jaws. (Photograph R. Kirby.)

(Courtesy Director of the Institute of Oceanographic Sciences)

Figure 256. Van Veen type of grab sampler being hauled on board. Note the inspection hatch cut into the top of the sample jaws. (Photograph J.R. Hails.)

(Photograph: Author)

Figure 257. 'Cocked' Shipek grab about to be lowered to seabed. Note the helically wound springs and triggering weight. This slides down the vertical piston and activates the springs when the grab hits the seabed.

(Photograph: Author)

Figure 258. Simple drop corer being hauled on board. Note that each corer weight is approximately 50 kg.

(Photograph: Author)

Figure 259. Umel drop corer being deployed. Note the bevelled core cutter at the bottom of the core barrel. Also the shaped fins at the top of the core which are added to improve stability during free fall.

(Photograph: Author)

(a) Piston Corer being deployed. (Note the trip mechanism attached to the top of the corer.)

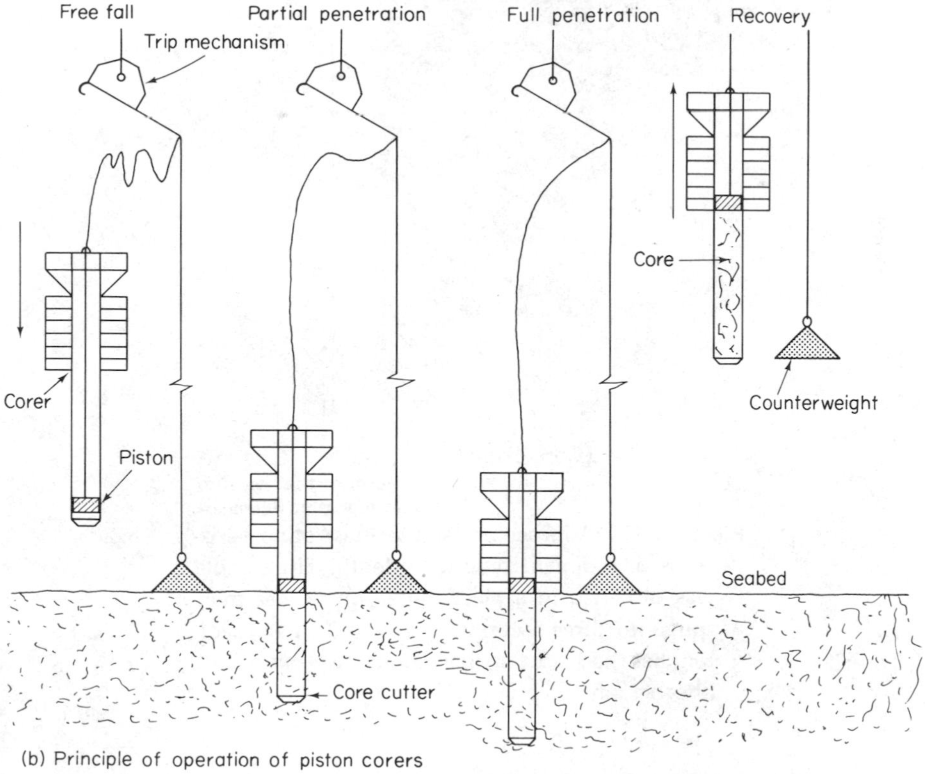

(b) Principle of operation of piston corers.

Figure 260. Piston corer being deployed and principle of operation.

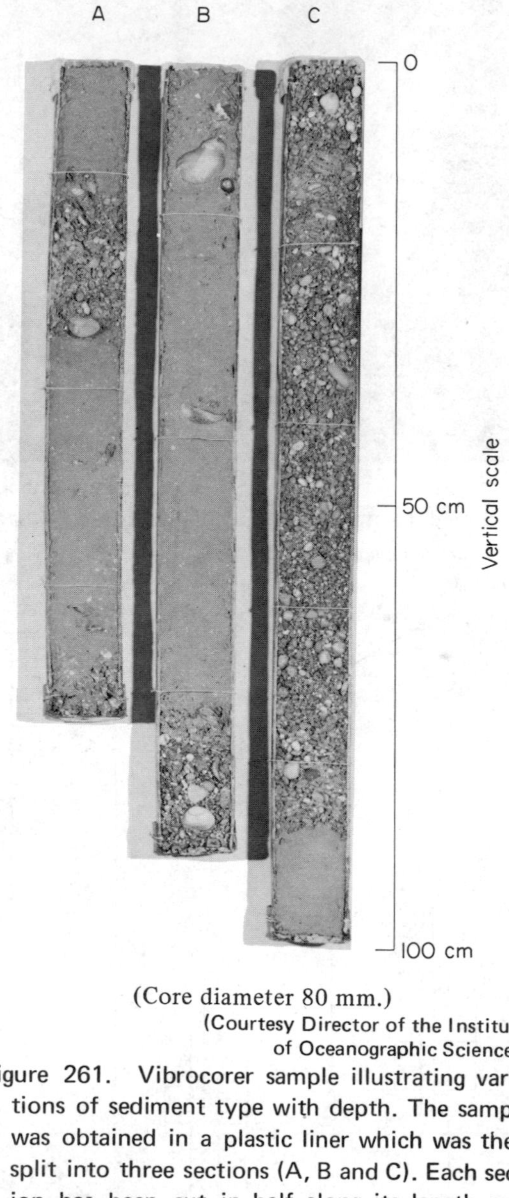

(Core diameter 80 mm.)

(Courtesy Director of the Institute of Oceanographic Sciences)

Figure 261. Vibrocorer sample illustrating variations of sediment type with depth. The sample was obtained in a plastic liner which was then split into three sections (A, B and C). Each section has been cut in half along its length and photographed.

(Courtesy Director of the Institute of Oceanographic Sciences and the Rijkswaterstaat)

Figure 262. Institute of Oceanographic Sciences vibrocorer being lowered into its sample handling position. Note the relocation pinger and orientation compass mounted on the raised part of the frame.

(Courtesy Director of the Institute of Oceanographic Sciences)

Figure 263. Lashing the Zenkovitch vibrocorer alongside prior to installing a core barrel. Note the fins on the bottom part of the frame. These orientate the vibrocorer in the current direction.

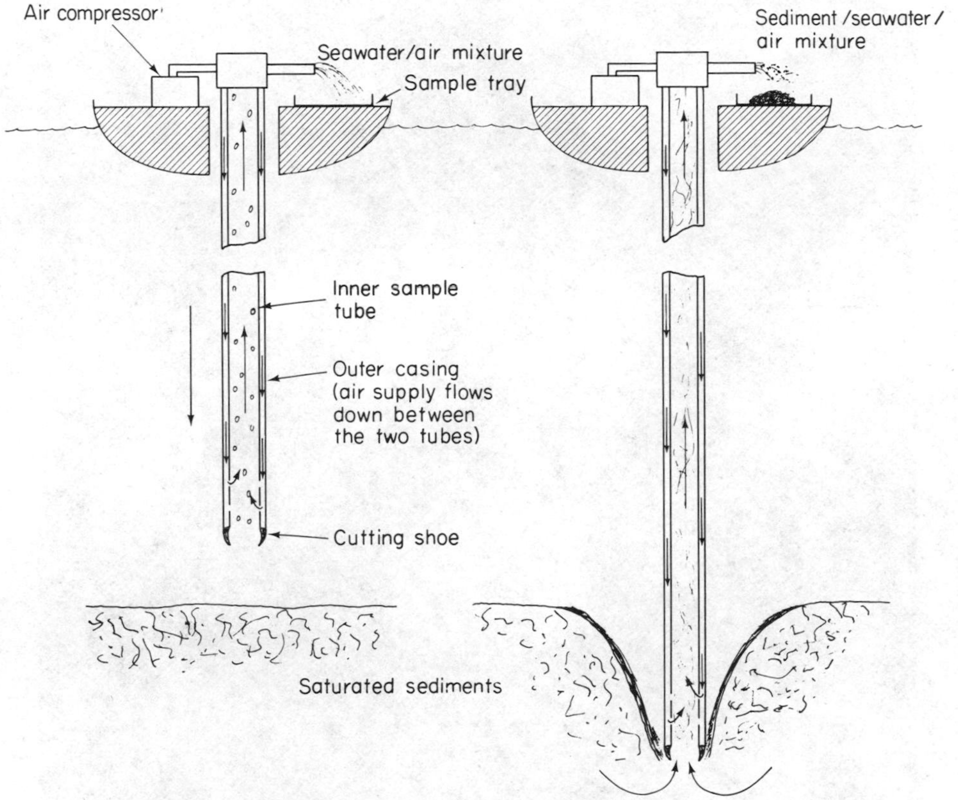

Figure 264. Principle of operation of airlift drills.

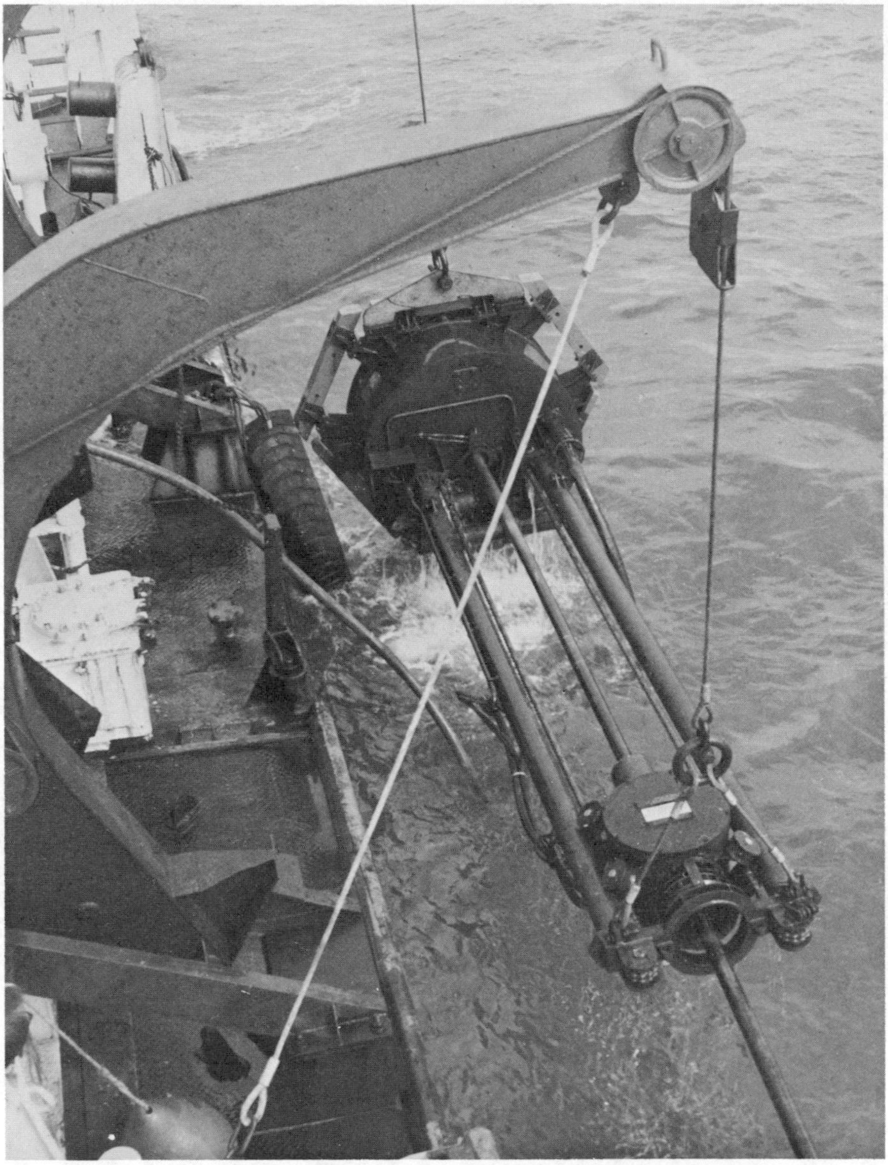

(Photograph: Author; Courtesy Director of the Institute of Oceanographic Sciences and Rijkswaterstaat)

Figure 265. Geodoff vibrocorer being deployed. The vibrocorer is swung outboard lying horizontal, as shown, and is lowered into a vertical position once in the water. Note the corer tube running through the vibrating head at the top of the drilling frame. This clamps on to the core tube during drilling or extraction and rides down or up the frame.

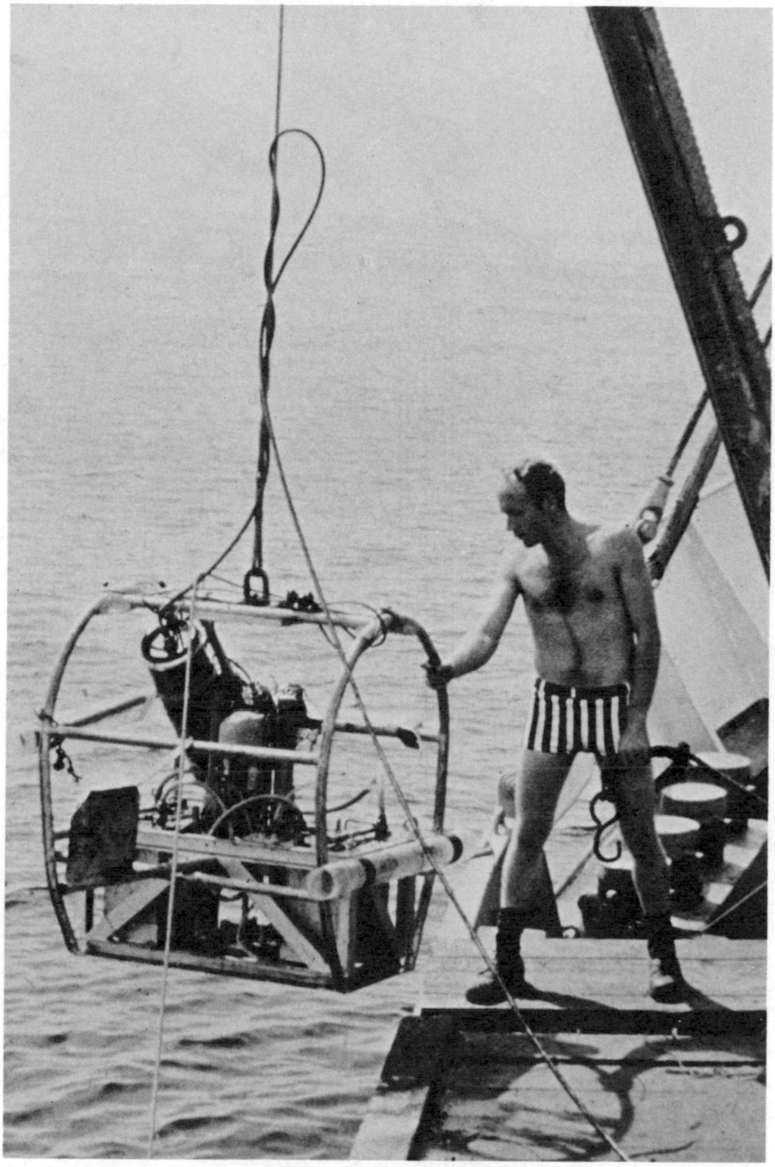

(Courtesy Director of the Institute of Geological Sciences)

Figure 266. Harrison Hardrock corer being deployed. Note the monitor underwater television camera mounted in the top left of the frame and the orientation fin. (Photograph: R. Kirby)

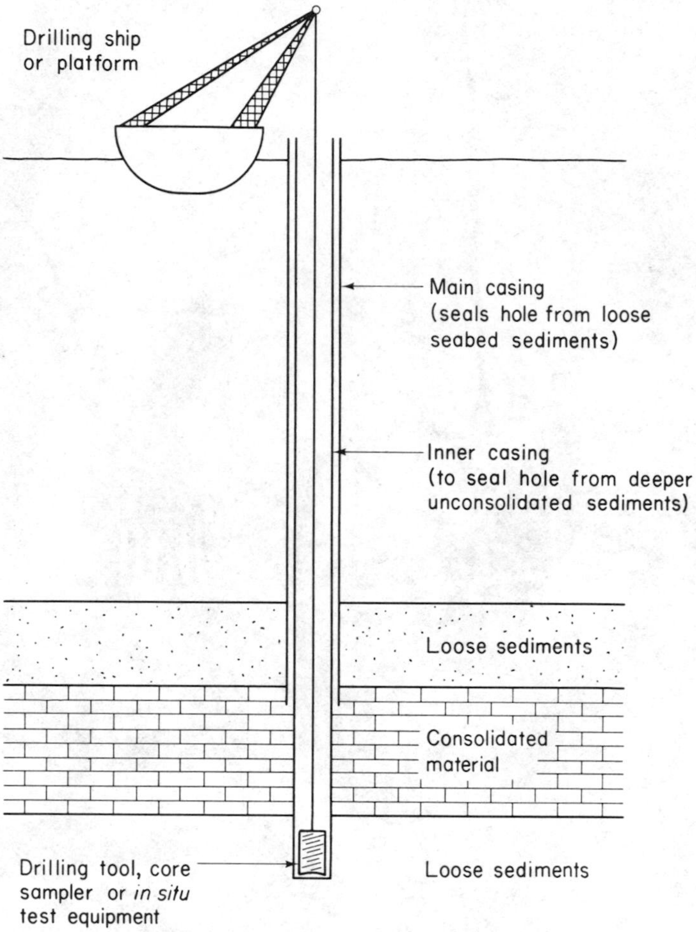

Figure 267. Principle of cased drilling technique.

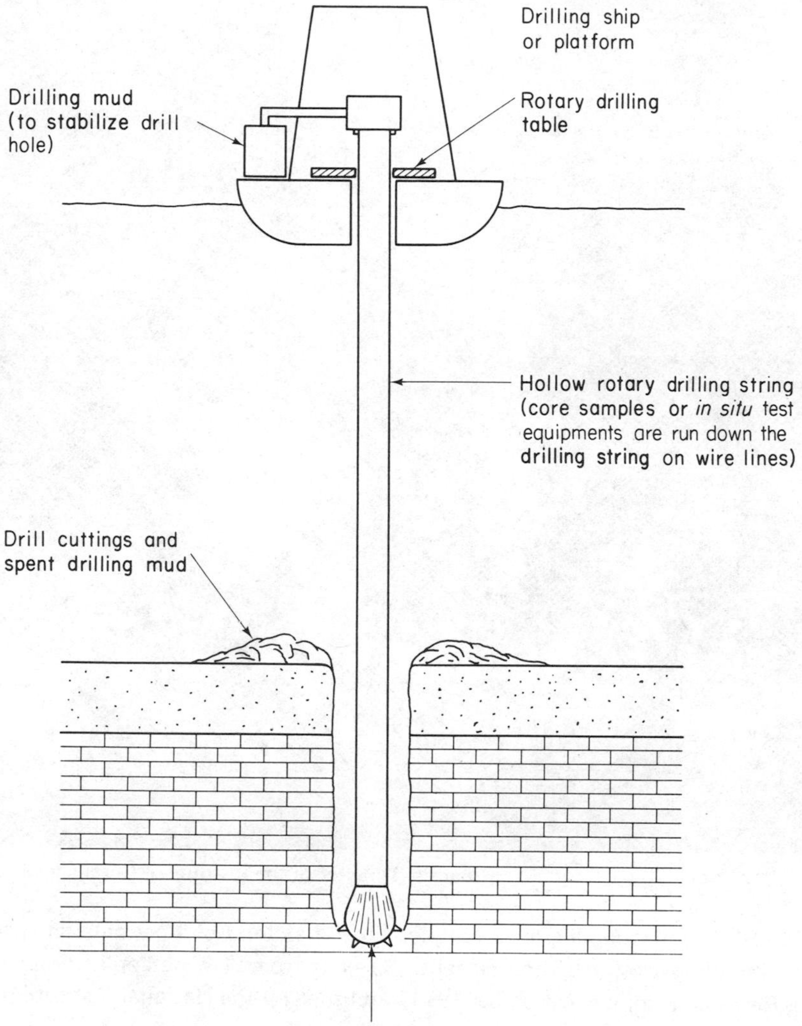

Figure 268. Principle of rotary drilling technique.

(Courtesy Director of the Institute of Oceanographic Sciences)

Figure 269. Diver collecting a seabed sample. Note the cloud of fine material swept away from the sampling trowel by the current. (Taken from 'Diving applications in Marine Sciences Research Seminar', December 1971, published by the National Institute of Oceanography.) (Photograph: E.J. Moore)

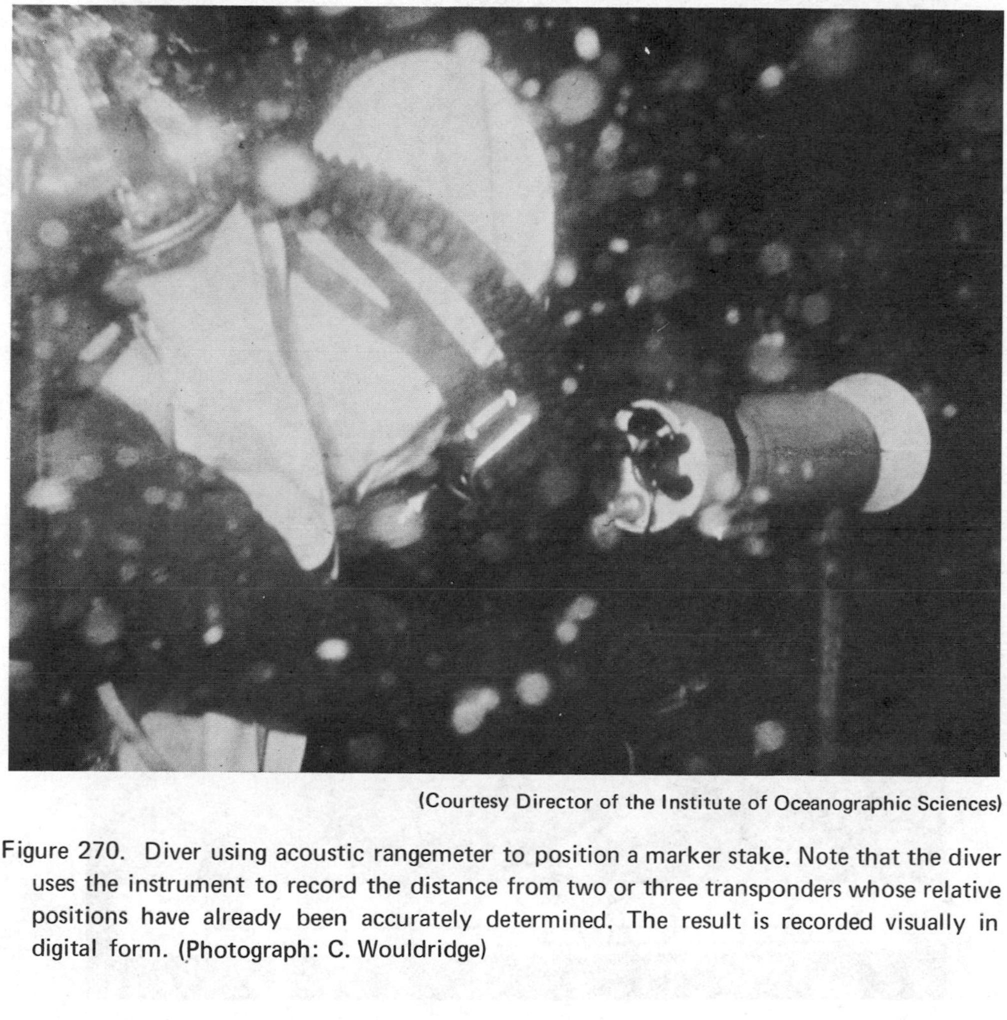

(Courtesy Director of the Institute of Oceanographic Sciences)

Figure 270. Diver using acoustic rangemeter to position a marker stake. Note that the diver uses the instrument to record the distance from two or three transponders whose relative positions have already been accurately determined. The result is recorded visually in digital form. (Photograph: C. Wouldridge)

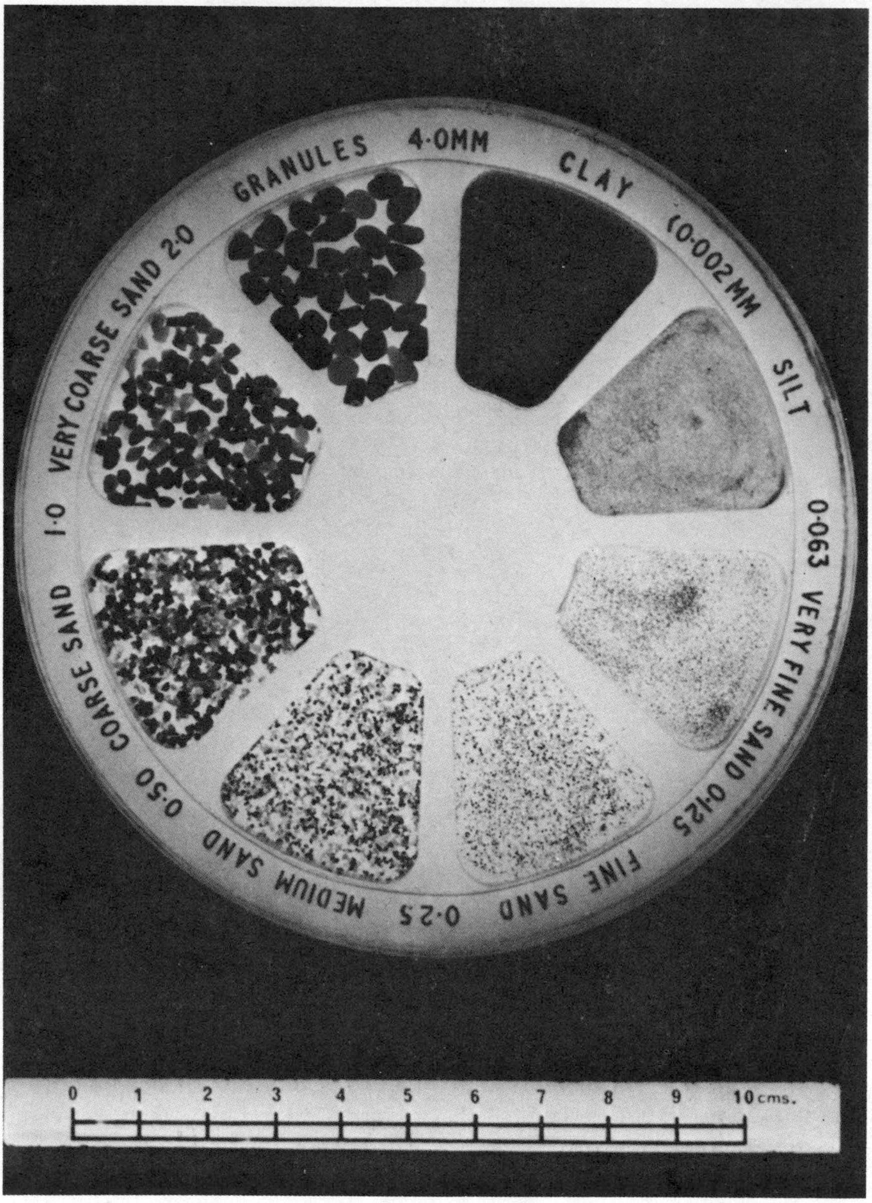

(Courtesy R. Kirby)

Figure 271. Top view of a grain-size comparator disc. Note the size limits of each sediment sample in millimeters. The class terms for each grade is after Wentworth. (Taken from 'The UCS grain size comparator disc', published in *Marine Geology,* 14/1973, M11-M14.)

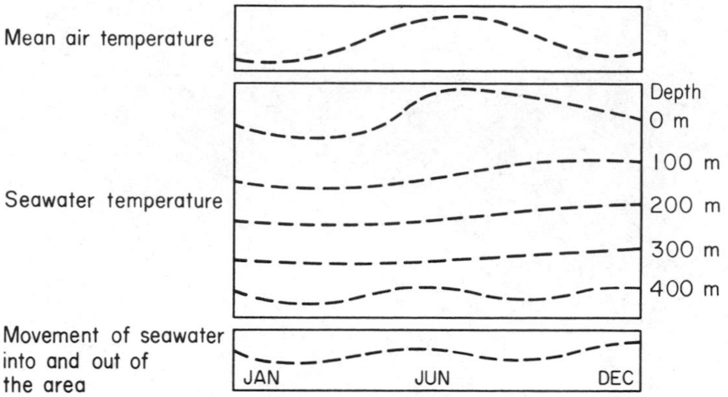

Figure 272. Typical time–depth–temperature profiles for mid-latitudes over one year, with similar profiles for the mean air temperature and seawater migration for comparison.

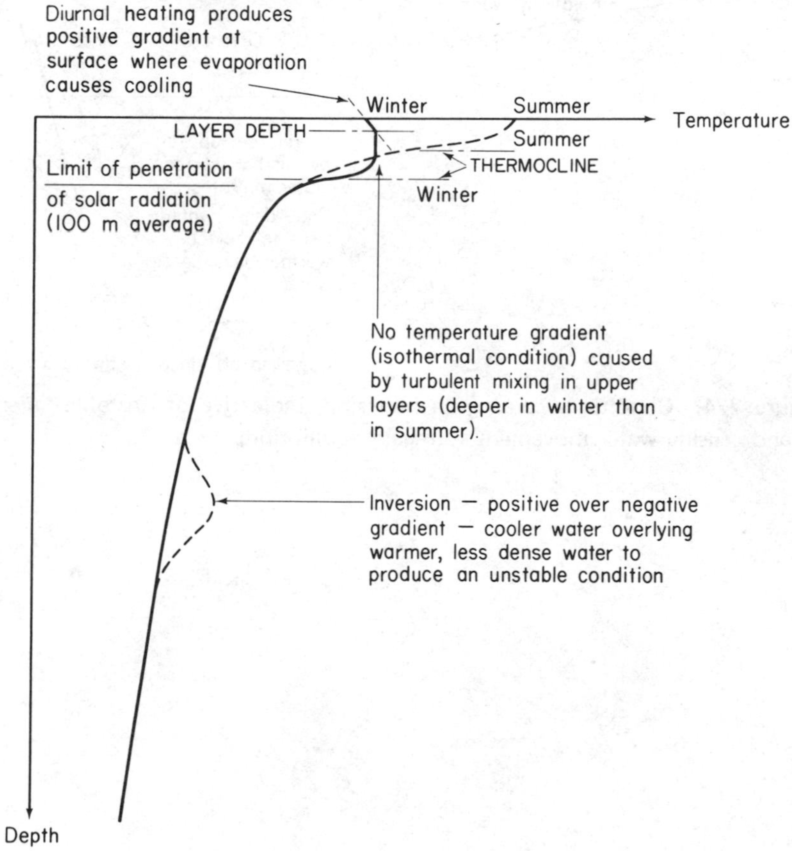

Figure 273. A typical temperature-depth profile.

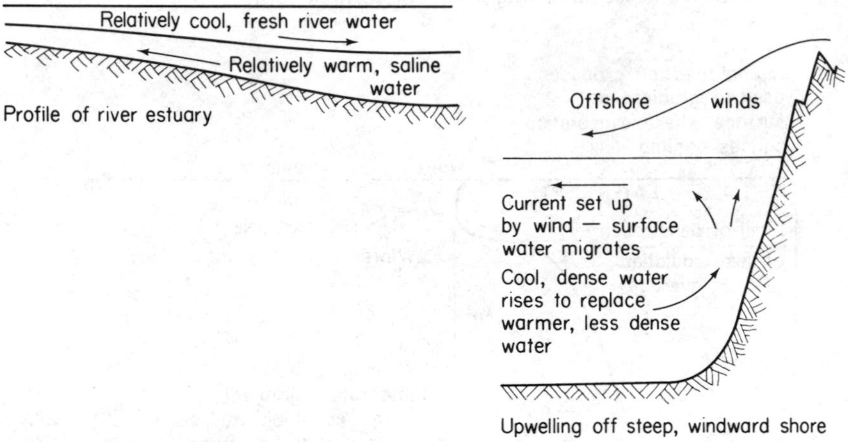

Figure 274. Conditions causing an inversion, indicative of unstable layering and causing water movement to regain equilibrium.

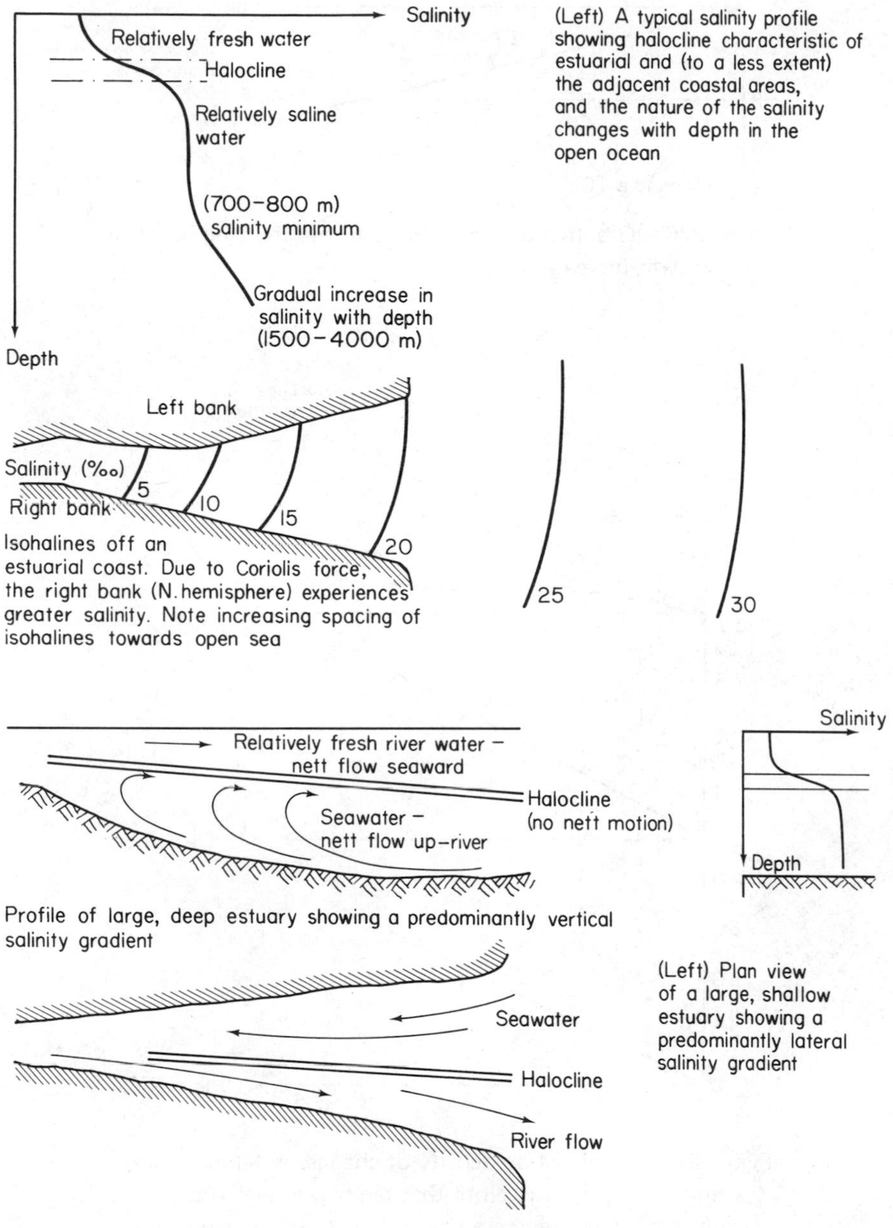
Salinity
Relatively fresh water
Halocline
Relatively saline water
(700–800 m) salinity minimum
Gradual increase in salinity with depth (1500–4000 m)
Depth
(Left) A typical salinity profile showing halocline characteristic of estuarial and (to a less extent) the adjacent coastal areas, and the nature of the salinity changes with depth in the open ocean
Left bank
Salinity (‰)
Right bank
5
10
15
20
25
30
Isohalines off an estuarial coast. Due to Coriolis force, the right bank (N. hemisphere) experiences greater salinity. Note increasing spacing of isohalines towards open sea
Relatively fresh river water – nett flow seaward
Halocline (no nett motion)
Seawater – nett flow up-river
Salinity
Depth
Profile of large, deep estuary showing a predominantly vertical salinity gradient
Seawater
Halocline
River flow
(Left) Plan view of a large, shallow estuary showing a predominantly lateral salinity gradient

Figure 275. The halocline and characteristic salinity conditions of estuarial areas.

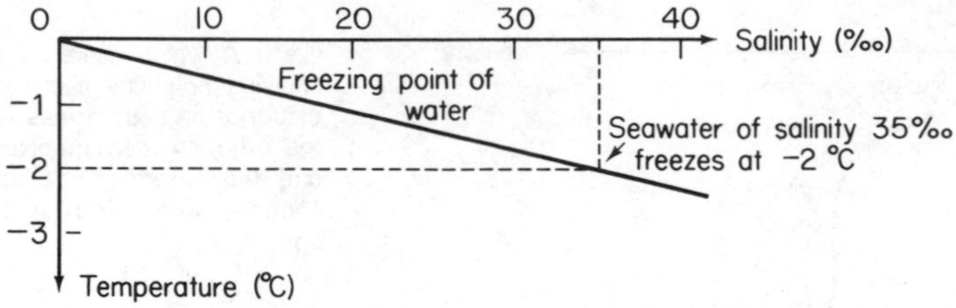

Figure 276. Graph showing changes in the freezing point of water with increasing salinity.

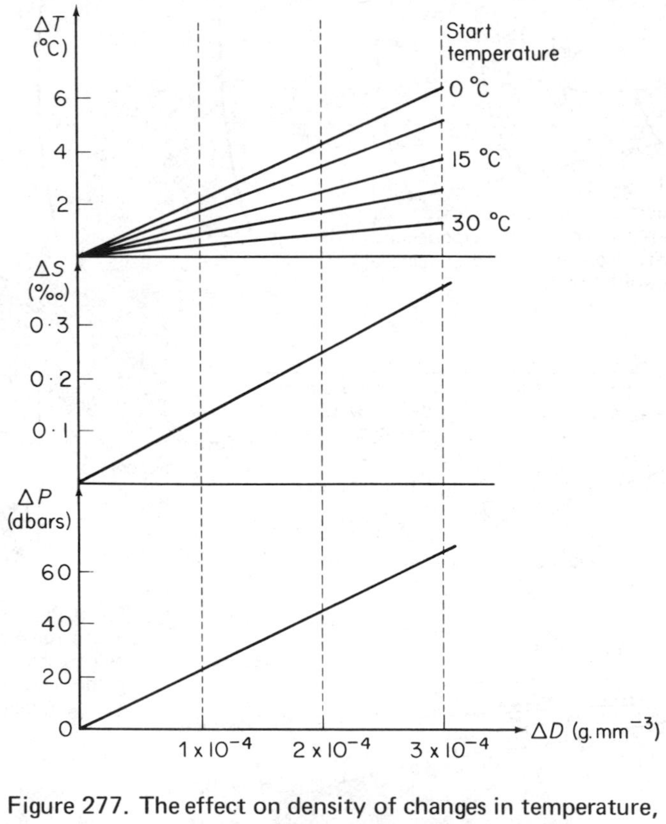

Figure 277. The effect on density of changes in temperature, salinity and pressure. Note that temperature effects vary with the 'start' temperature, salinity and pressure; for example at atmospheric pressure and a temperature of 0°C, seawater of 35‰ salinity increases in density at a rate of 5×10^{-5} g.mm^{-3} per 1°C change in temperature.

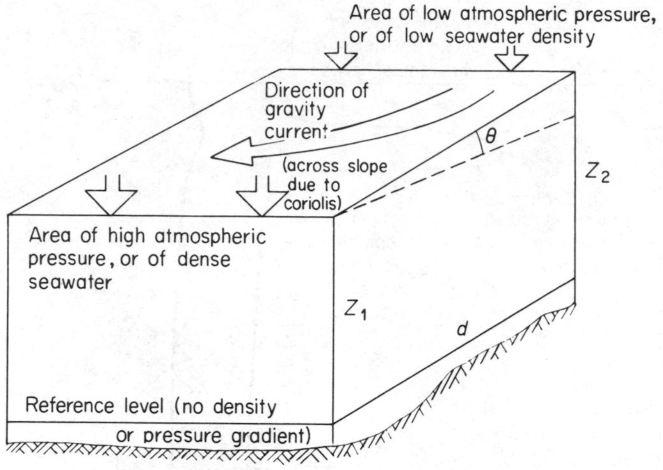

Figure 278. Gravity currents set up, either by atmospheric pressure or by density differentials.

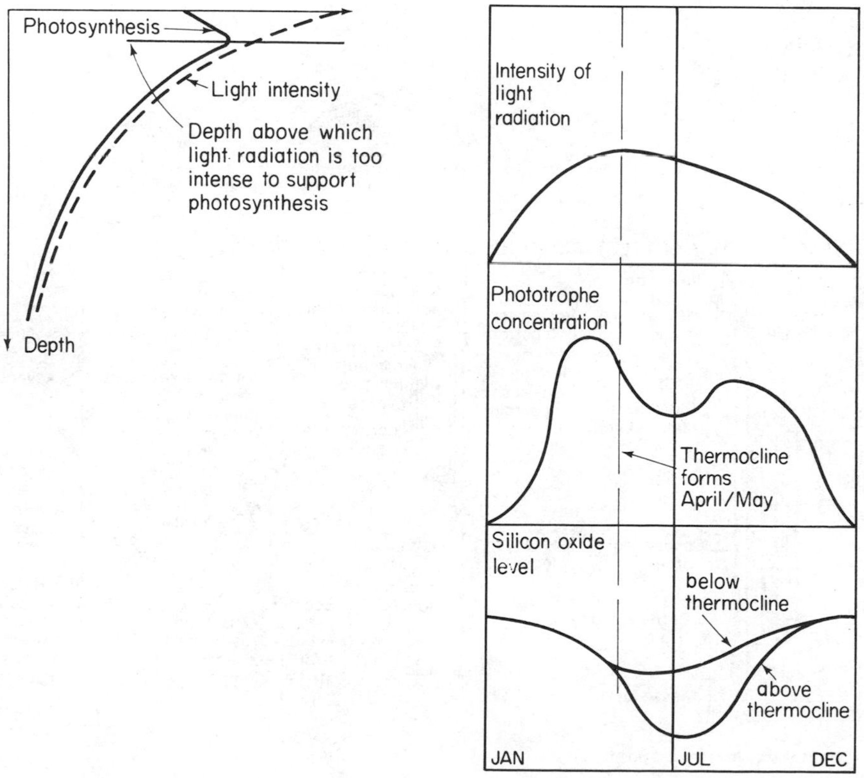

Figure 279. The relationship of photosynthesis to seasonal light intensity and the consequent level of silicon concentration in seawater.

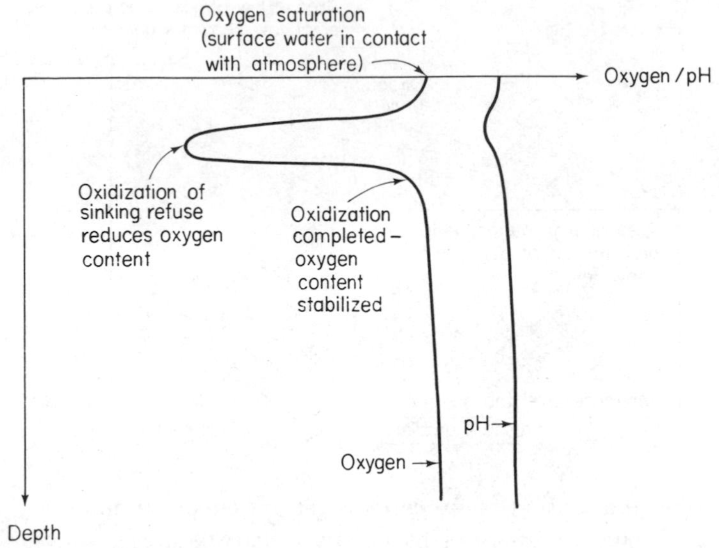

Figure 280. Typical variations in oxygen content at an oceanographic station and the damping action of carbon dioxide on the pH value.

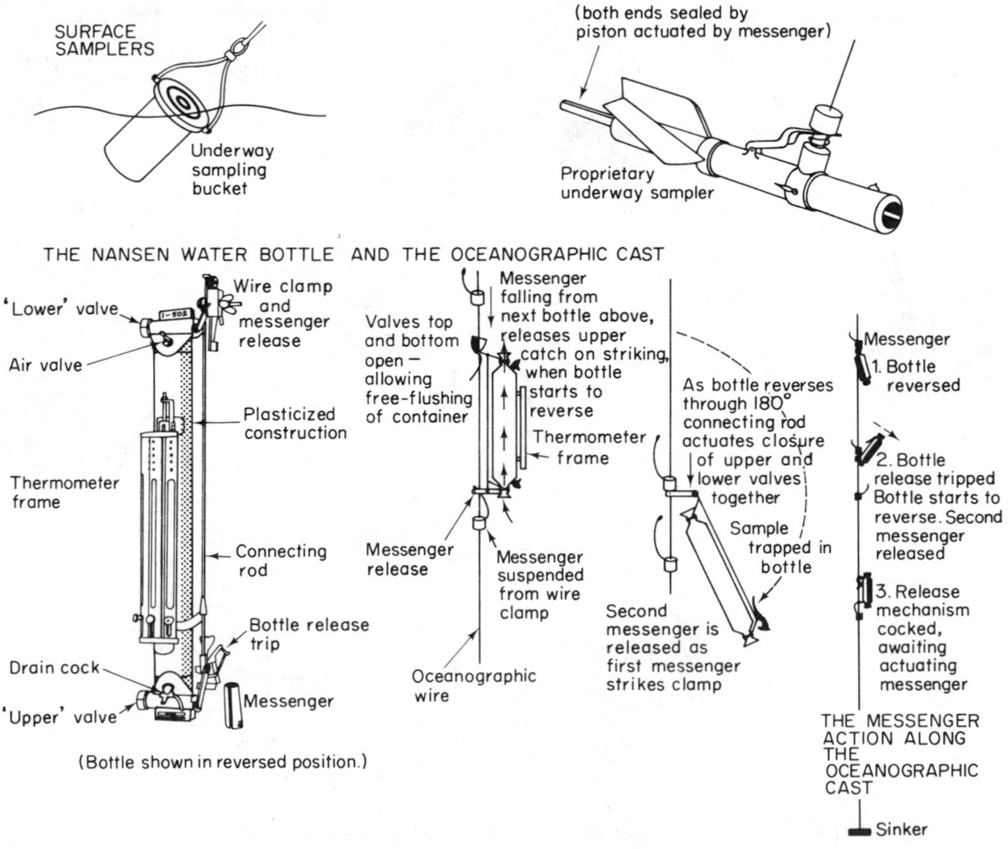

Figure 281. Water samplers.

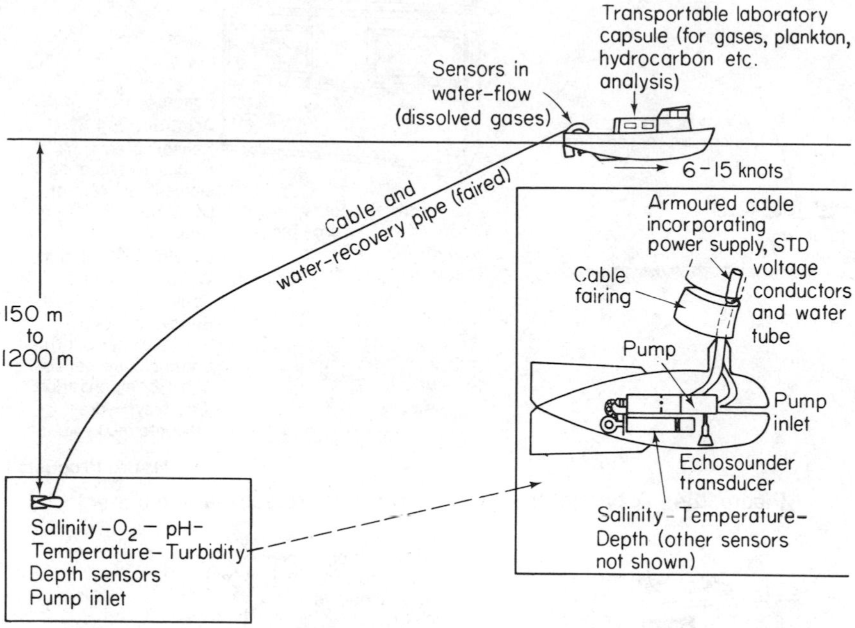

Figure 282. The InterOcean Systems Inc. Submersible Pumping Systems.

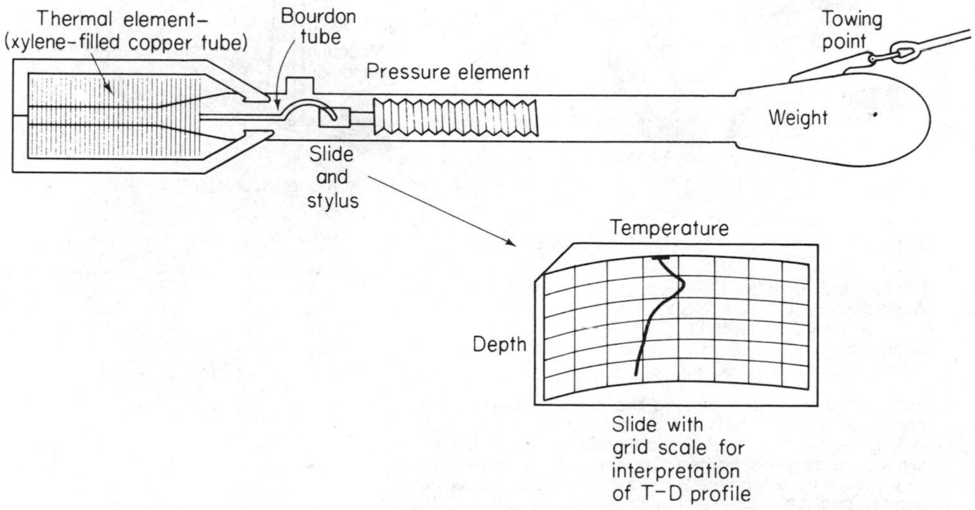

Figure 283. The bathythermograph.

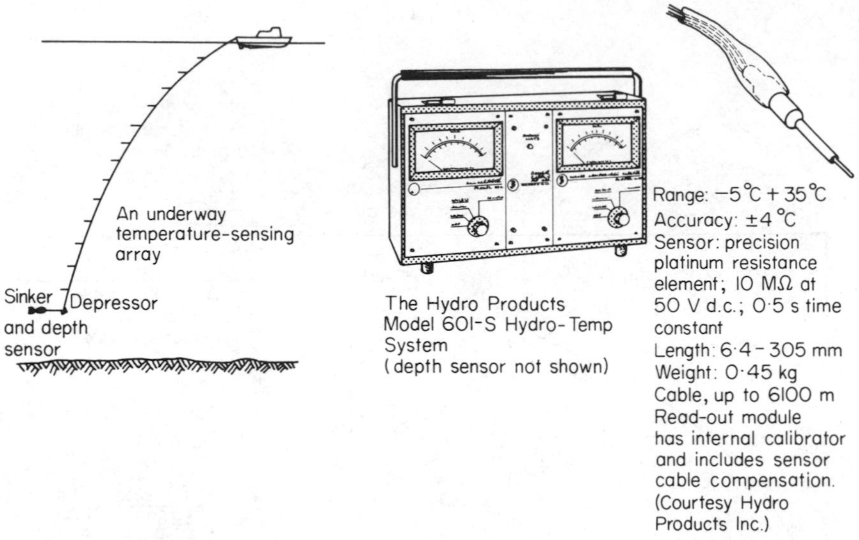

(Courtesy Hydro Products Inc.)

Figure 284. A typical thermistor probe with direct reading meter.

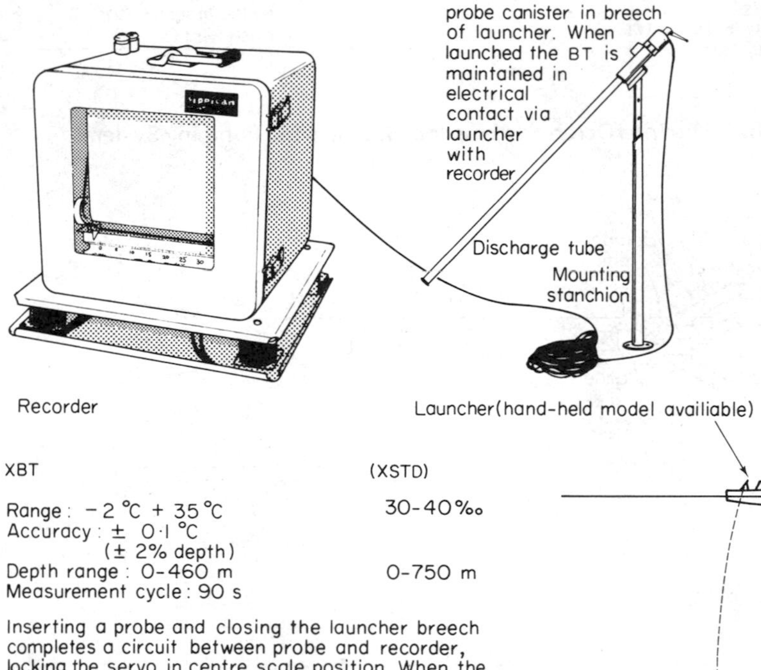

XBT	(XSTD)
Range: −2 °C + 35 °C	30–40 ‰
Accuracy: ± 0·1 °C (± 2% depth)	
Depth range: 0–460 m	0–750 m
Measurement cycle: 90 s	

Inserting a probe and closing the launcher breech completes a circuit between probe and recorder, locking the servo in centre scale position. When the probe is gravity-discharged over the side, a seawater trigger circuit is completed, the recorder chart-drive is started and the measurement cycle commences.

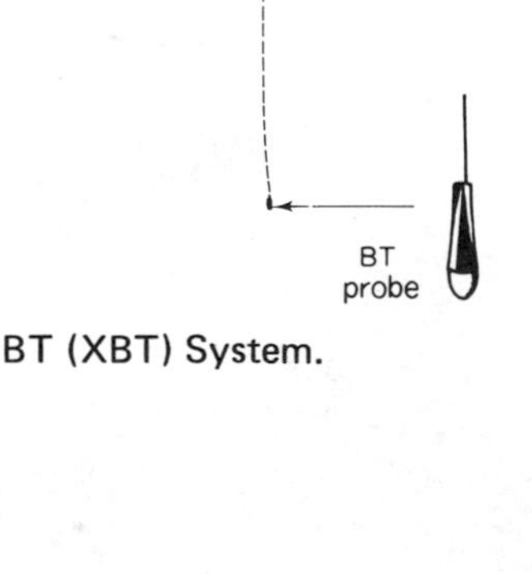

(Courtesy The Plessey Co. Ltd.)

Figure 285. The Sippican Expendable BT (XBT) System.

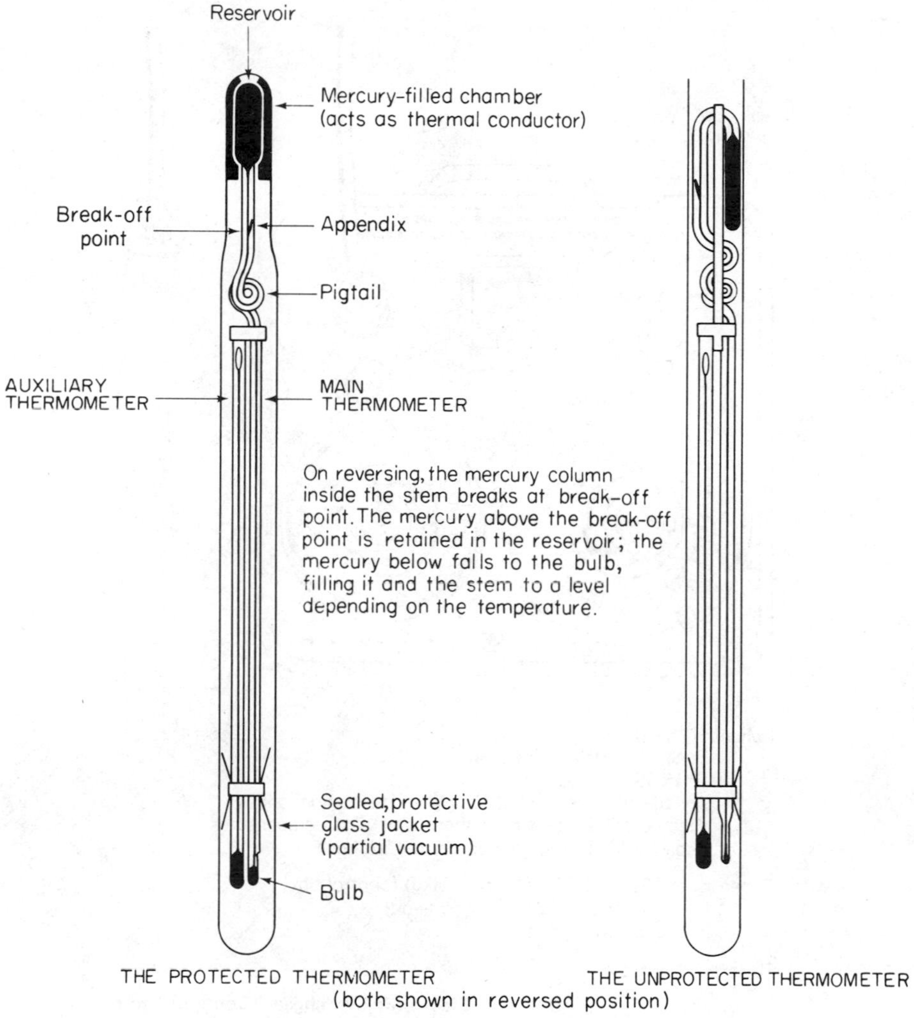
Reservoir
Mercury-filled chamber (acts as thermal conductor)
Break-off point
Appendix
Pigtail
AUXILIARY THERMOMETER
MAIN THERMOMETER
On reversing, the mercury column inside the stem breaks at break-off point. The mercury above the break-off point is retained in the reservoir; the mercury below falls to the bulb, filling it and the stem to a level depending on the temperature.
Sealed, protective glass jacket (partial vacuum)
Bulb
THE PROTECTED THERMOMETER
THE UNPROTECTED THERMOMETER
(both shown in reversed position)

Figure 286. Protected and unprotected thermometers.

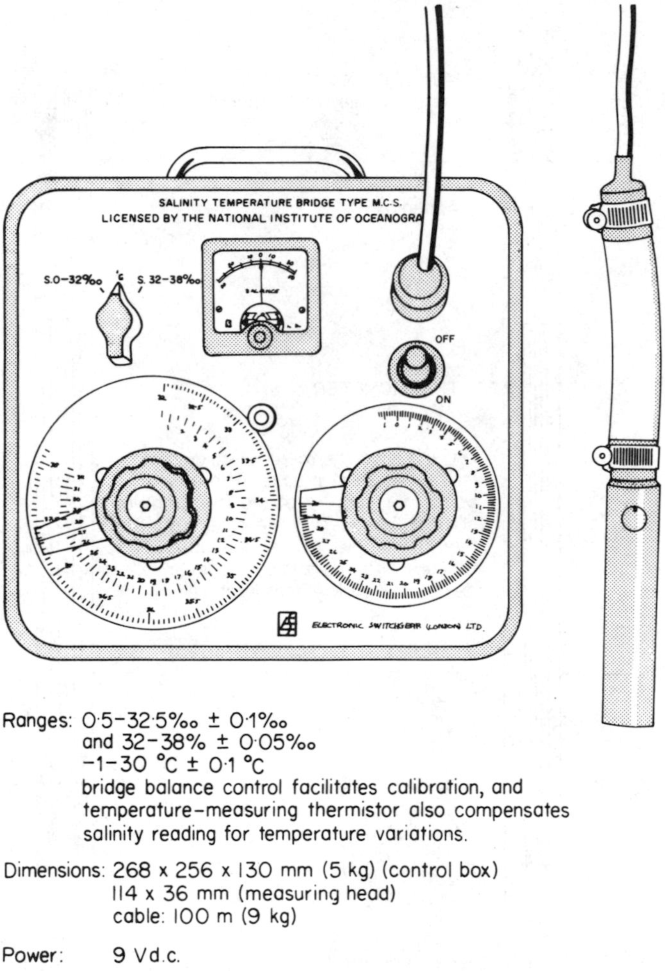

Ranges: 0·5–32·5‰ ± 0·1‰
and 32–38% ± 0·05‰
−1–30 °C ± 0·1 °C
bridge balance control facilitates calibration, and temperature-measuring thermistor also compensates salinity reading for temperature variations.

Dimensions: 268 x 256 x 130 mm (5 kg) (control box)
114 x 36 mm (measuring head)
cable: 100 m (9 kg)

Power: 9 V d.c.

(Courtesy Electronic Switchgear (London) Ltd.)

Figure 287. The MC5 *in situ* salinity and temperature bridge.

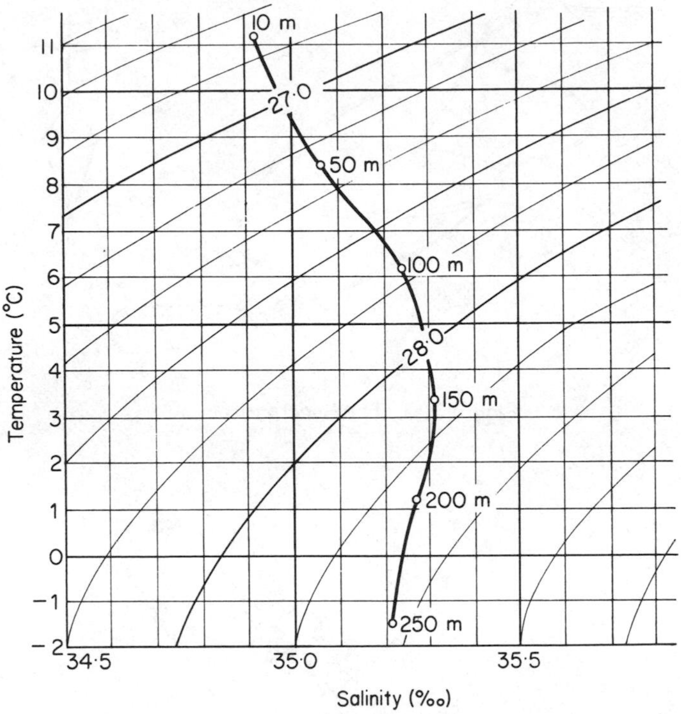

Figure 288. A temperature-salinity graph with σ_t *'isopycnals'*. Water masses may be identified within an area of interest by their characteristic T–S curves and their stability by the slope of the curve relative to the slope of the isopycnals.

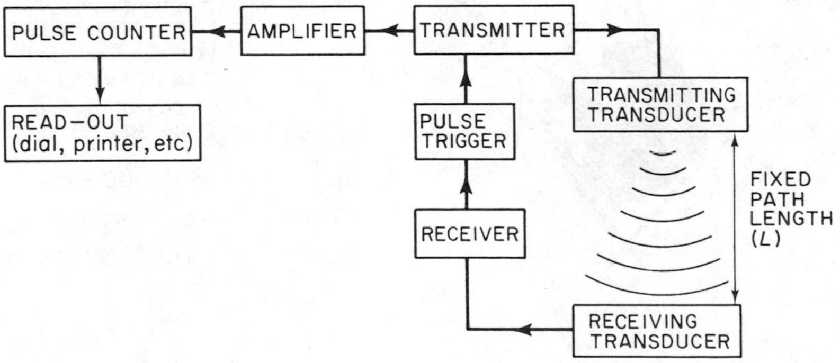

Figure 289. The 'sing-around' principle – block diagram.

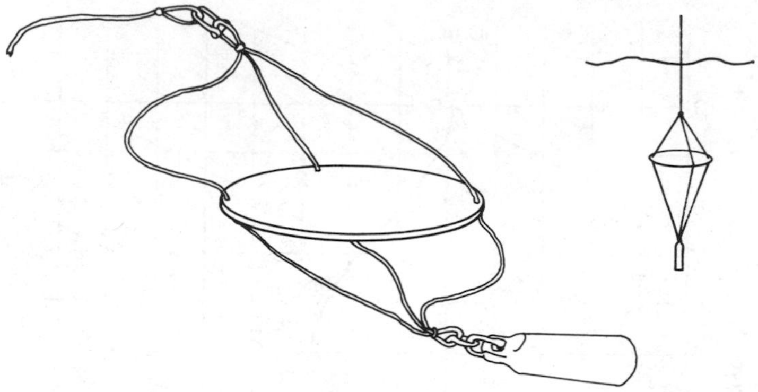

Figure 290. The Secchi disc.

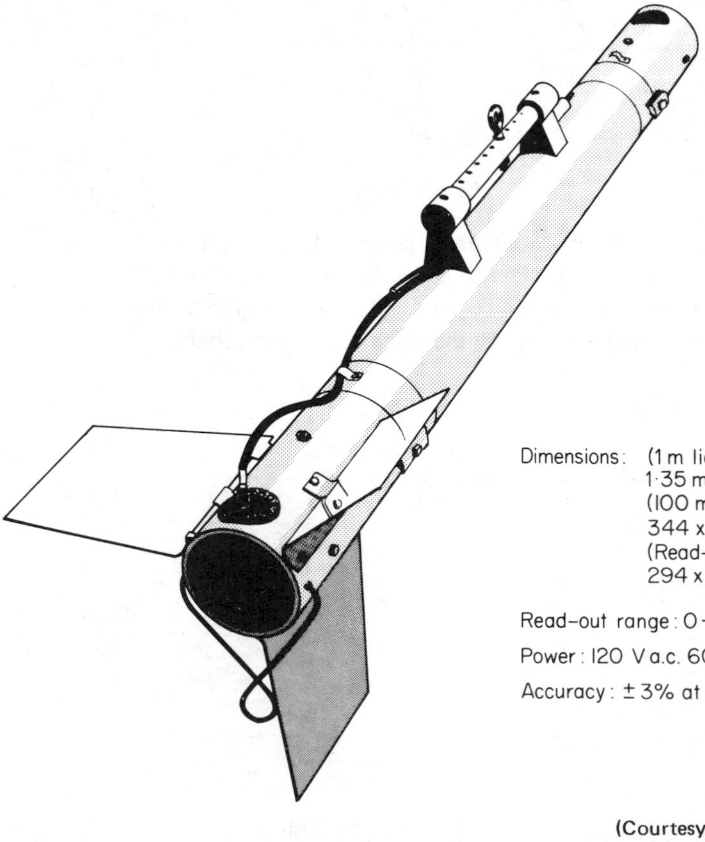

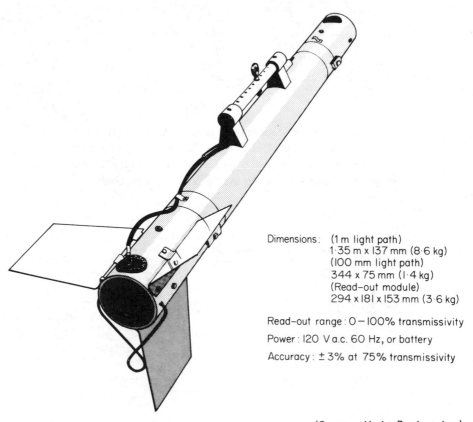

(Courtesy Hydro Products Inc.)

Figure 291. The Hydro Products Inc. Model 612S Transmissometer System.

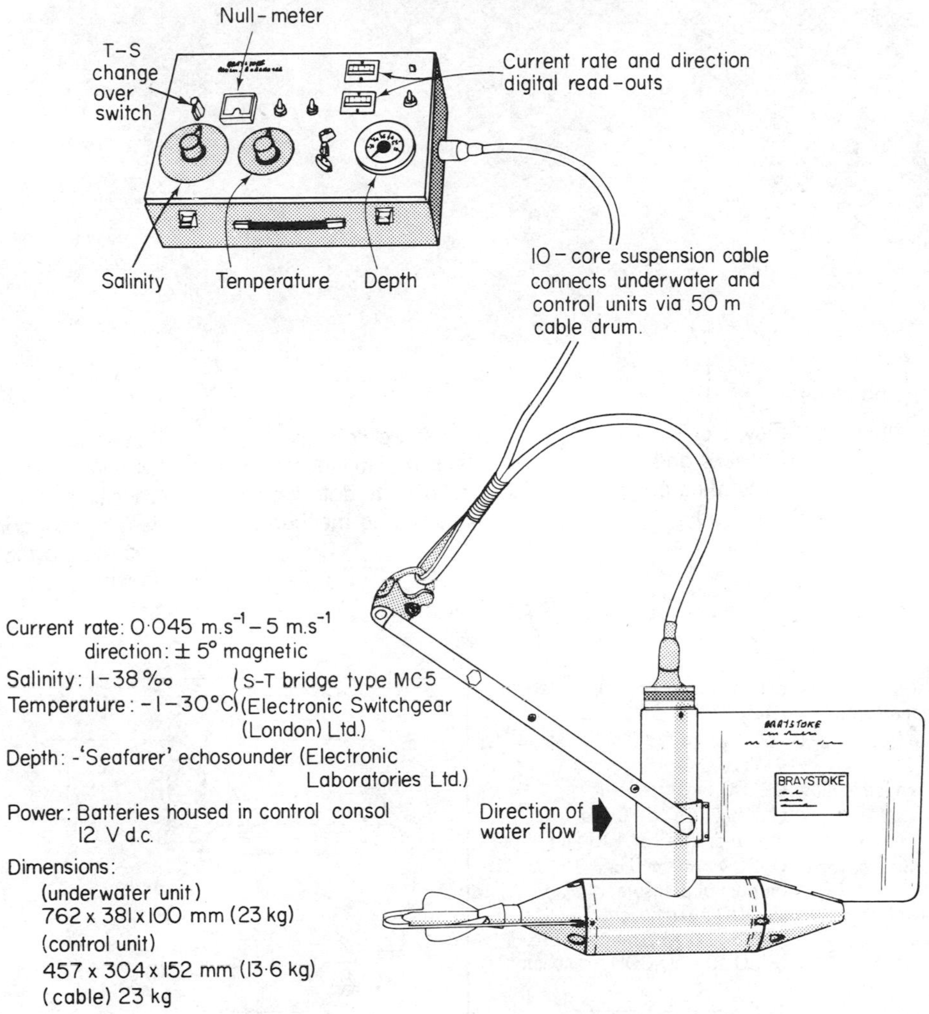

(Courtesy Valeport Engineering Ltd.)

Figure 292. The Braystoke Multi-Parameter Current Flowmeter.

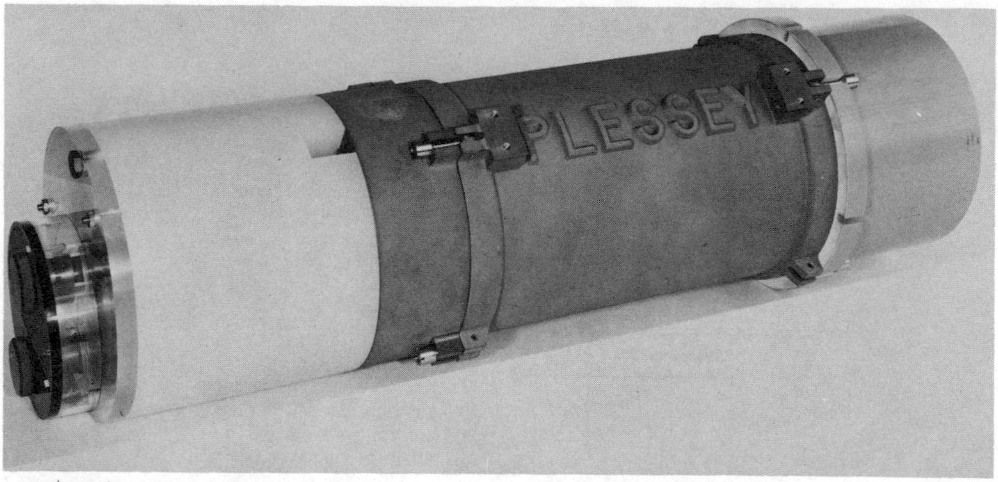

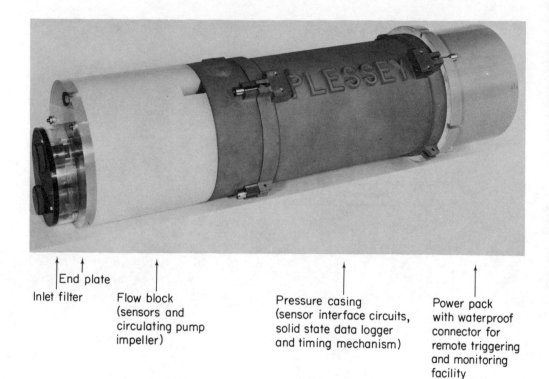

Sensors:	(all outputs reference voltages)
Temperature:	0–30 °C(±0·3 °C)
Depth (pressure)	0–500 kN.m^{-2}(±2% full scale)
pH:	4–10 (±0·2 pH)
Suspended solids:	0–100 mg. litre^{-1}
	(±1 mg. litre^{-1} at 0 mg. litre^{-1},
	± 10 mg. litre^{-1} at 100 mg. litre^{-1})
Dissolved oxygen:	0–200 mm mercury
	(in partial pressure)(±5%)
Conductivity:	10–1000 μmho. cm^{-1}
	or 50–3000 μmho. cm^{-1}
	or 3000–50,000 μmho. cm^{-1}
Specification:	
Dimensions:	760 mm x 190 mm dia.
	25 kg
Working depth:	33 m
Power supply:	Encapsulated rechargeable
	Ni–Cd batteries ± 18 V

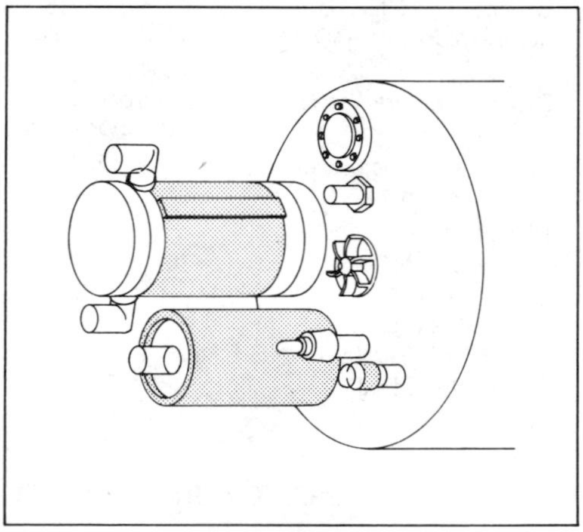

(Courtesy The Plessey Co. Ltd.)

Figure 293. The Plessey Type MM4 Submersible Water Quality Station.

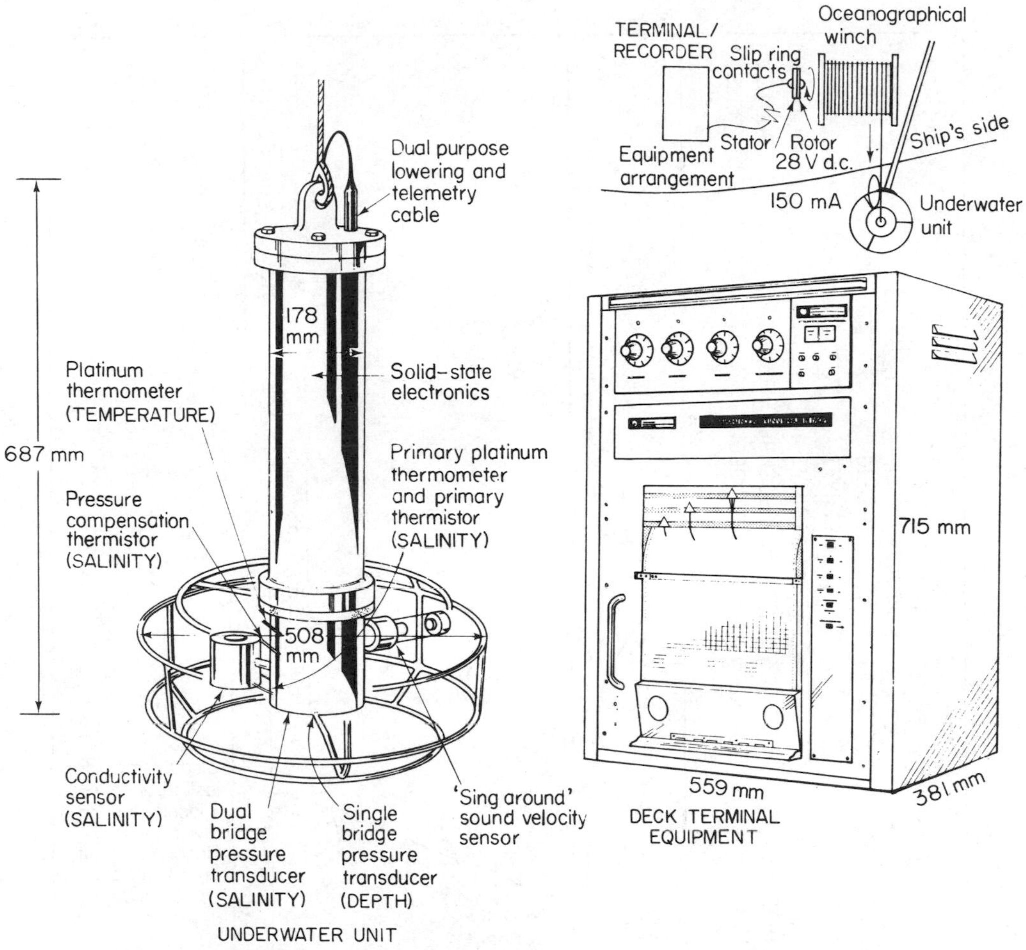

Parameters measured:

Salinity	30–40‰	to	±0·02‰	Output 4995–7901 Hz
Temperature	−2 °C–36 °C	to	±0·02 °C	Output 2127–4193 Hz
Depth	0–6000 m	to	±4 m from 0–500 m	
			±6 m from 500–6000 m	Output 9712–11,288 Hz
Sound velocity	1400–1600 m/s	to	±0·3 m/s	Output 14,000–16,000 Hz

(Depth is recorded by the paper drive roller, and other parameters are recorded across the graph as a function of depth.)

The Bissett–Berman Corporation (USA) Salinity/Temperature/Depth/Sound velocity measuring system is an example of a sophisticated electronic seawater parameter measuring device for oceanographical observations to a depth of 6000 m throughout the water column.

(Courtesy The Plessey Co. Ltd.)

Figure 294. The Bissett-Berman Model 9040 Environmental Profiling System.

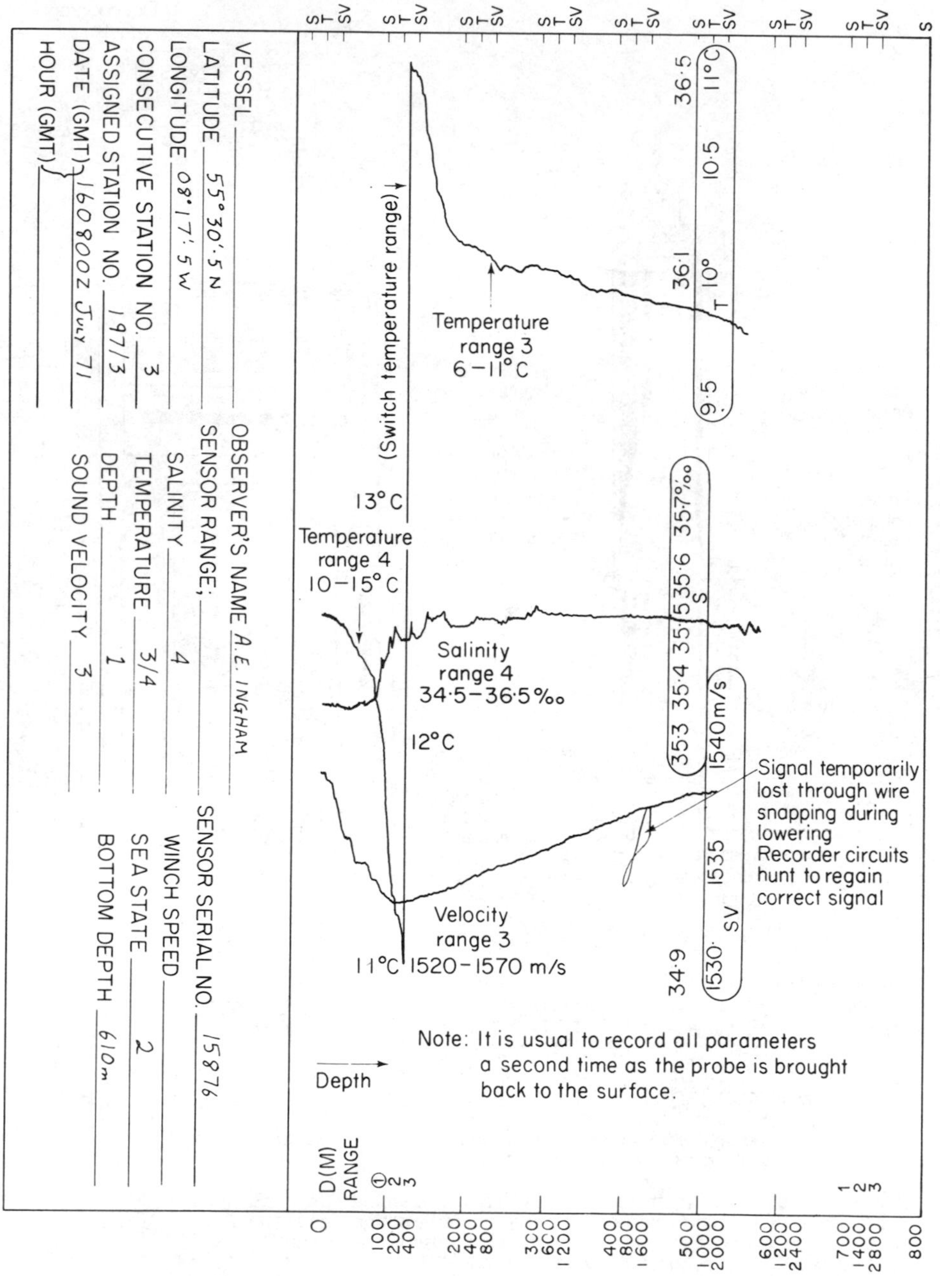

(Courtesy The Plessey Co. Ltd.)

Figure 295. Part of the record from the Bissett-Berman STDV Environmental Profiling System, Model 9040.

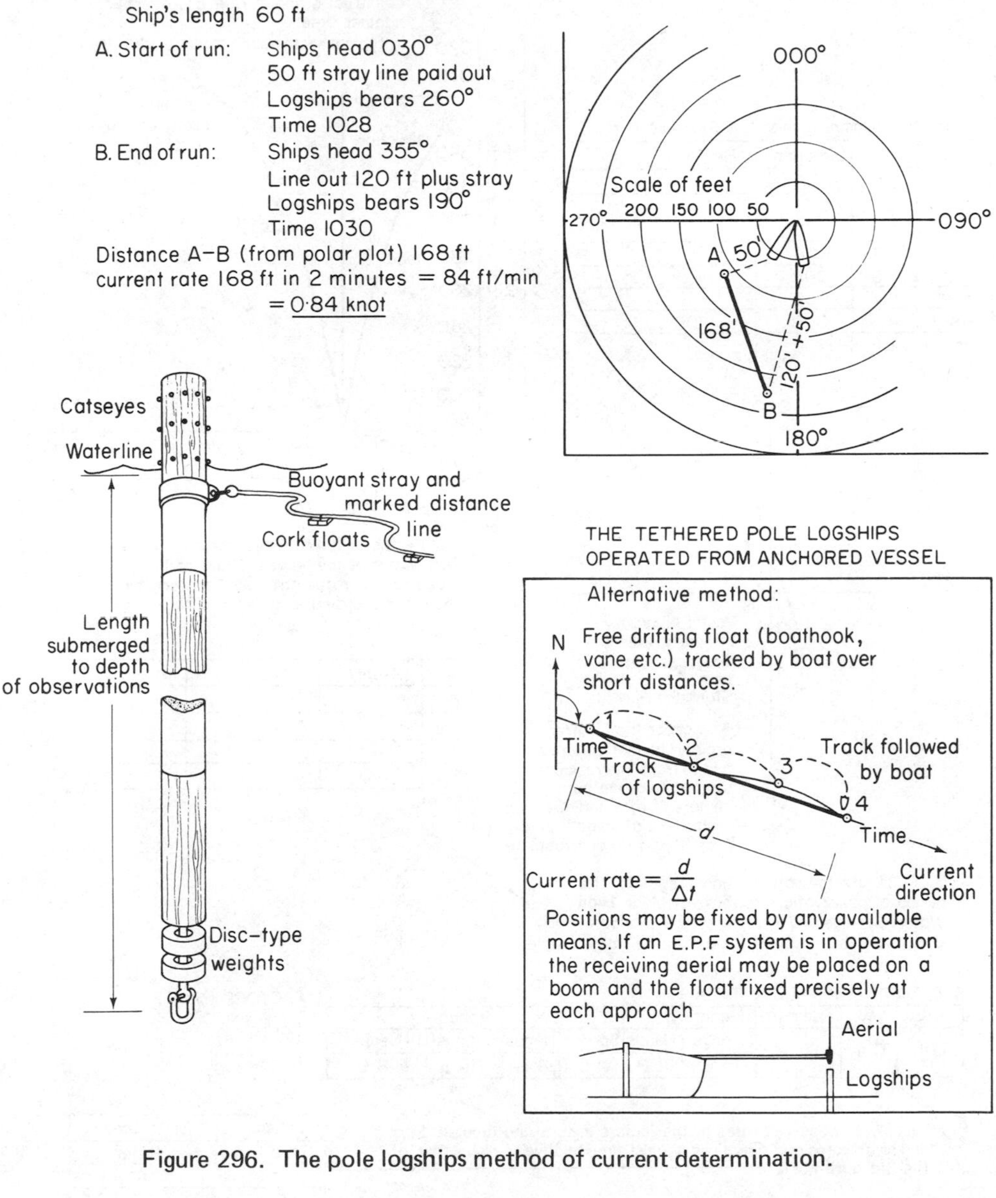
Ship's length 60 ft
A. Start of run: Ships head 030°
50 ft stray line paid out
Logships bears 260°
Time 1028
B. End of run: Ships head 355°
Line out 120 ft plus stray
Logships bears 190°
Time 1030
Distance A–B (from polar plot) 168 ft
current rate 168 ft in 2 minutes = 84 ft/min
= 0·84 knot
000°
090°
270°
180°
Scale of feet
200 150 100 50
A
B
50'
168'
120' + 50'
Catseyes
Waterline
Buoyant stray and
marked distance
line
Cork floats
Length
submerged
to depth
of observations
Disc-type
weights
THE TETHERED POLE LOGSHIPS
OPERATED FROM ANCHORED VESSEL
Alternative method:
N
Free drifting float (boathook,
vane etc.) tracked by boat over
short distances.
1
2
3
4
Time
Track
of logships
Track followed
by boat
d
Time
Current
direction
Current rate = d/Δt
Positions may be fixed by any available
means. If an E.P.F system is in operation
the receiving aerial may be placed on a
boom and the float fixed precisely at
each approach
Aerial
Logships

Figure 296. The pole logships method of current determination.

Time		Ship's head		Bearing of pole		Line out	Rate	Direction
start	end	start	end	start	end		knots	deg. true

① Log readings for each run

② Construct graph of rate and direction against time
(a typical rectilinear stream is depicted)

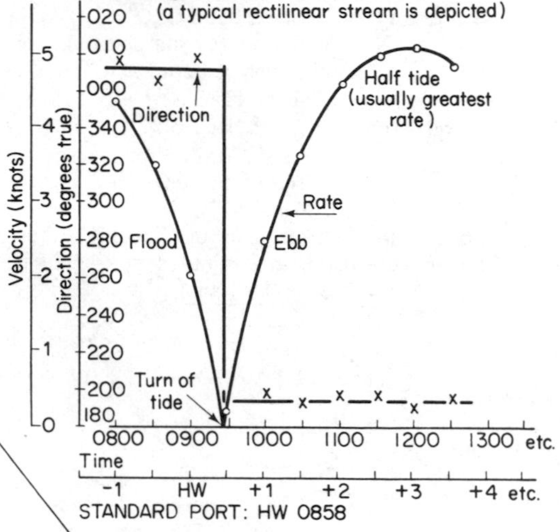

Rates/Directions relative to HW Standard Port

Time	Rate	Direction	Remarks
−6	0·8	008	
−5	2·2	008	
−4	3·9	008	
−3	4·8	008	
−2	4·9	008	
−1	4·7	008	
HW	4·3	008	
+1	2·6	192	
etc.			

③ Relate the graph to times before and after high water at the nearest Standard Port, abstract the rates and directions and plot in polar co-ordinate vectors

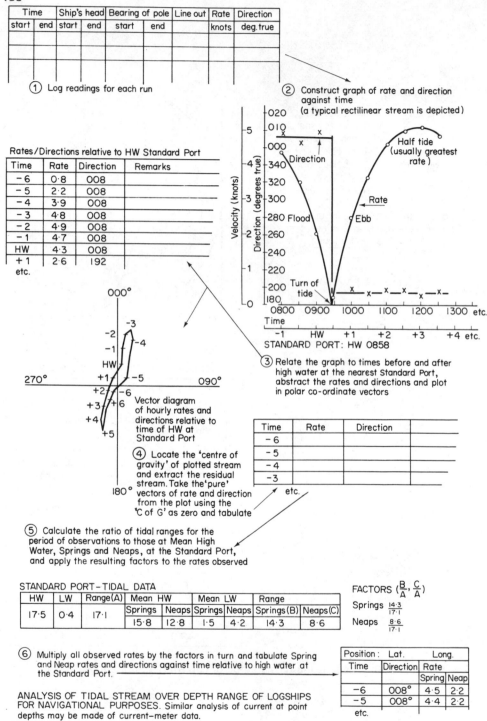

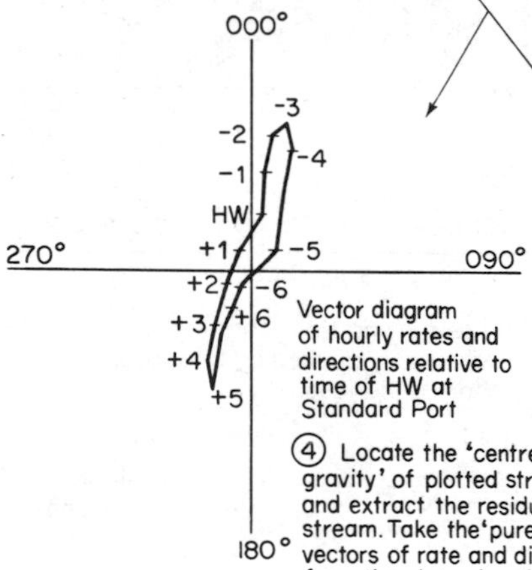

Vector diagram of hourly rates and directions relative to time of HW at Standard Port

④ Locate the 'centre of gravity' of plotted stream and extract the residual stream. Take the 'pure' vectors of rate and direction from the plot using the 'C of G' as zero and tabulate

Time	Rate	Direction	
−6			
−5			
−4			
−3			
etc.			

⑤ Calculate the ratio of tidal ranges for the period of observations to those at Mean High Water, Springs and Neaps, at the Standard Port, and apply the resulting factors to the rates observed

STANDARD PORT–TIDAL DATA

HW	LW	Range(A)	Mean HW		Mean LW		Range	
			Springs	Neaps	Springs	Neaps	Springs (B)	Neaps (C)
17·5	0·4	17·1	15·8	12·8	1·5	4·2	14·3	8·6

FACTORS $(\frac{B}{A}, \frac{C}{A})$

Springs $\frac{14{\cdot}3}{17{\cdot}1}$

Neaps $\frac{8{\cdot}6}{17{\cdot}1}$

⑥ Multiply all observed rates by the factors in turn and tabulate Spring and Neap rates and directions against time relative to high water at the Standard Port.

Position:	Lat.	Long.	
Time	Direction	Rate	
		Spring	Neap
−6	008°	4·5	2·2
−5	008°	4·4	2·2
etc.			

ANALYSIS OF TIDAL STREAM OVER DEPTH RANGE OF LOGSHIPS FOR NAVIGATIONAL PURPOSES. Similar analysis of current at point depths may be made of current-meter data.

Figure 297. The pole logships method of current determination (contd.) (Note: For full details, refer to the *Admiralty Manual of Hydrographic Surveying,* Vol. II.)

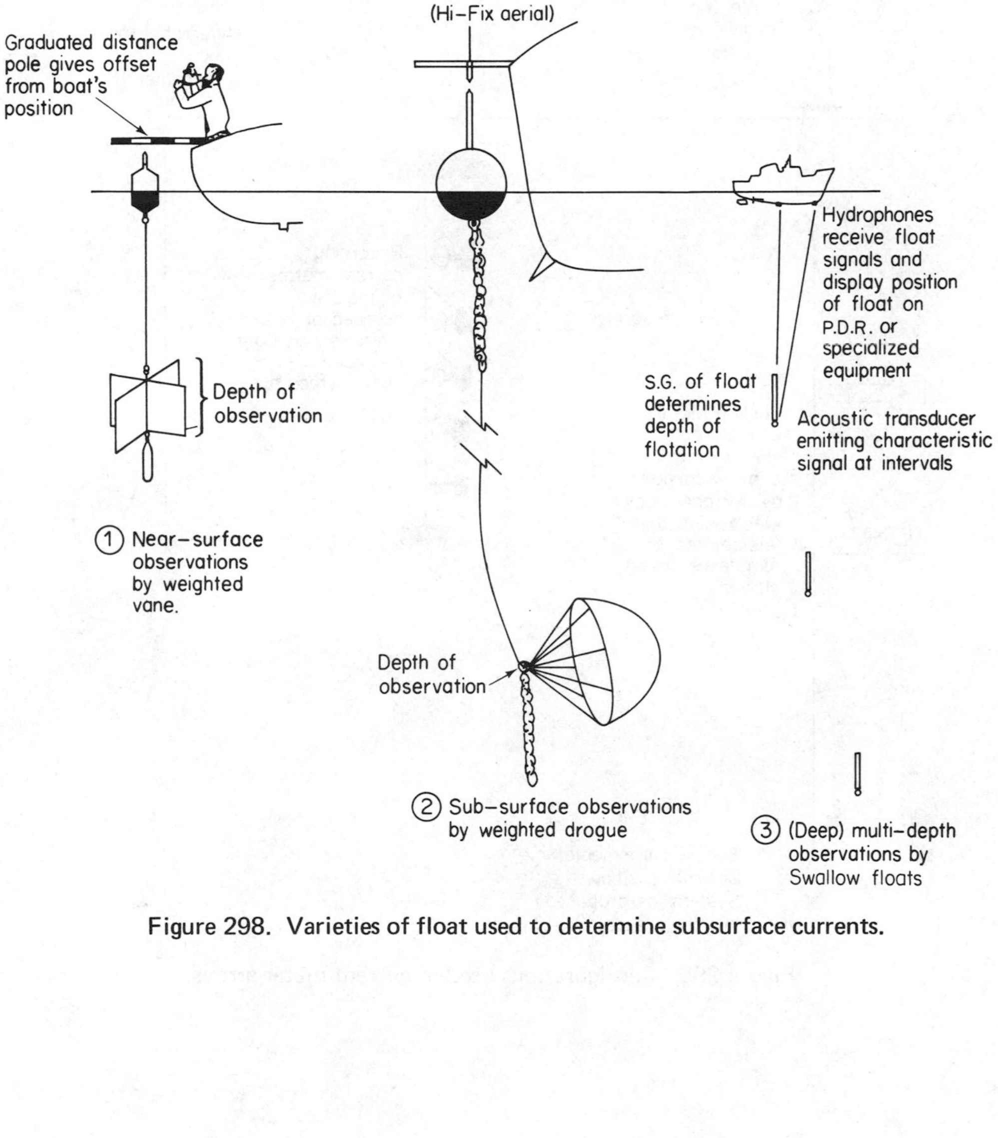

Figure 298. Varieties of float used to determine subsurface currents.

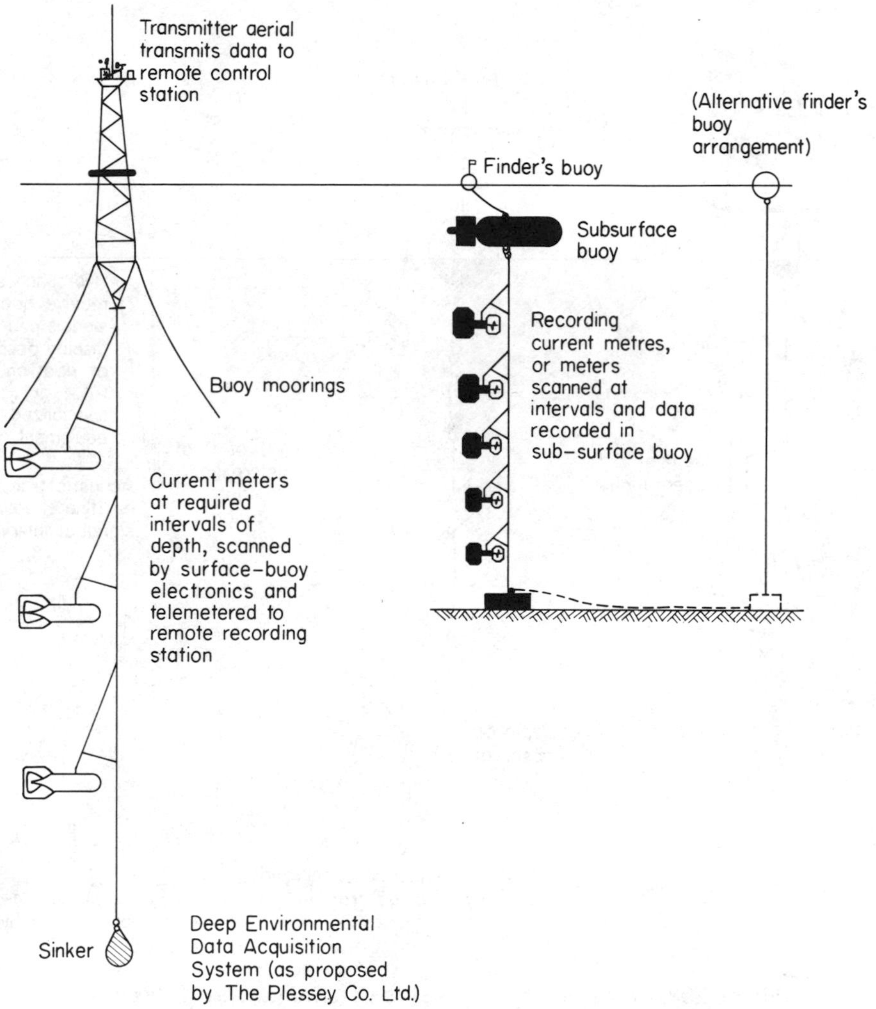

Figure 299. Configurations used in current meter arrays.

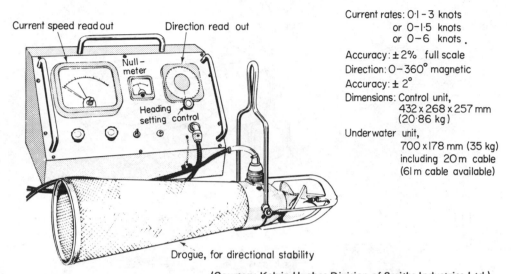

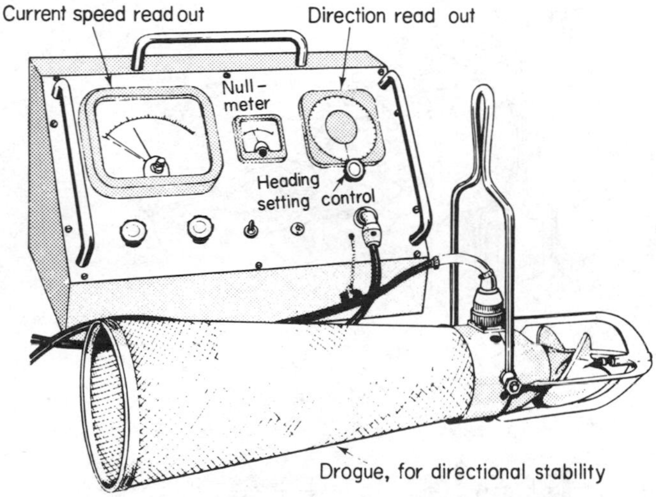

Current rates: 0·1 – 3 knots
or 0–1·5 knots
or 0–6 knots.
Accuracy: ± 2% full scale
Direction: 0–360° magnetic
Accuracy: ± 2°
Dimensions: Control unit,
432 x 268 x 257 mm
(20·86 kg)
Underwater unit,
700 x 178 mm (35 kg)
including 20 m cable
(61 m cable available)

(Courtesy Kelvin Hughes Division of Smiths Industries Ltd.)

The impeller rotates in the current, at a speed proportional to the current rate. A magnetic coupling drives a hexagonal cam inside the sealed, oil-filled body, making six electrical contacts per revolution of the impeller. A succession of pulses is sent via the umbilical cable to the control unit, where the current rate is displayed direct in knots as a voltage dependent on the pulse repetition frequency.

Direction is sensed by an aircraft-type magnetic compass inside the body of the underwater unit. Compensation for residual deviation is factory-arranged to within 1°. Pressure is equalized by a bellows device. The compass forms two arms of a bridge circuit, its two outputs being connected to a null-meter and to a resolver. The resolver controls the 'heading setting control' which indicates the current direction when operated to null the meter.

Figure 300. The Kelvin Hughes Direct Reading Current Meter. A direct reading (analogue) meter using magnetic direction finding, with corrector magnets for deviation and a pressure-compensated, oil-filled body. (A gyro compass giving direction relative to true North is rare in current meters, such a unit costing well in excess of £500.)

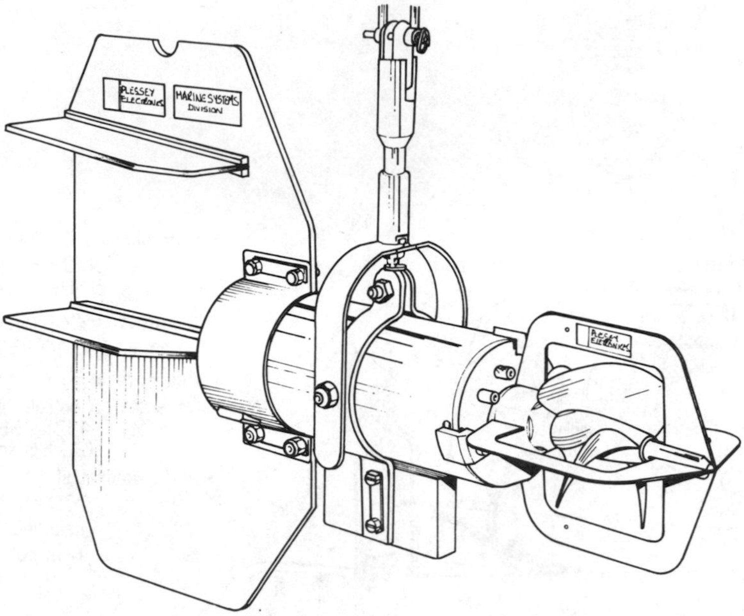

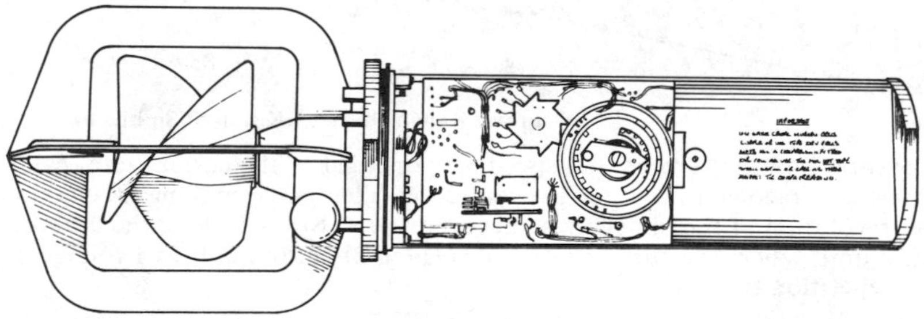

The velocity and direction instrumentation (tape-deck and circuitry otherside)

Body: 127 mm dia. x 610 mm
Complete: 610 x 915 mm (21·5 kg)

Speed: 50 – 2500 $mm.s^{-1}$ (0·1–5 knots approx.)

Accuracy: ±20 $mm.s^{-1}$ or 2%

Direction: 0–360° magnetic (continuous)

Accuracy: ±5°

Recording system: Magnetic tape, digital (55,200 measurements) processed and printed out by manufacturer.

Clock: Battery powered $1\frac{1}{2}$ V d.c. (150 days)

Timing interval: 5 – 60 min at intervals of 5, 10, 15, 20, 30 or 60 min

Maximum operating depth: 2000 m

(Courtesy The Plessey Co. Ltd.)

Figure 301. The Plessey Model MO21 Recording Current Meter.

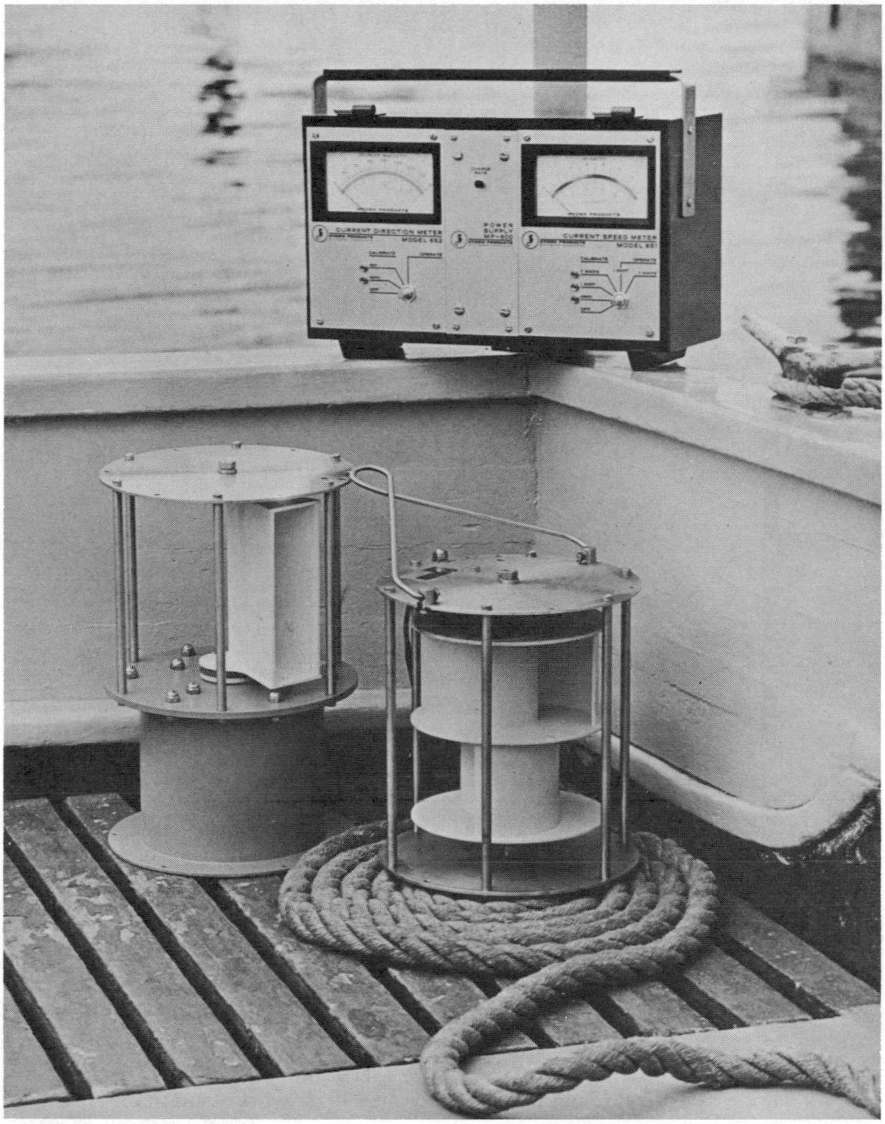

(Courtesy Hydro Products Inc.)

The vane is coupled to a magnetic compass and potentiometer take-off, giving a proportionate voltage dependent on the compass needle's travel away from Magnetic North.
The Savonius rotor has ten magnets mounted around its top. A magnetic reed switch is energized by a d.c. source and a circuit is completed each time a magnet passes by, i.e. ten times per rotor revolution. The pulsed output is therefore at a frequency dependent on rotor speed and, thus, current rate.

Range: 0.05–7.0 knots (± 3% full scale) 0–360° magnetic, ± 5°
Maximum operating depth: 12,000 m. Battery life: 200 h continuous.

Figure 302. The Hydro Products Inc. Savonius Rotor Current Meter with Graphic Recording Facility.

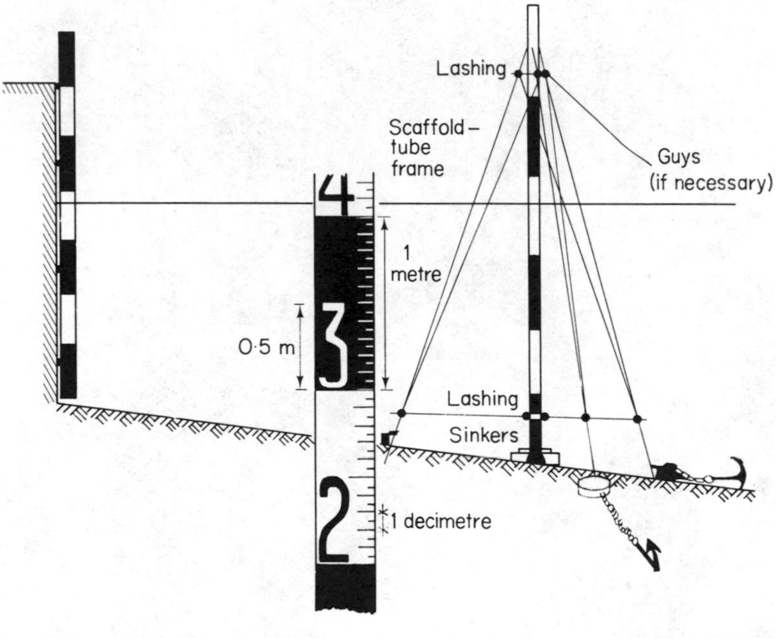

Figure 303. The visual tide scale: typical installations and markings.

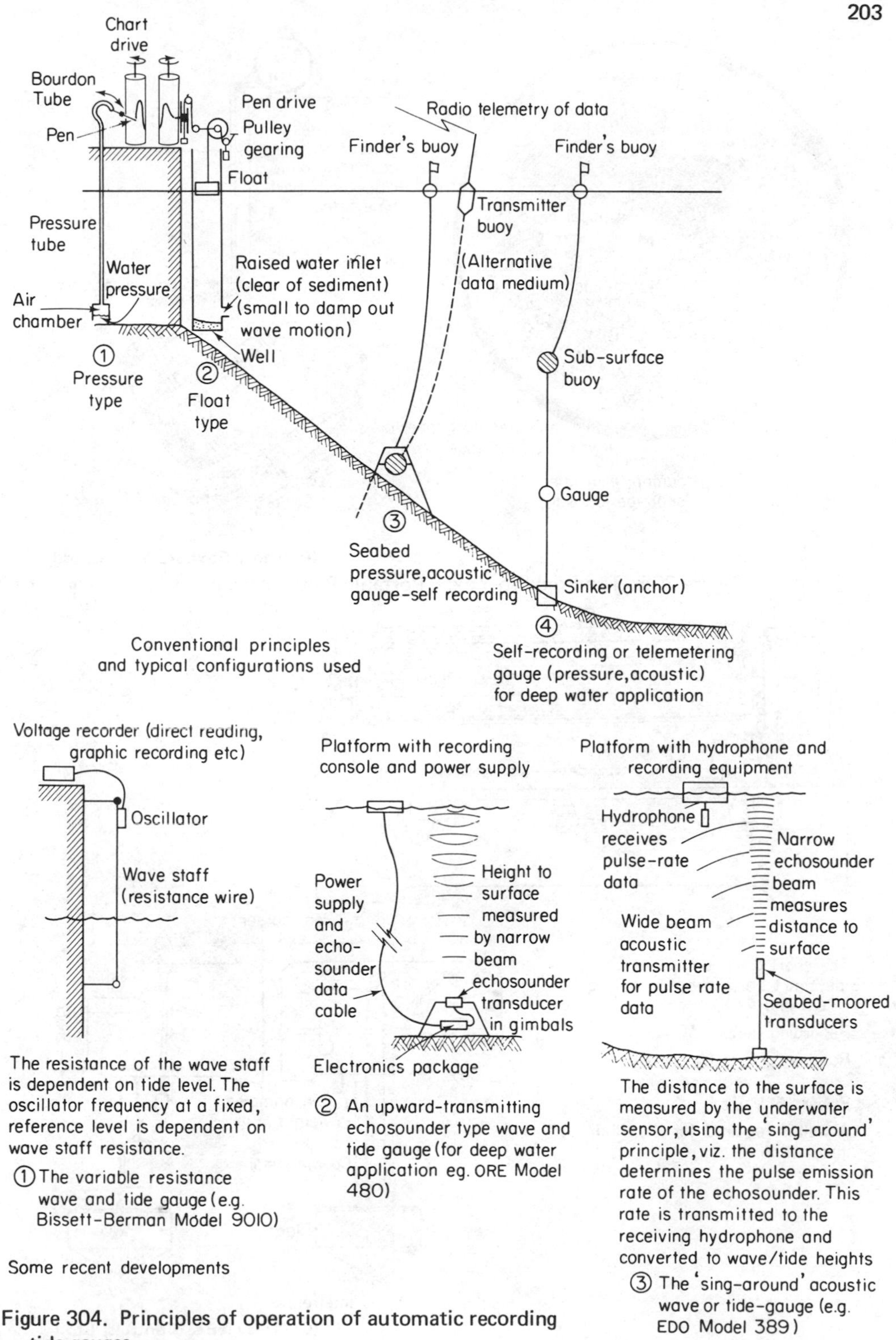
Chart drive
Bourdon Tube
Pen
Pen drive
Pulley gearing
Float
Radio telemetry of data
Finder's buoy
Finder's buoy
Transmitter buoy
Pressure tube
Water pressure
Air chamber
Raised water inlet (clear of sediment) (small to damp out wave motion)
Well
(Alternative data medium)
① Pressure type
② Float type
Sub-surface buoy
Gauge
③ Seabed pressure, acoustic gauge-self recording
Sinker (anchor)
④ Self-recording or telemetering gauge (pressure, acoustic) for deep water application
Conventional principles and typical configurations used
Voltage recorder (direct reading, graphic recording etc)
Oscillator
Wave staff (resistance wire)
The resistance of the wave staff is dependent on tide level. The oscillator frequency at a fixed, reference level is dependent on wave staff resistance.
① The variable resistance wave and tide gauge (e.g. Bissett-Berman Model 9010)
Platform with recording console and power supply
Power supply and echo-sounder data cable
Height to surface measured by narrow beam echosounder transducer in gimbals
Electronics package
② An upward-transmitting echosounder type wave and tide gauge (for deep water application eg. ORE Model 480)
Platform with hydrophone and recording equipment
Hydrophone receives pulse-rate data
Narrow echosounder beam measures distance to surface
Wide beam acoustic transmitter for pulse rate data
Seabed-moored transducers
The distance to the surface is measured by the underwater sensor, using the 'sing-around' principle, viz. the distance determines the pulse emission rate of the echosounder. This rate is transmitted to the receiving hydrophone and converted to wave/tide heights
③ The 'sing-around' acoustic wave or tide-gauge (e.g. EDO Model 389)
Some recent developments

Figure 304. Principles of operation of automatic recording tide-gauges.

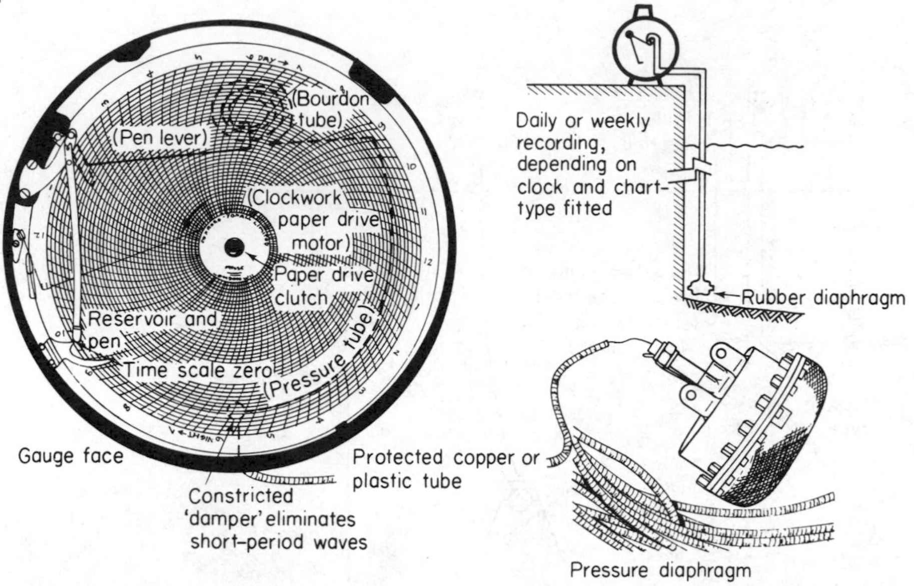

(Courtesy Foxboro-Yoxall Ltd.)

Figure 305. The Foxboro-Yoxall Portable Pressure Gauge.

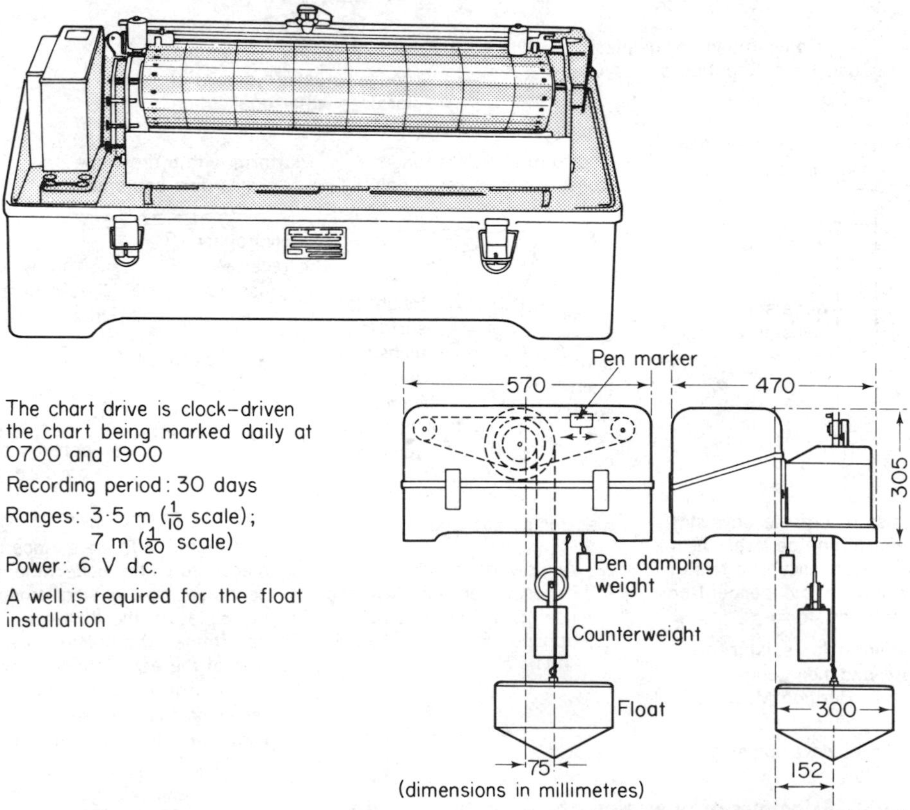

(Courtesy NBA (Controls) Ltd.)

Figure 306. The NBA (Controls) Ltd. Foose Type Float Gauge.

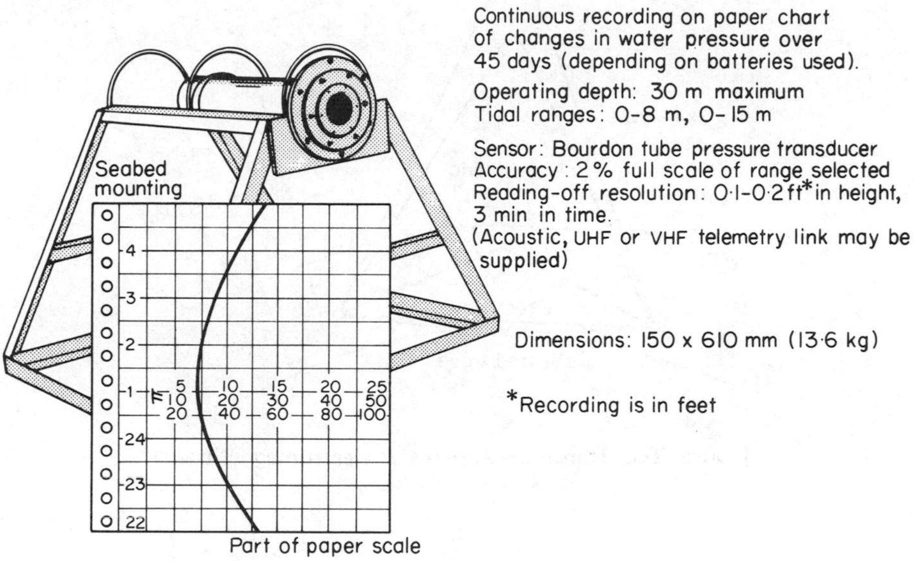
Continuous recording on paper chart of changes in water pressure over 45 days (depending on batteries used).
Operating depth: 30 m maximum
Tidal ranges: 0-8 m, 0-15 m
Sensor: Bourdon tube pressure transducer
Accuracy: 2 % full scale of range selected
Reading-off resolution: 0·1-0·2 ft* in height, 3 min in time.
(Acoustic, UHF or VHF telemetry link may be supplied)
Dimensions: 150 x 610 mm (13·6 kg)
*Recording is in feet
Seabed mounting
Part of paper scale
ft
5 10 15 20 25
10 20 30 40 50
20 40 60 80 100
4
3
2
1
24
23
22

(Courtesy NBA (Controls) Ltd.)

Figure 307. The NBA (Controls) Ltd. Model DNT-1 bottom-mounted Tide Recorder.

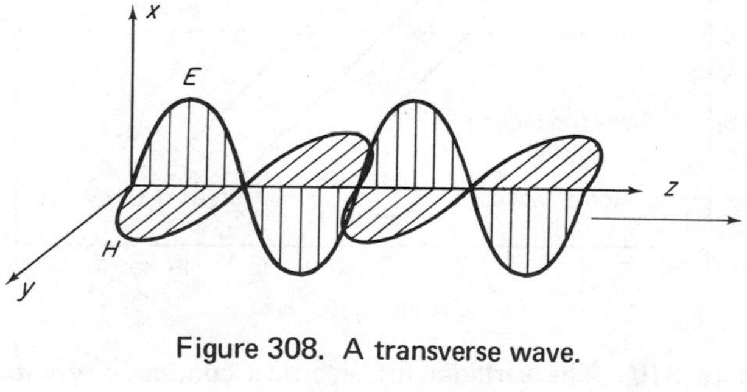
x
E
z
H
y

Figure 308. A transverse wave.

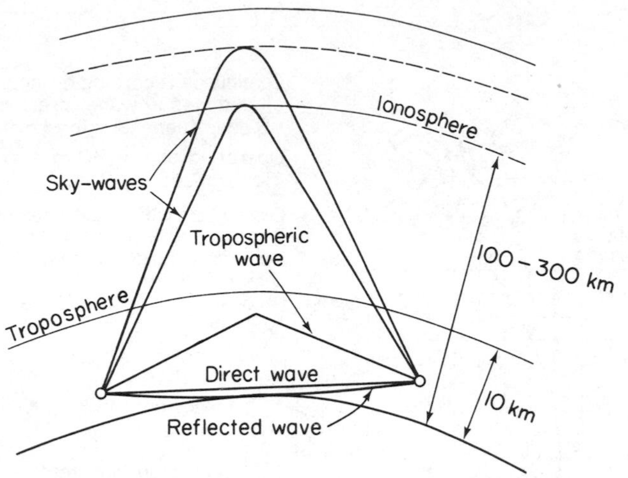

Figure 309. Transmission paths of electromagnetic waves.

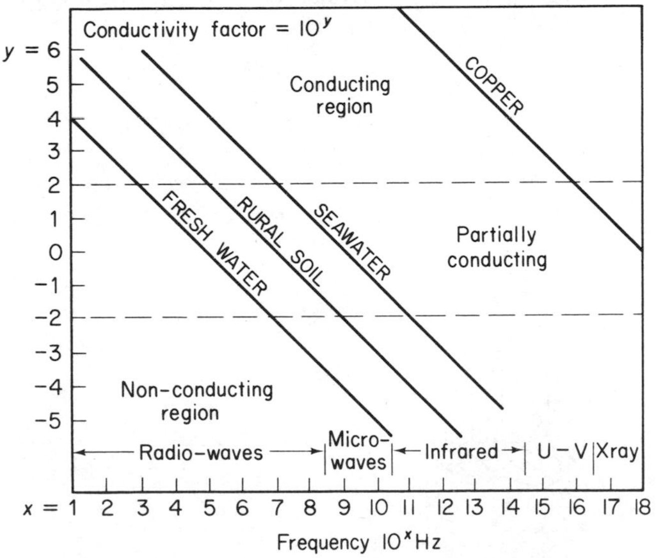

Figure 310. The variation of electrical conductivity with frequency for some materials.

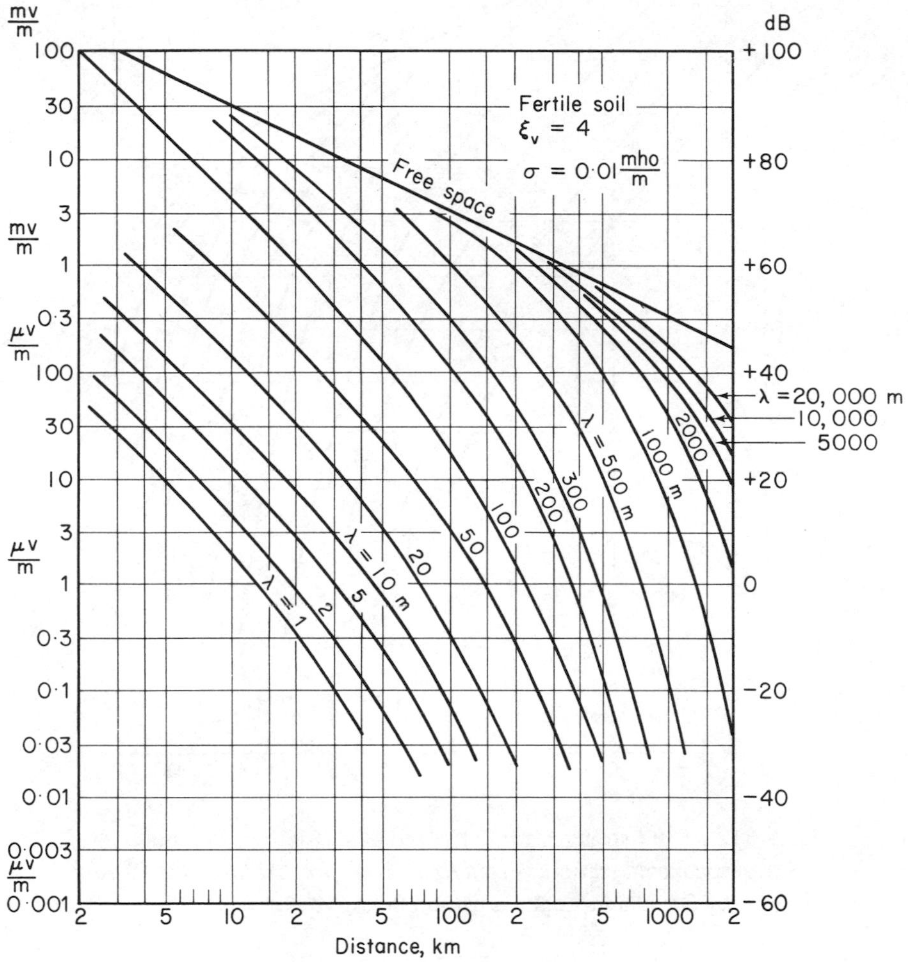

Figure 311. Field strength for 1 kW radiated. Stations at ground level. Land: vertical polarization. (From *Radiowave propagation in the atmosphere,* by David and Vogl, Pergamon Press, 1969.)

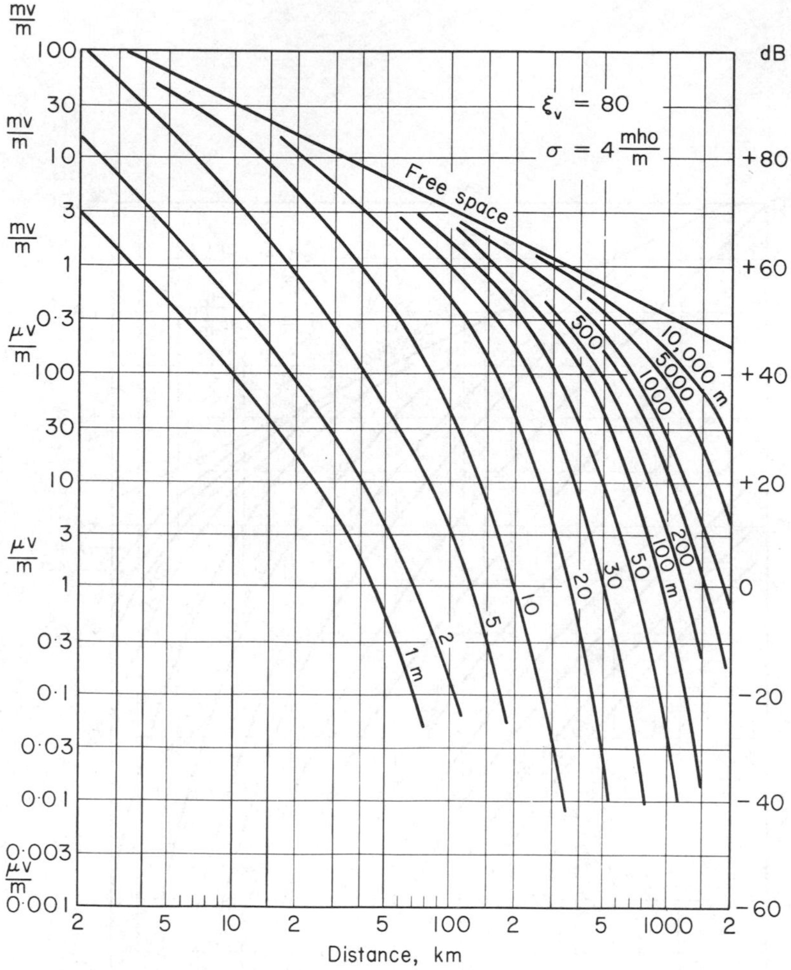

Figure 312. Field strength for 1 kW radiated. Stations at ground level. Sea: vertical polarization. (From *Radiowave propagation in the atmosphere,* by David and Vogl, Pergamon Press, 1969.)

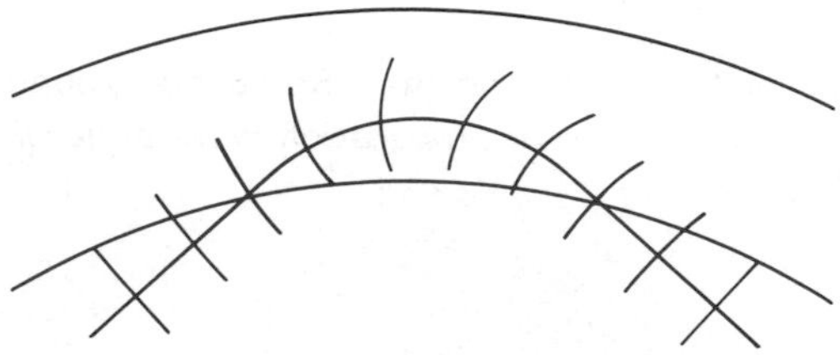

Figure 313. The refraction of an electromagnetic transmission by the layering of charged particles present in the ionosphere.

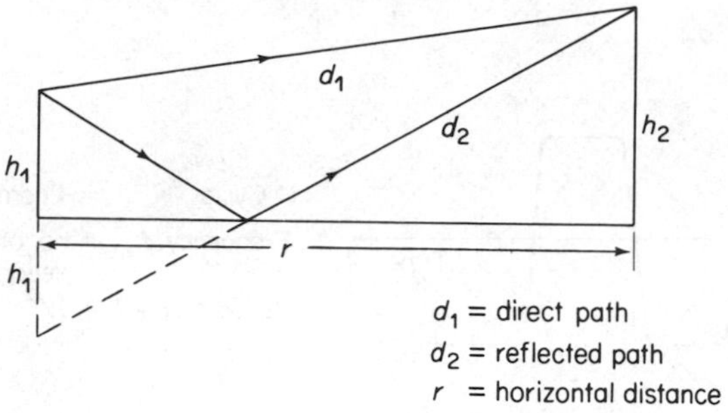

Figure 314. The phase interference between direct and reflected paths and the effect of differences in antenna heights.

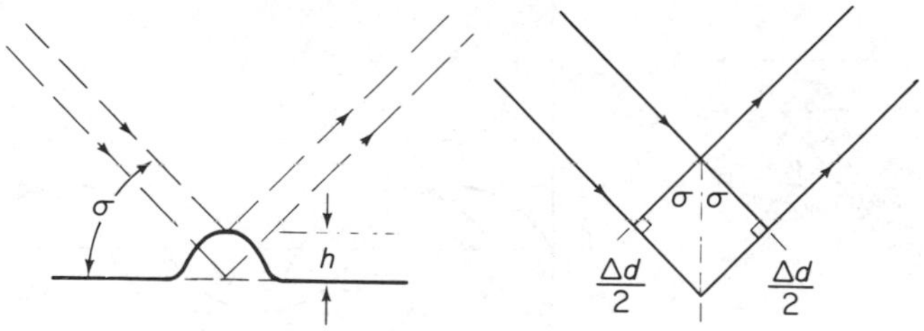

Figure 315. The 'smooth reflecting surface. $\Delta d = 2h \sin \sigma$ = difference in path length, i.e. $\lambda/8 > 2h \sin \sigma$ for specular reflection.

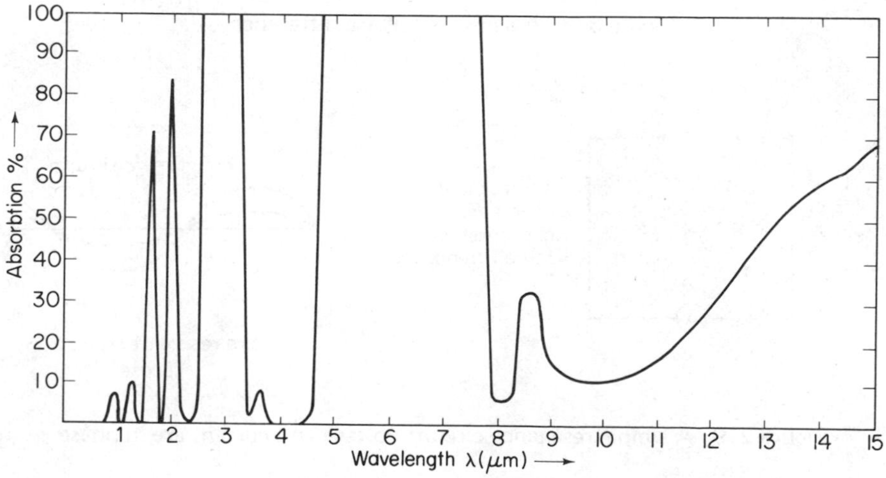

Figure 316. Infrared absorption bands in the atmosphere.

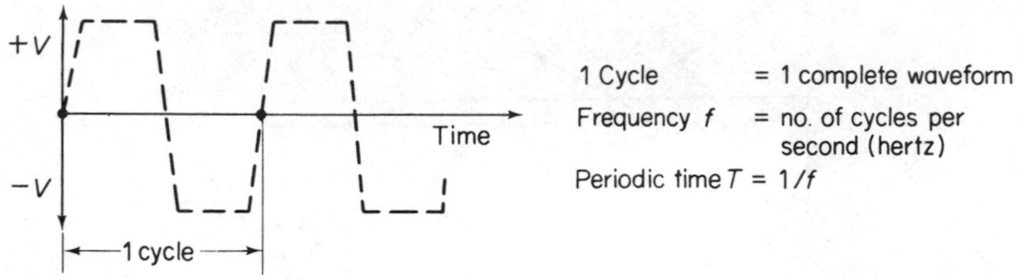

Figure 317. The square waveform produced by a switching device from a d.c. source.

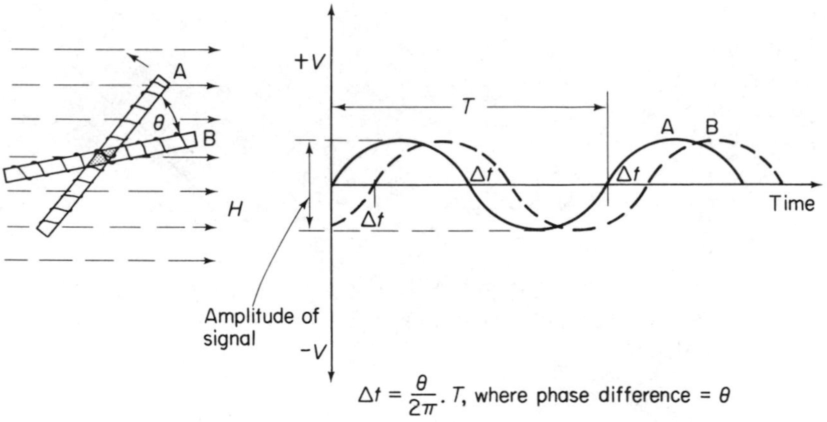

Figure 318. A generated phase difference.

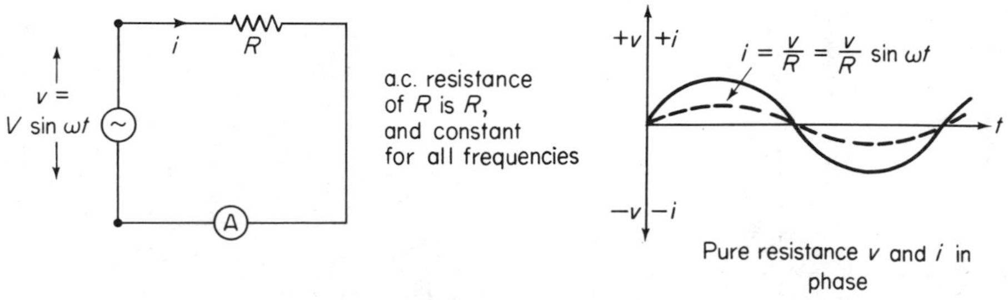

Figure 319. A simple resistance circuit: voltage and current are in phase.

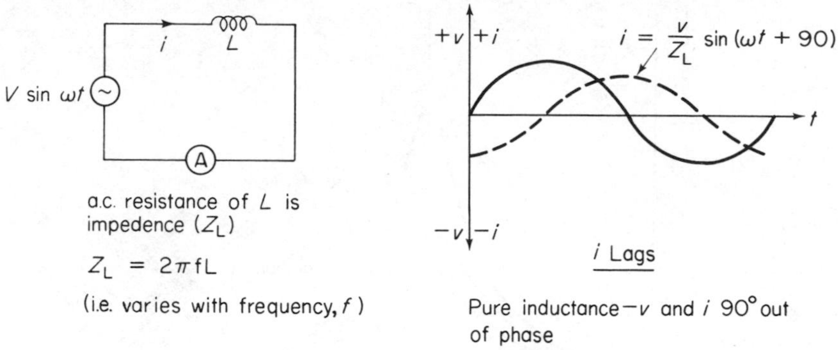

Figure 320. An inductance circuit: current lagging 90° in phase on voltage.

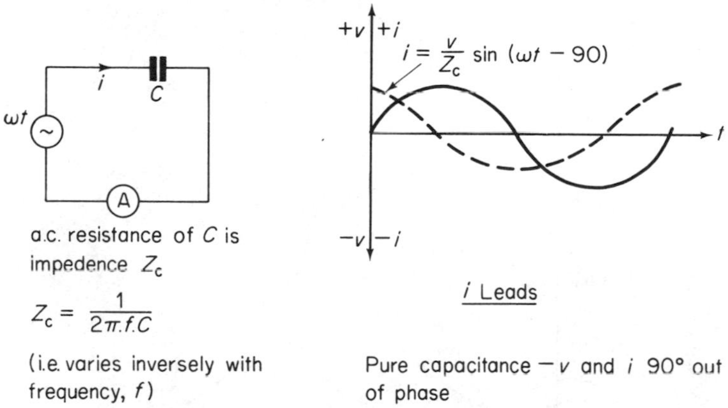

Figure 321. A capacitance circuit: current leading in phase by 90° on voltage.

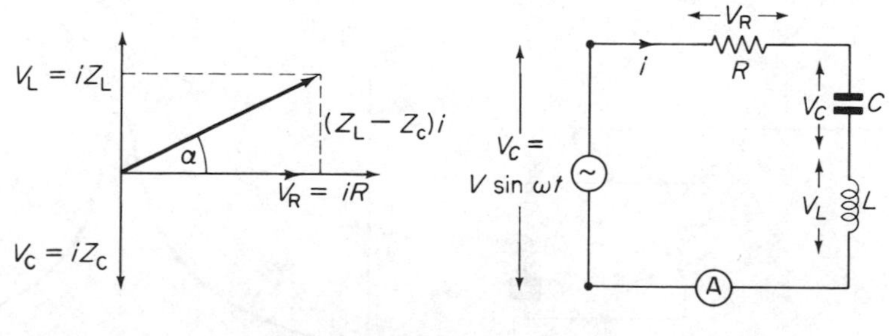

Figure 322. A combined resistance, inductance and capacitance circuit. The vector sum of the voltages across the three components is equal to the applied voltage.

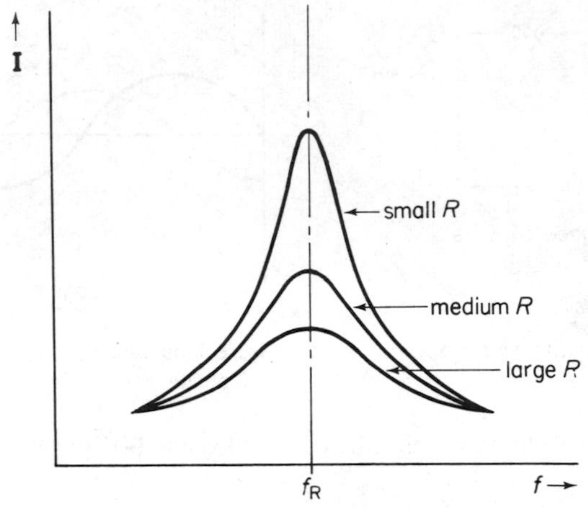

Figure 323. Characteristic response curves for a tuned circuit.

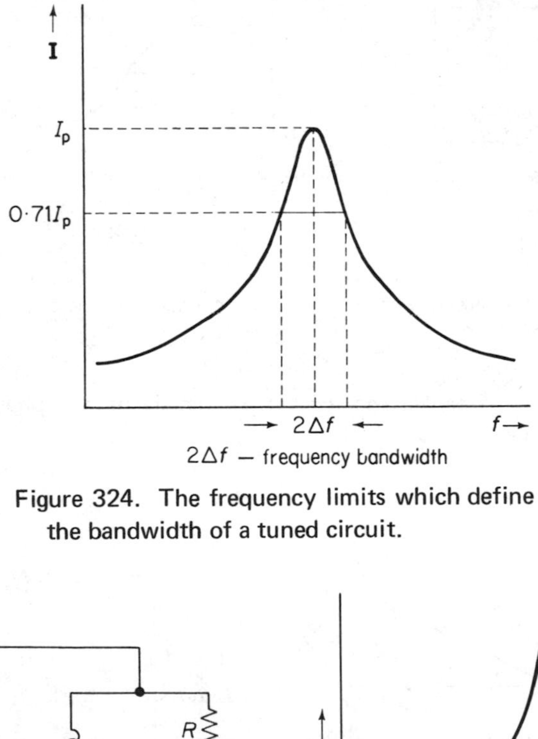

Figure 324. The frequency limits which define the bandwidth of a tuned circuit.

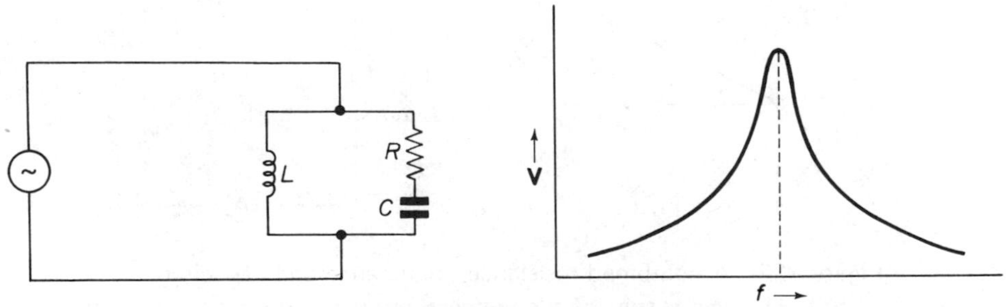

Figure 325. The parallel connected resistance, inductance, capacitance circuit.

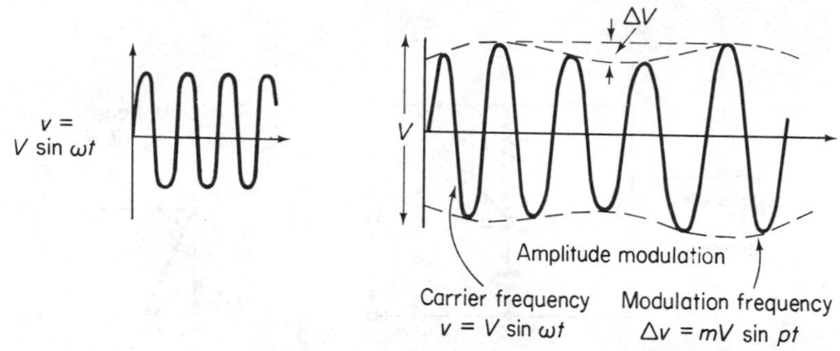

Figure 326. Amplitude modulation.

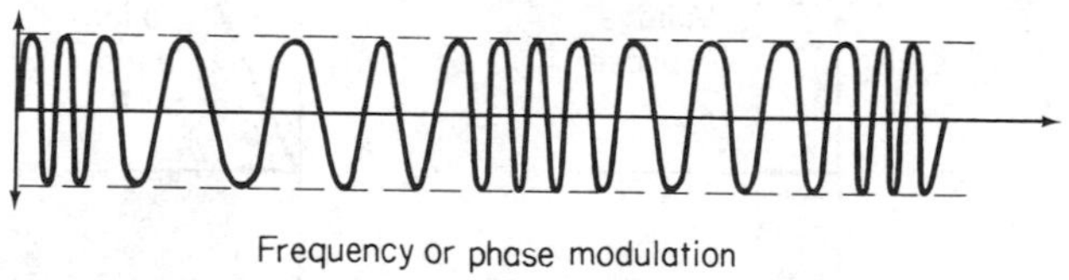

Figure 327. Frequency or phase modulation.

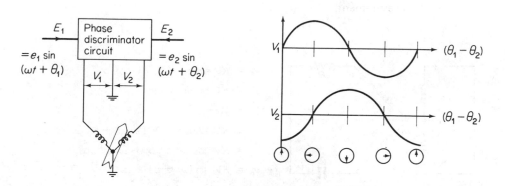

Figure 328. The principles of the Decca Navigator Decometer phasemeter.

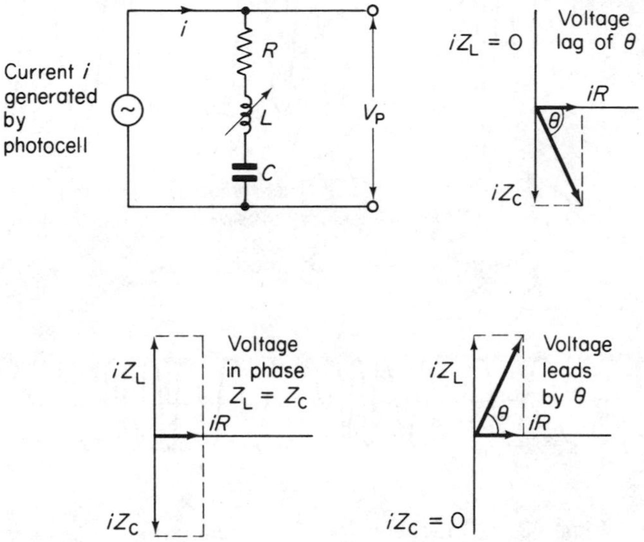

Figure 329. The principles of the null-point method of phase measurement.

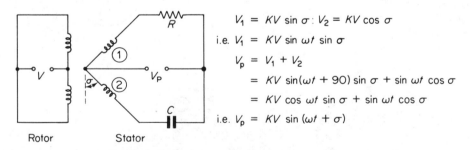

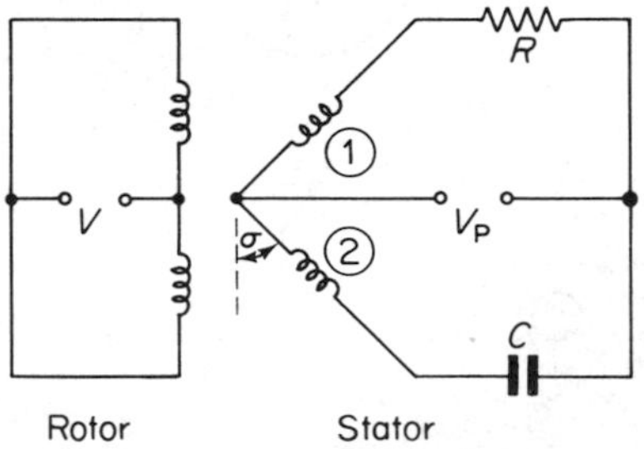

$$V_1 = KV \sin \sigma : V_2 = KV \cos \sigma$$

$$\text{i.e. } V_1 = KV \sin \omega t \sin \sigma$$

$$V_p = V_1 + V_2$$

$$= KV \sin(\omega t + 90) \sin \sigma + \sin \omega t \cos \sigma$$

$$= KV \cos \omega t \sin \sigma + \sin \omega t \cos \sigma$$

$$\text{i.e. } V_p = KV \sin(\omega t + \sigma)$$

Figure 330. The resolver principle of phase measurement.

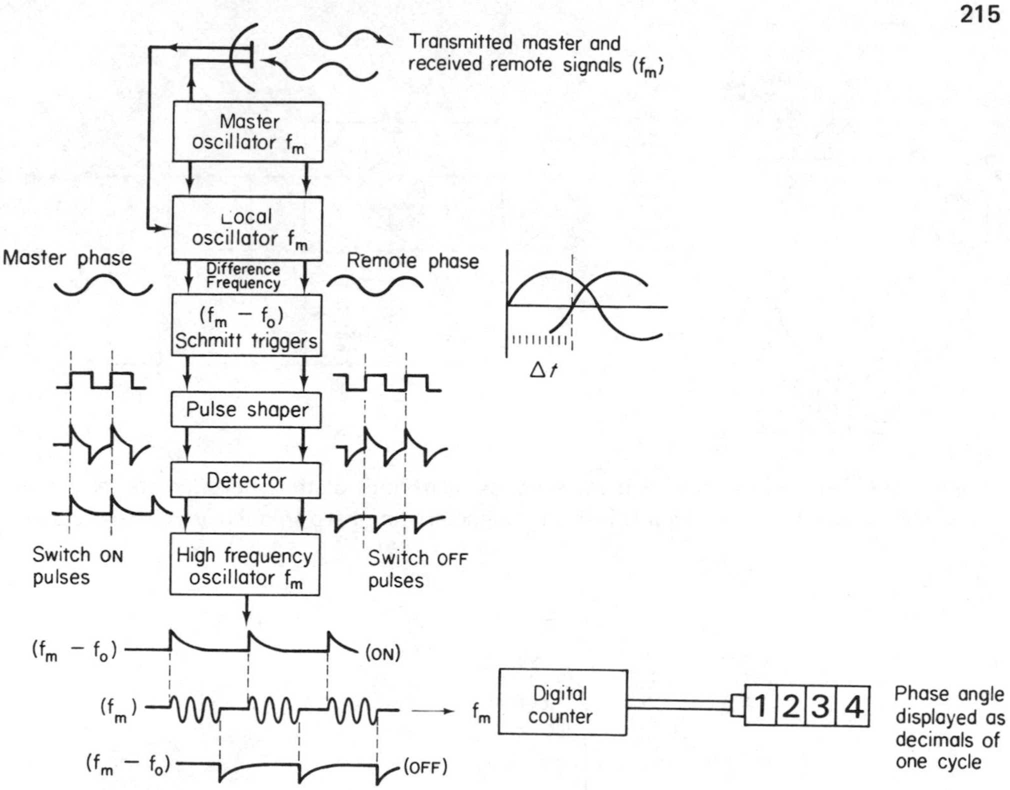

Figure 331. A digital principle of phase measurement.

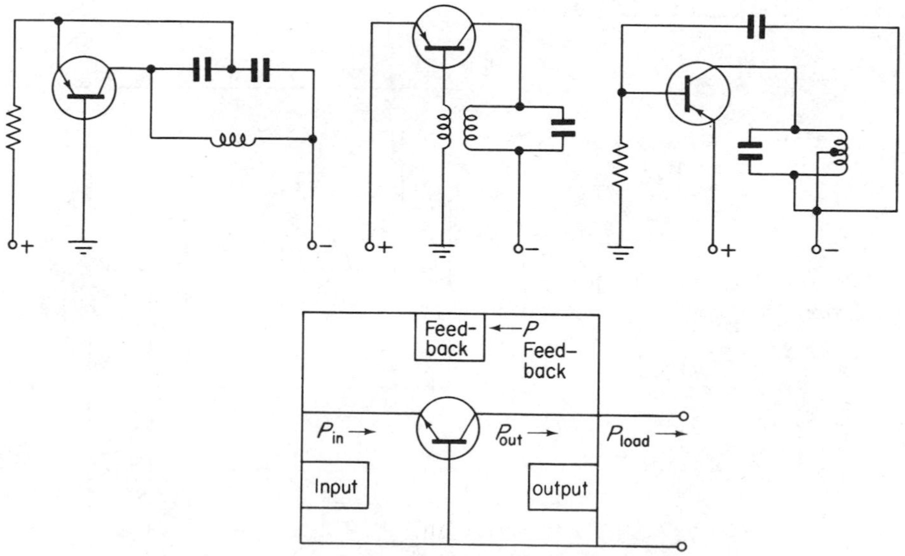

Figure 332. Typical oscillator circuits and principal networks.

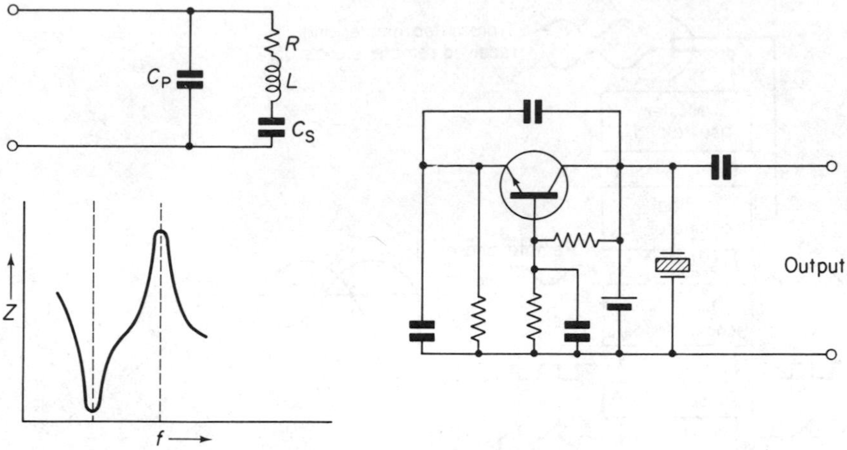

Figure 333. An equivalent circuit for electrical behaviour of the quartz crystal and a typical crystal-oscillator circuit, with the characteristic, sharp frequency response curve.

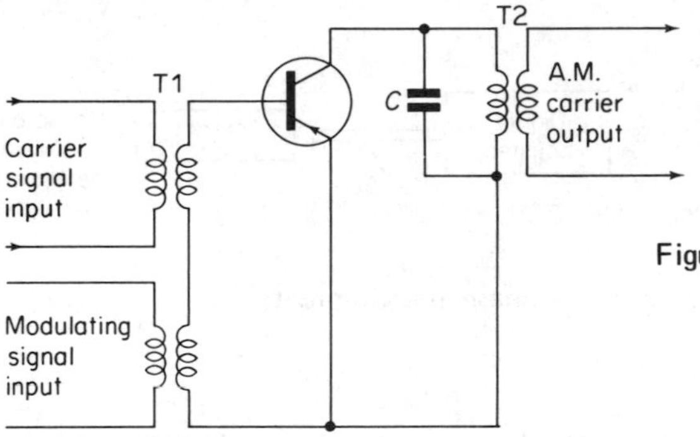

Figure 334. Direct amplitude modulation.

Figure 335. A circuit for frequency modulation.

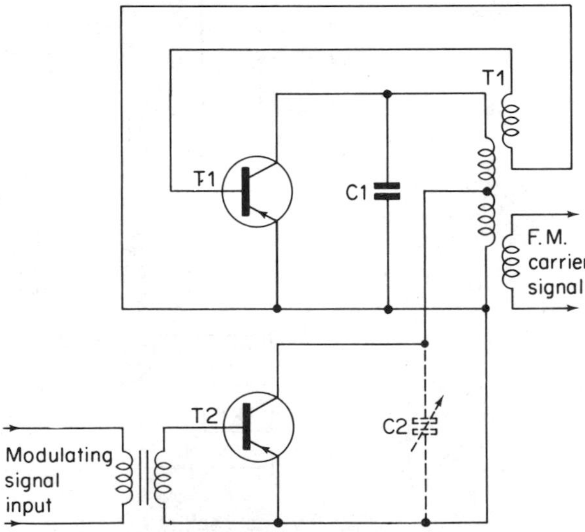

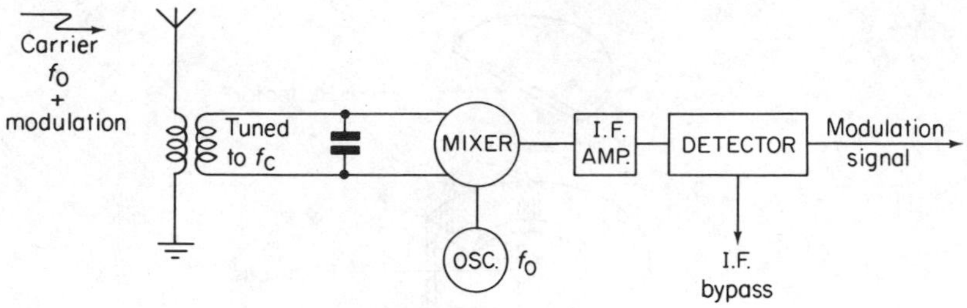

Figure 336. The function of a local oscillator in producing an intermediate frequency from the carrier wave.

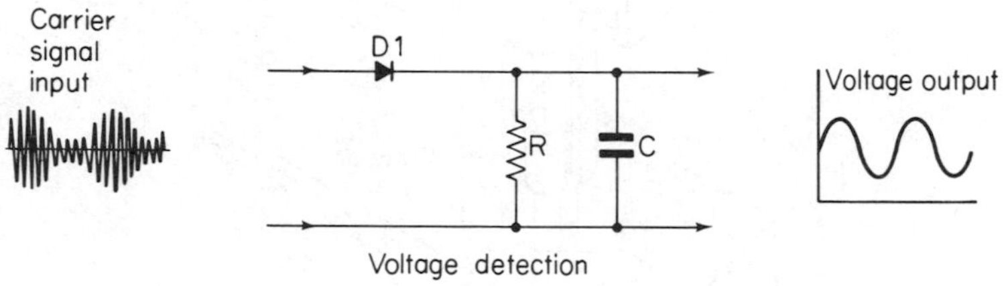

Figure 337. The diode used to demodulate an amplitude modulated carrier signal.

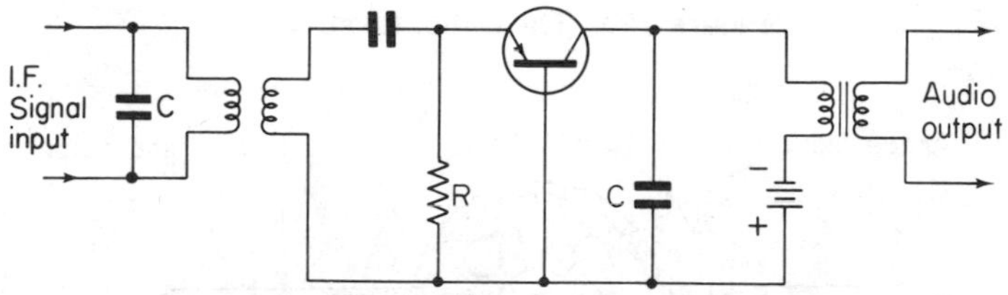

Figure 338. A transistor detector-amplifier.

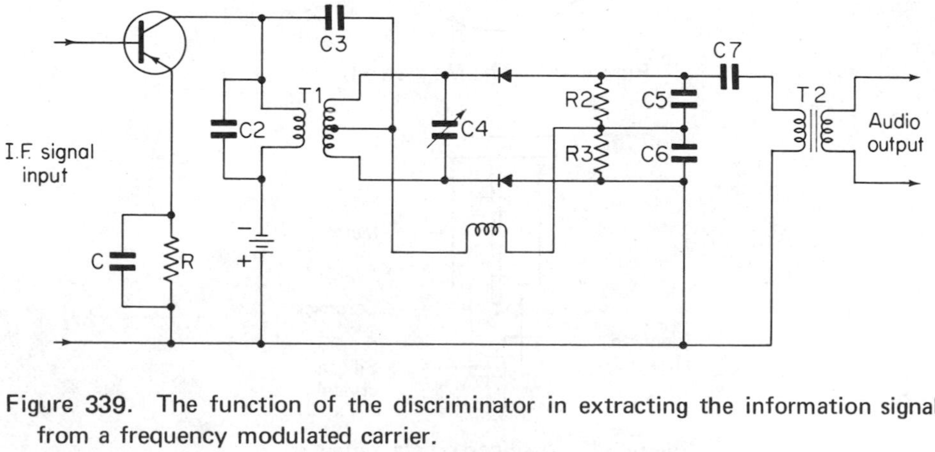

Figure 339. The function of the discriminator in extracting the information signal from a frequency modulated carrier.

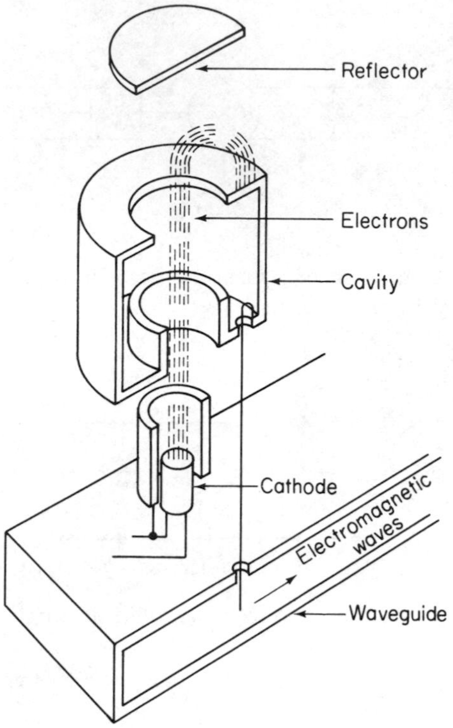

Figure 340. The reflex klystron.

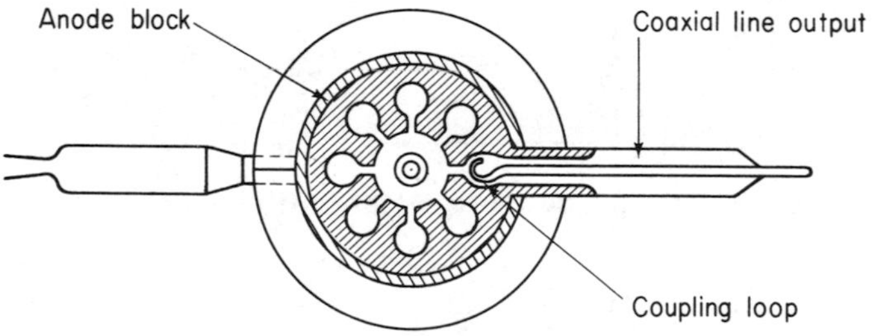

Figure 341. The Magnetron.

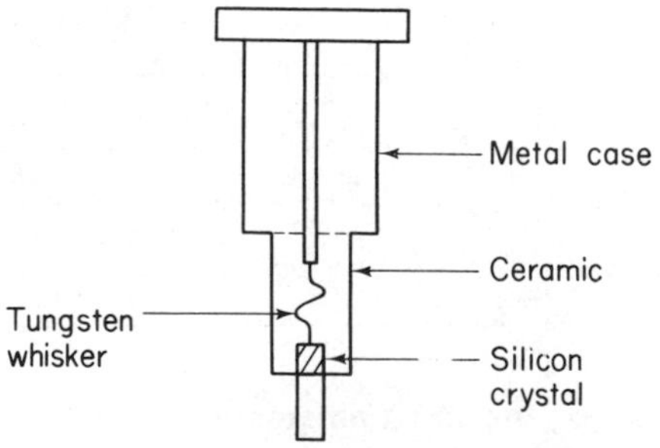

Figure 342. A silicon crystal detector.

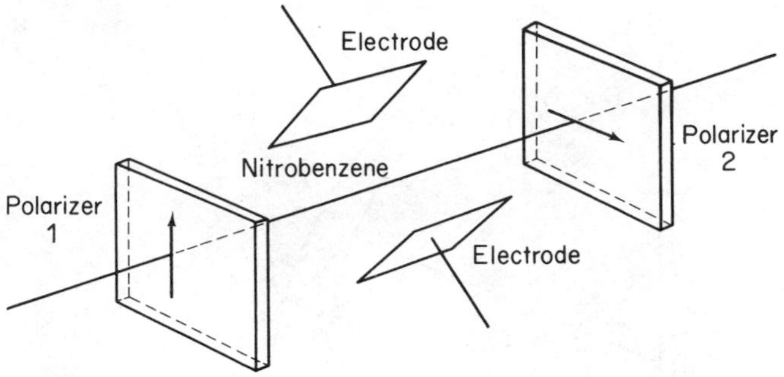

Figure 343. The modulation of visible light by polarizers and a Kerr cell.

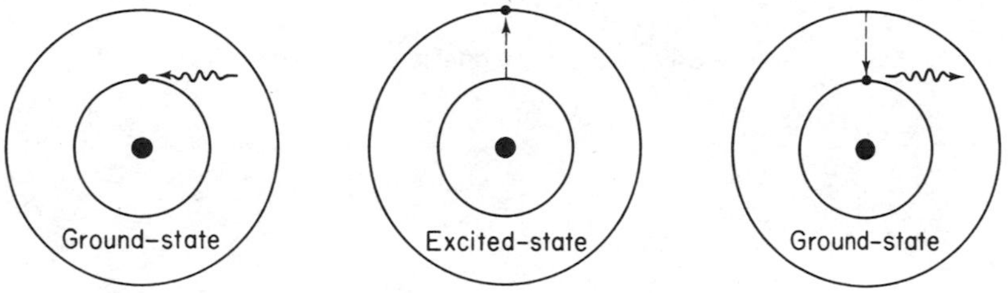

Figure 344. Absorption and spontaneous emission of light by atoms of a gas.

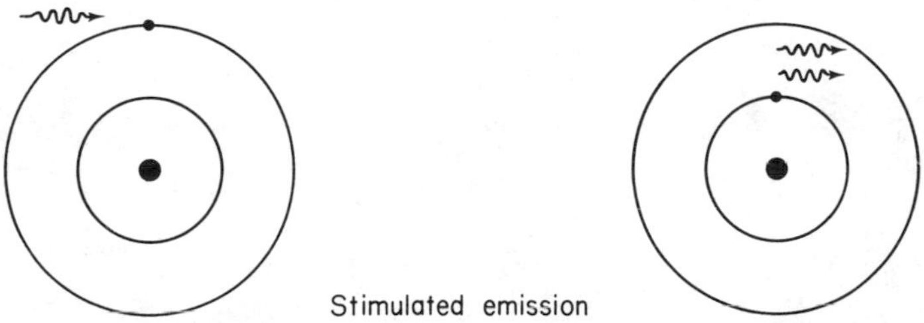

Figure 345. The principle of stimulated emission.

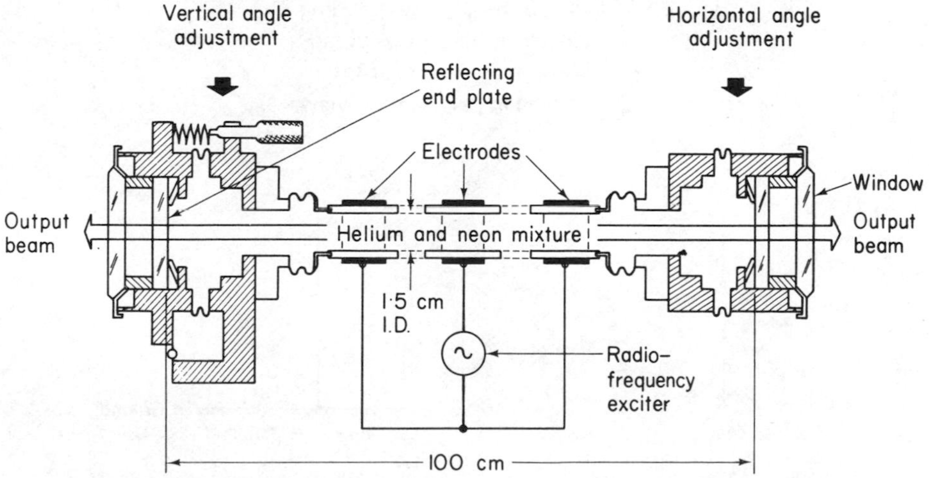

Figure 346. The helium-neon gas laser; cross-section (From *Lasers and their Applications,* M.J. Beasley, Taylor and Francis Ltd., 1971, by permission).

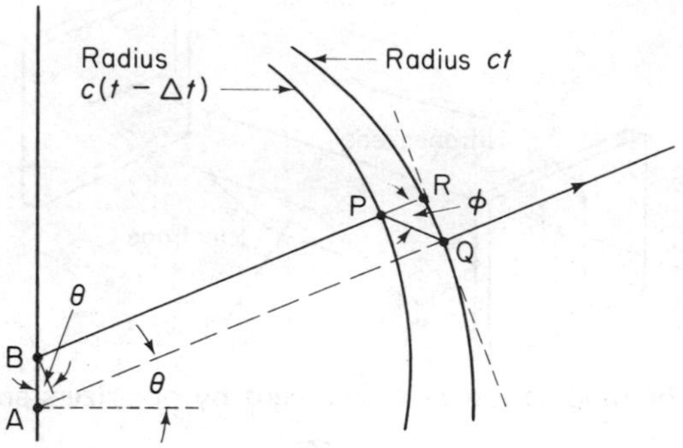

Figure 347. Electromagnetic wave production.

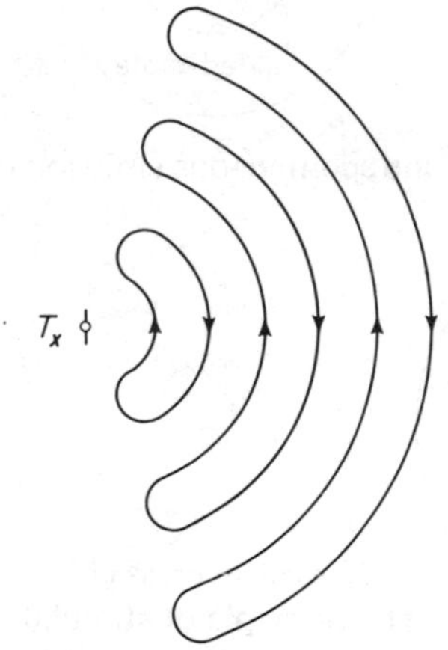

Figure 348. The expanding electrical field loops of radiation from a vertical antenna seen in part-section view.

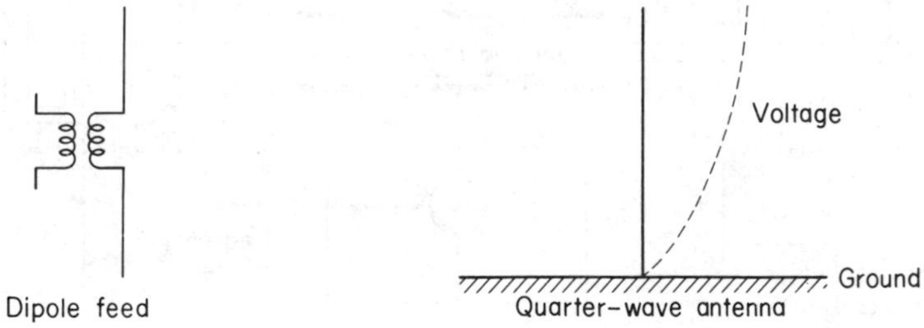

Figure 349. Dipole feed and a quarter-wave antenna.

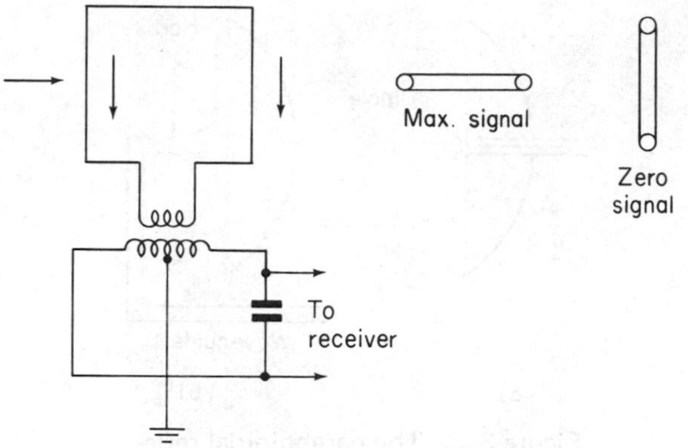

Figure 350. The frame or loop directional antenna.

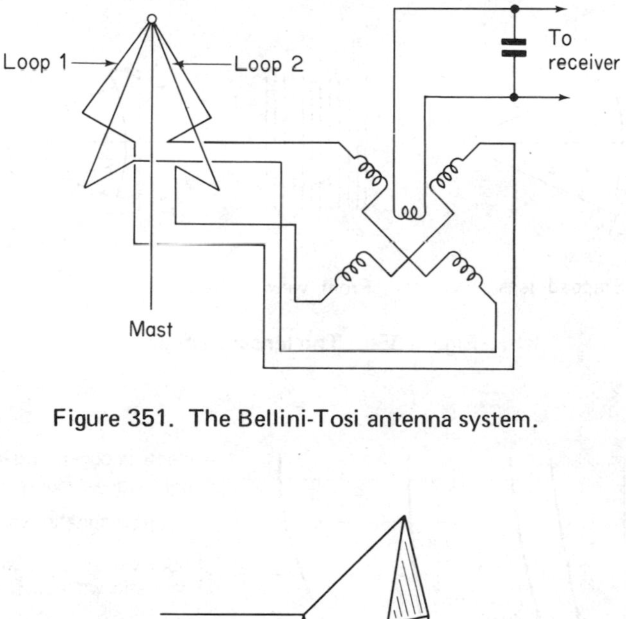

Figure 351. The Bellini-Tosi antenna system.

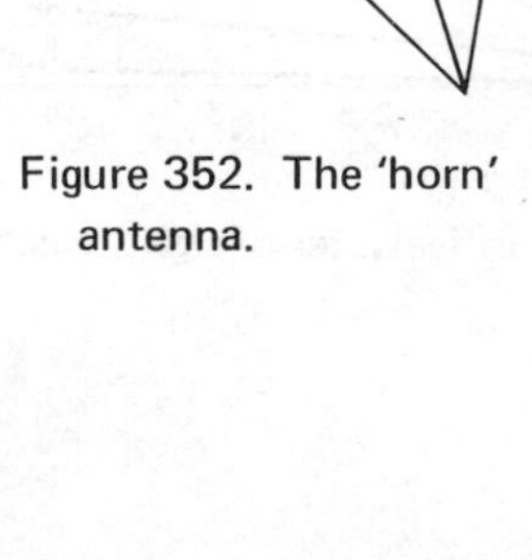

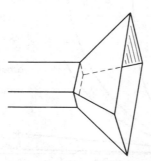

Figure 352. The 'horn' antenna.

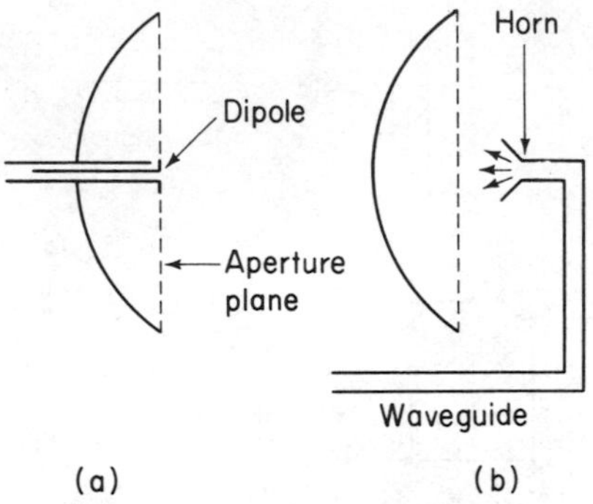

Figure 353. The paraboloidal reflector with (a) dipole and (b) horn feeds.

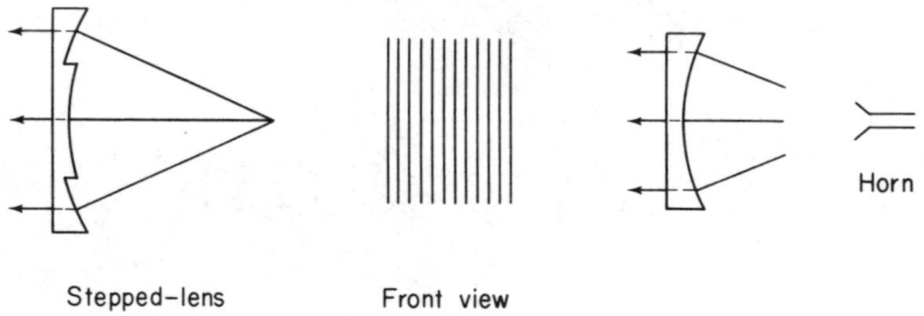

Figure 354. The lens antenna.

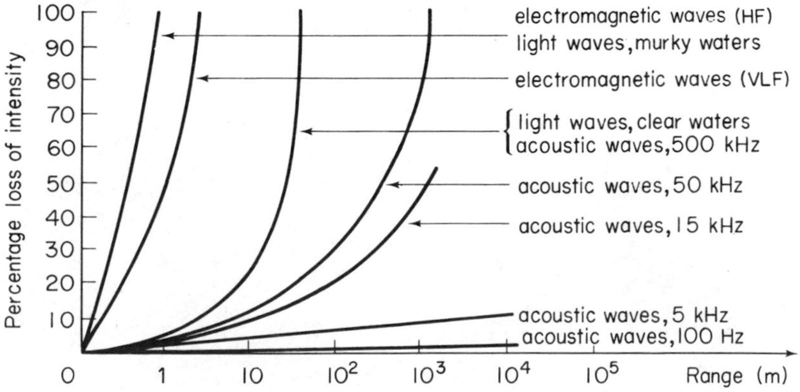

Figure 355. Propagation losses in seawater of light, electromagnetic and sound waves.

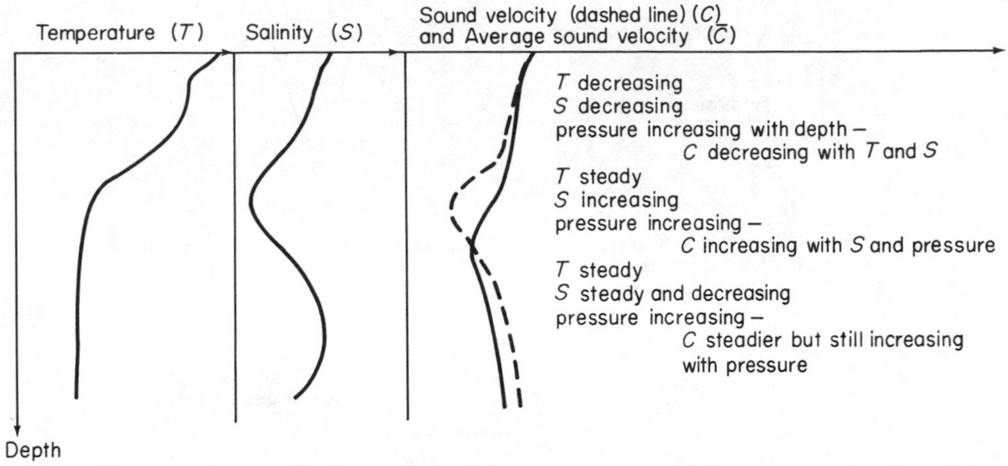

Figure 356. **Typical curves for temperature, salinity, sound velocity and average sound velocity against depth.**

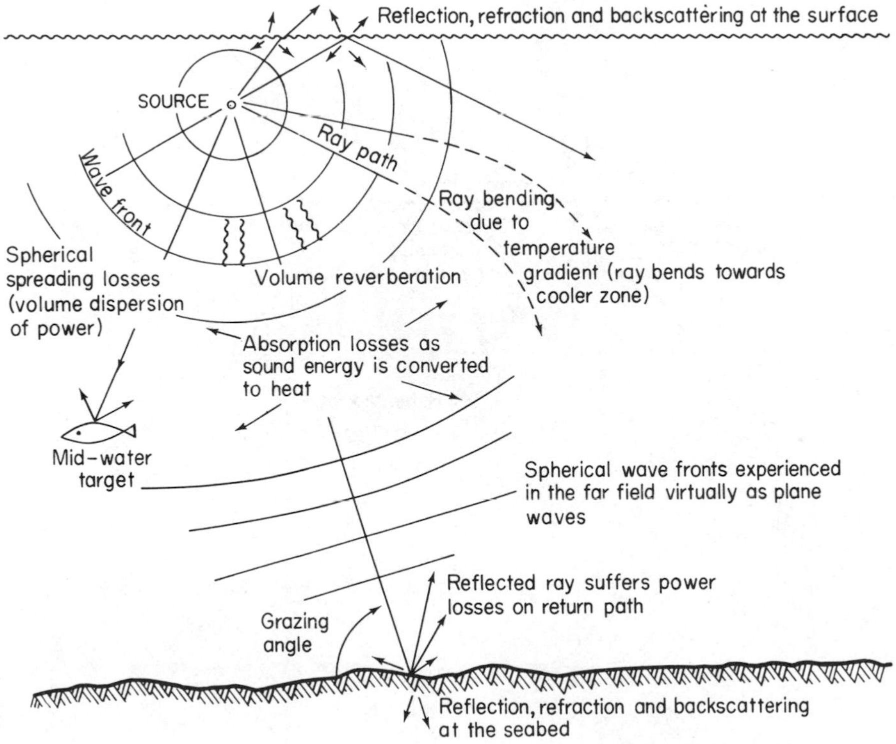

Figure 357. **A summary of the characteristics of acoustic wave propagation.**

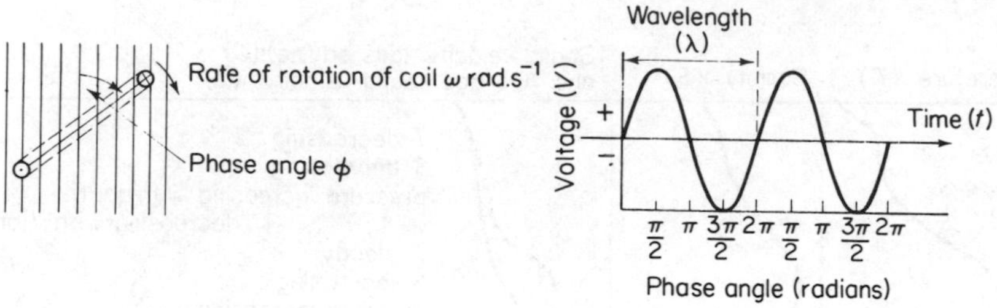

Coil rotating in magnetic field produces alternating current of sinusoidal waveform. The number of wavelengths, or cycles, per second gives the frequency (f) $\lambda f = c$, the velocity of propagation.

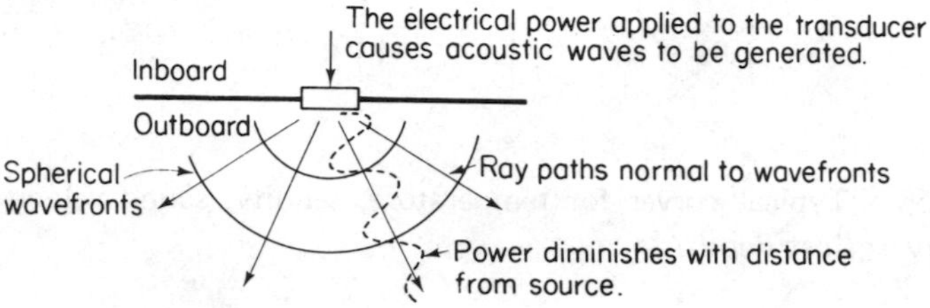

Figure 358. The electrical and acoustic parameters and terminology.

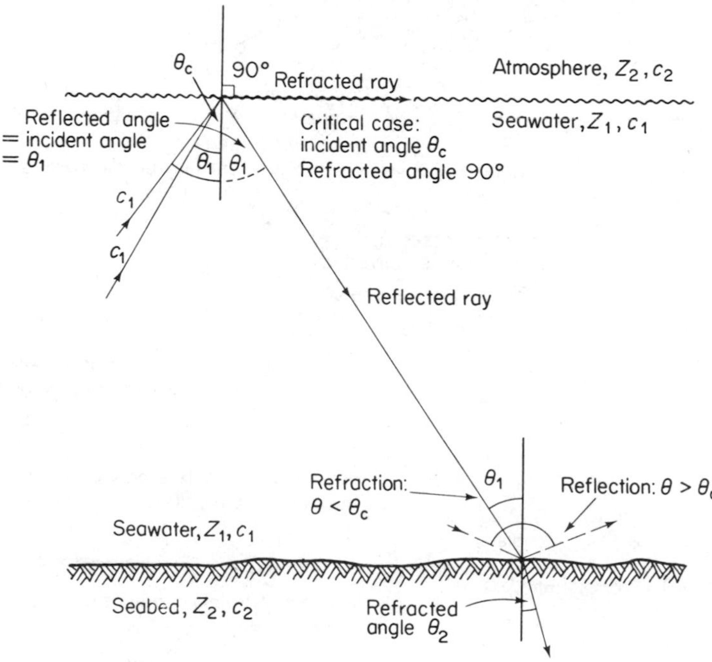

Figure 359. The characteristics of reflection and refraction at the surface and seabed.

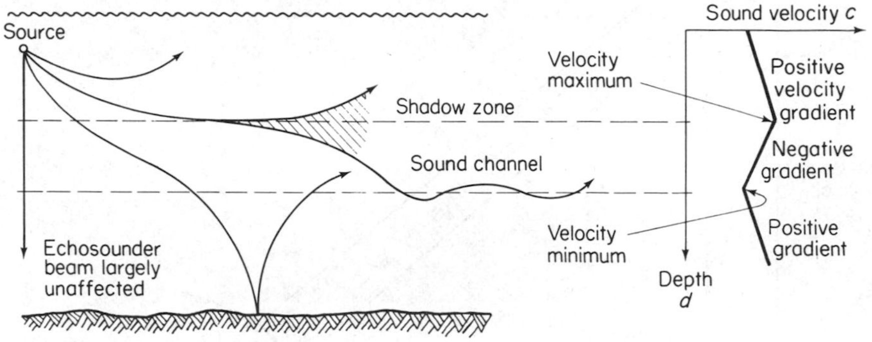

Figure 360. Ray path bending in deep water.

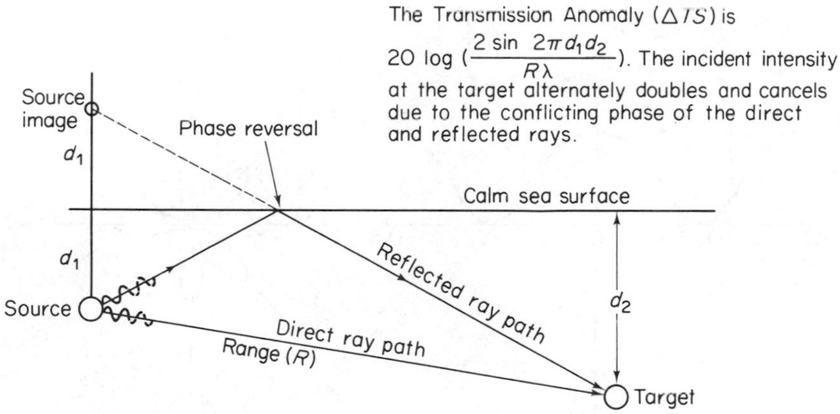

Figure 361. The geometry of the Lloyd Mirror Effect.

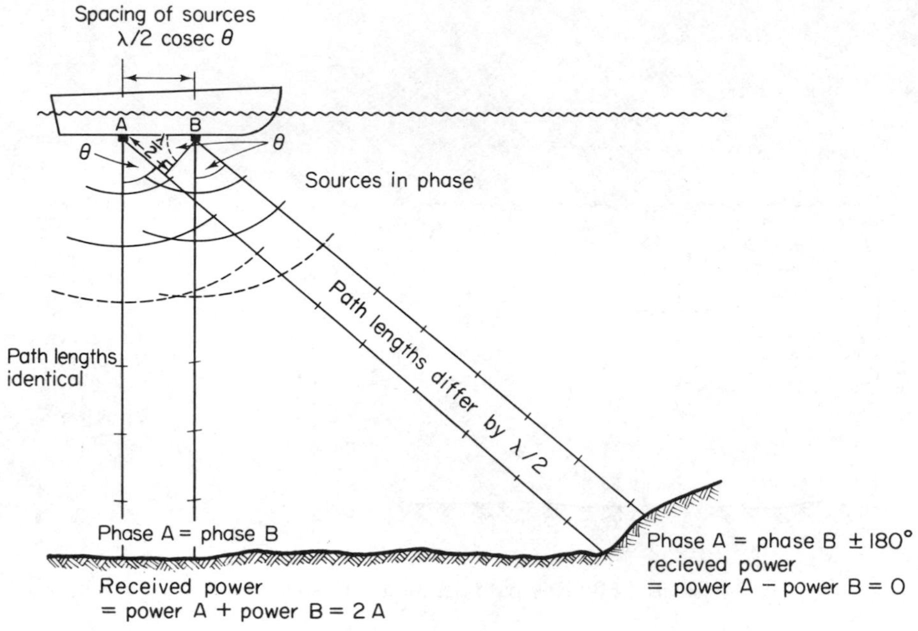

Figure 362. Phase interference due to differing path lengths between source and target.

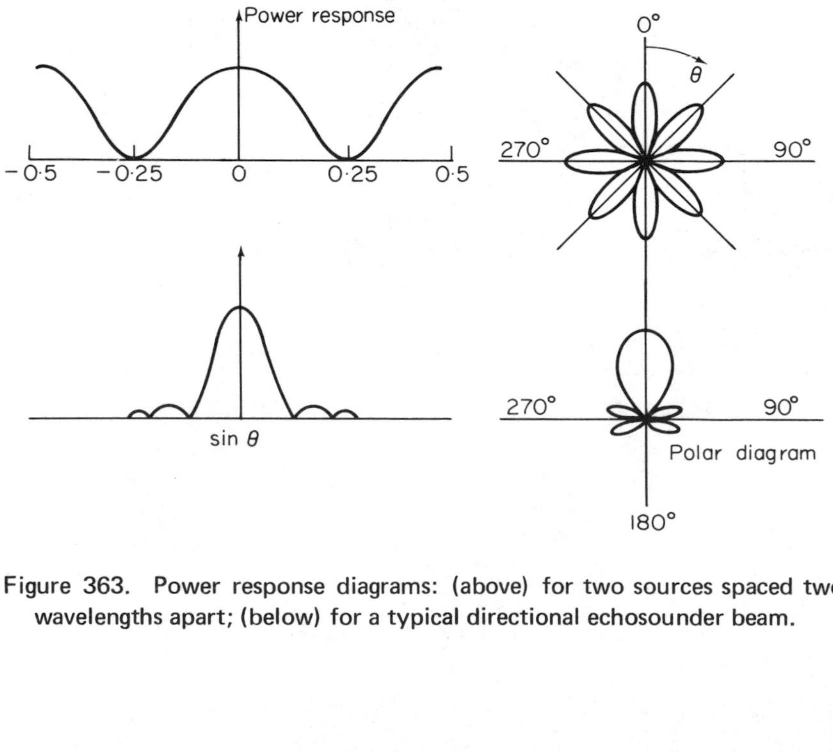

Figure 363. Power response diagrams: (above) for two sources spaced two wavelengths apart; (below) for a typical directional echosounder beam.

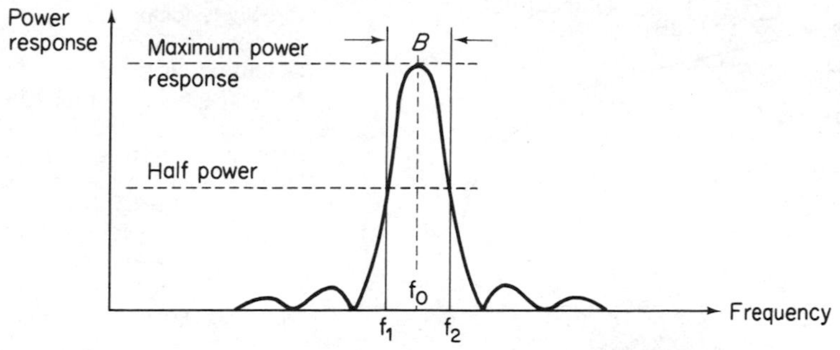

Figure 364. The power response of a resonant system.

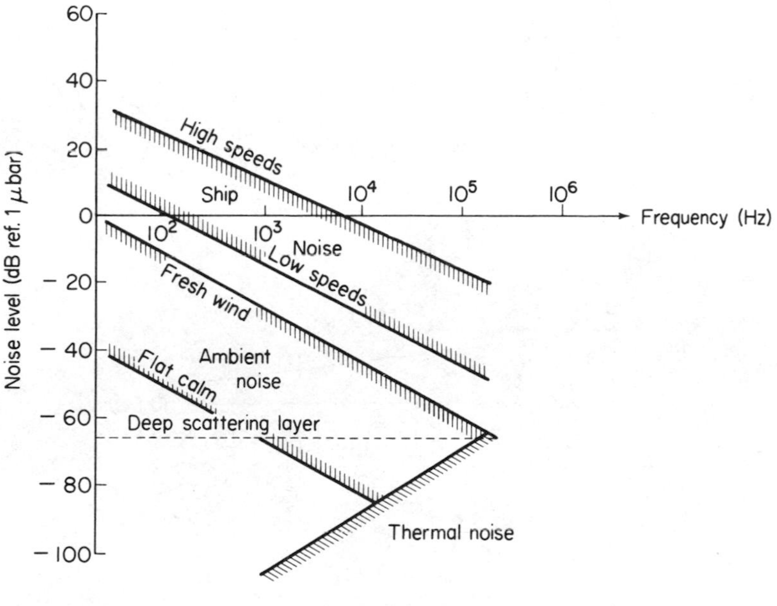

Figure 365. An indication of noise levels and their typical frequency ranges.

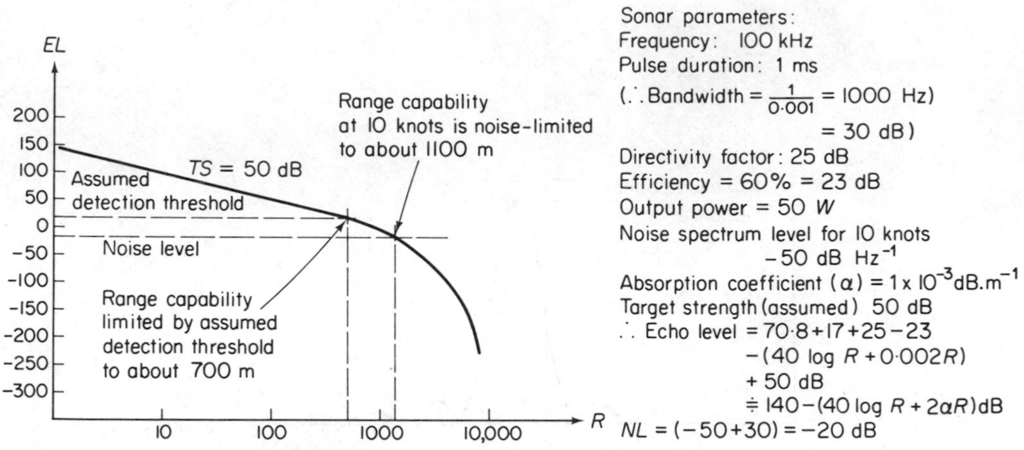

Figure 366. The interrelationship of sonar parameters.

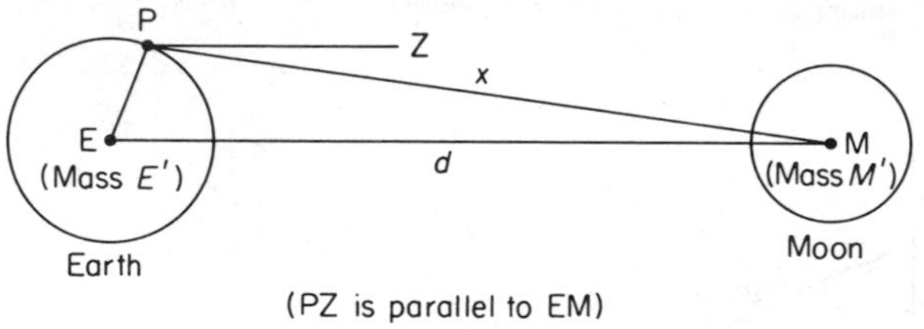

Figure 367

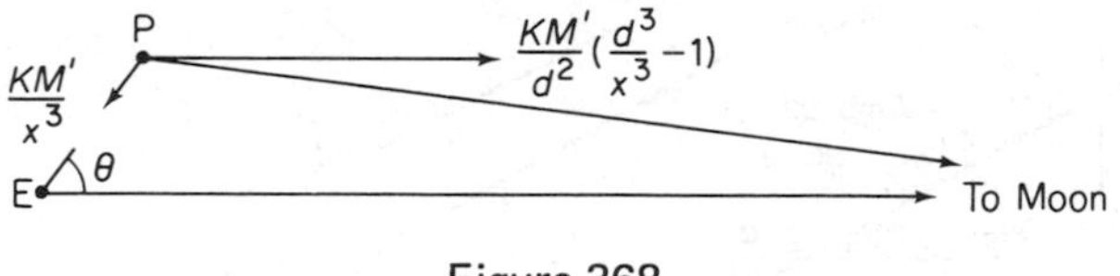

Figure 368

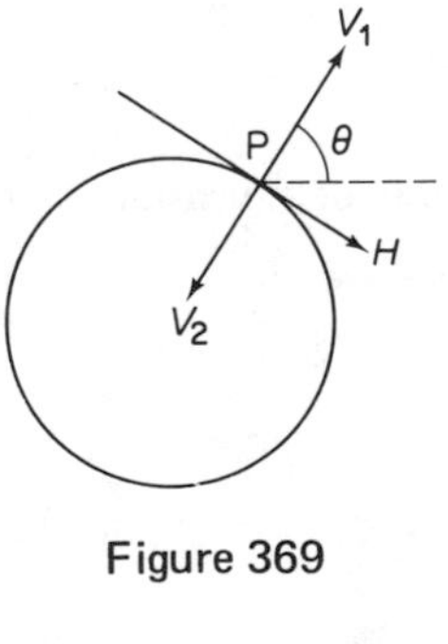

Figure 369

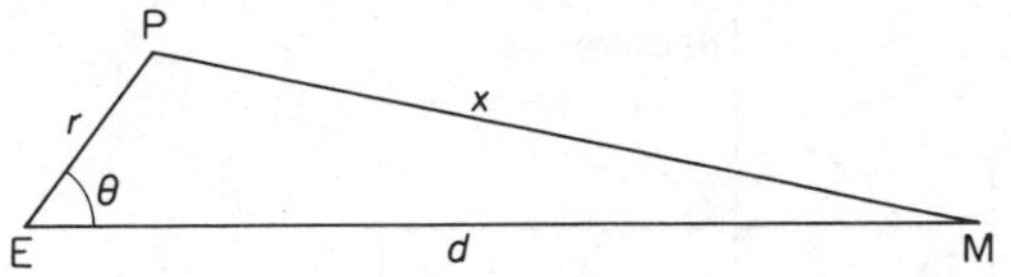

Figure 370

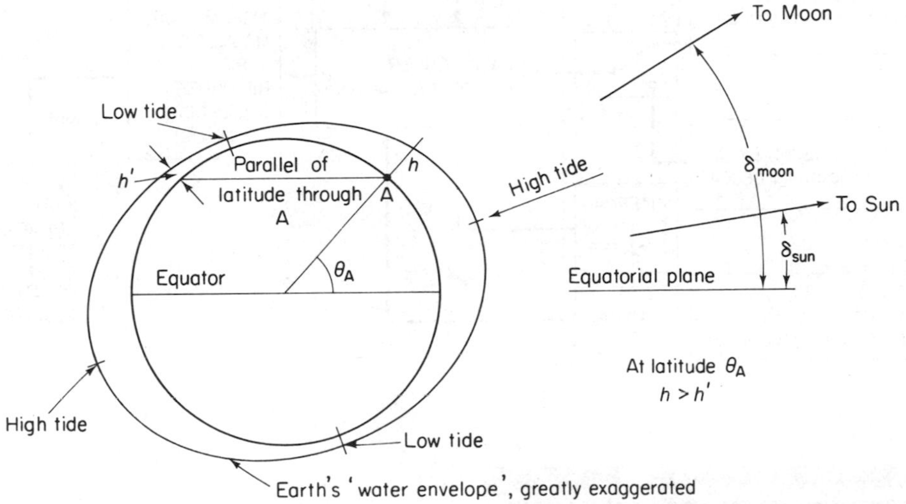

Figure 371

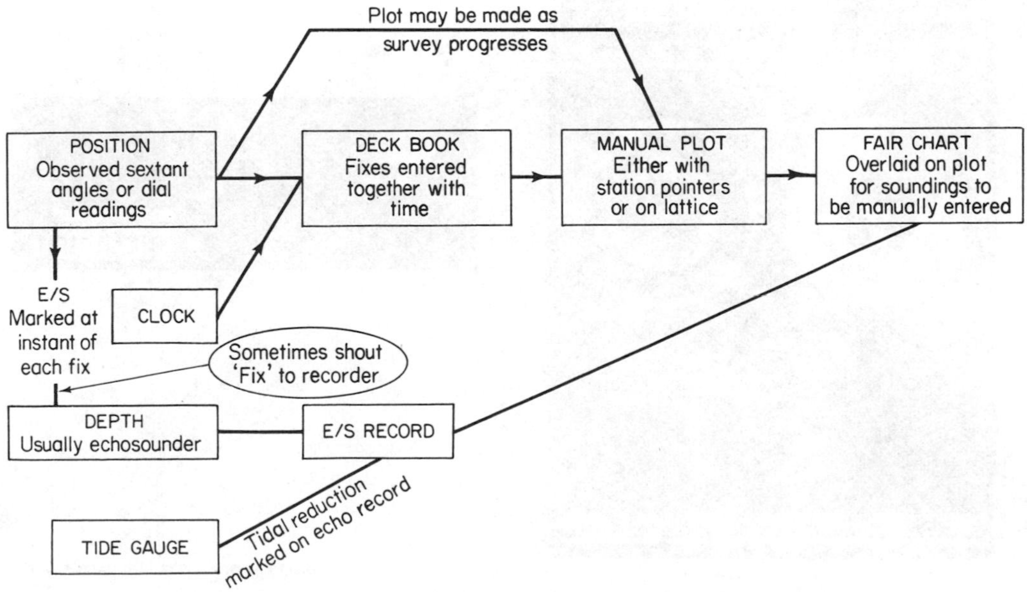

Figure 372. The processes of manual data processing.

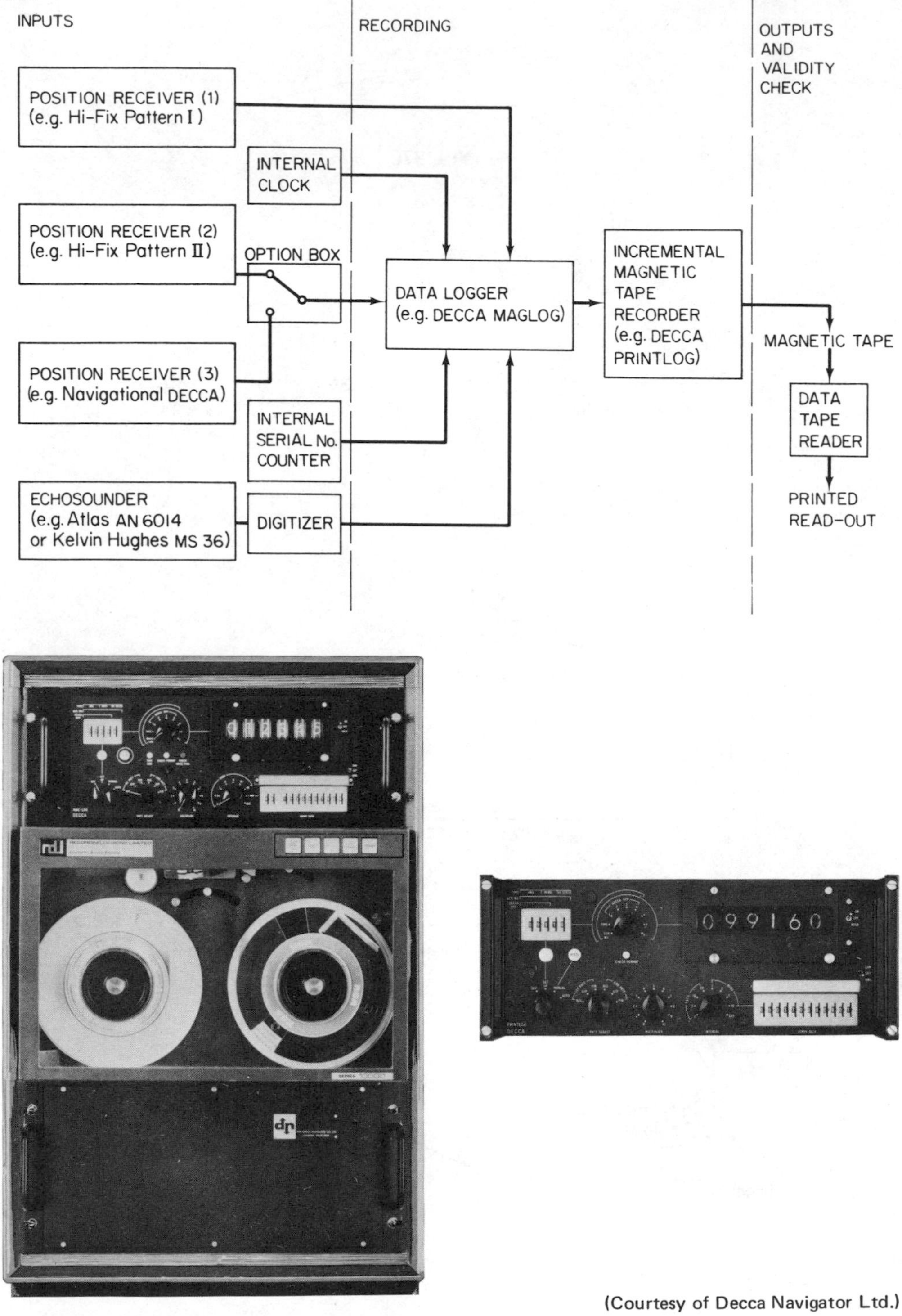

(Courtesy of Decca Navigator Ltd.)

Figure 373. The components of a typical ship-borne automated survey system.

```
N00721 T23.17.36 R711.25 G644.53 A061.99 B434.25
Z3835 Z3830 Z3830 Z3830 Z3830
//N00722 T23.17.37 R711.25 G644.53 A062.03 B434.29
Z3830 Z3830 Z3830 Z3830 Z3835
//N00723 T23.17.38 R711.25 G644.53 A062.06 B434.32
Z3835 Z3835 Z3835 Z3835 Z3835
//N00724 T23.17.39 R711.25 G644.54 A062.08 B434.34
Z3830 Z3830 Z3835 Z3835 Z3830
//N00725 T23.17.40 R711.24 G644.54 A062.10 B434.37
Z3835 Z3835 Z3835 Z3830 Z3830
//N00726 T23.17.41 R711.24 G644.54 A062.13 B434.41
Z3835 Z3835 Z3840 Z3840 Z3845
//N00727 T23.17.42 R711.24 G644.54 A062.15 B434.43
Z3850 Z3855 Z3855 Z3860 Z3865
//N00728 T23.17.43 R711.24 G644.54 A062.18 B434.45
Z3865 Z3870 Z3870 Z3870 Z3865
//N00729 T23.17.44 R711.24 G644.54 A062.21 B434.48
Z3870 Z3865 Z3860 Z3865 Z3860
//N00730 T23.17.45 R711.24 G644.54 A062.24 B434.51
Z3860 Z3860 Z3840 Z3835 Z3835
//N00731 T23.17.46 R711.24 G644.54 A062.26 B434.55
Z3830 Z3855 Z3850 Z3855 Z3850
//N00732 T23.17.47 R711.24 G644.54 A062.29 B434.58
Z3850 Z3850 Z3845 Z3840 Z3835
//N00733 T23.17.48 R711.24 G644.53 A062.32 B434.61
Z3830 Z3830 Z3825 Z3820 Z3820
//N00734 T23.17.49 R711.24 G644.53 A062.34 B434.64
Z3815 Z3815 Z3815 Z3810 Z3810
//N00735 T23.17.50 R711.24 G644.54 A062.37 B434.66
Z3810 Z3805 Z3805 Z3805 Z3805
//N00736 T23.17.51 R711.24 G644.54 A062.39 B434.69
Z3800 Z3805 Z3805 Z3805 Z3805
//N00737 T23.17.52 R711.24 G644.54 A062.41 B434.72
Z3805 Z3810 Z3810 Z3810 Z3805
//N00738 T23.17.53 R711.24 G644.54 A062.44 B434.75
Z3810 Z3815 Z3810 Z3815 Z3820
//N00739 T23.17.54 R711.25 G644.54 A062.47 B434.77
Z3820 Z3820 Z3820 Z3825 Z3825
//N00740 T23.17.55 R711.24 G644.54 A062.50 B434.79
Z3830 Z3830 Z3830 Z3835 Z3830
//N00741 T23.17.56 R711.24 G644.54 A062.52 B434.82
Z3830 Z3830 Z3830 Z3830 Z3830
//N00742 T23.17.57 R711.24 G644.54 A062.55 B434.86
Z3830 Z3830 Z3830 Z3830 Z3830
//N00743 T23.17.58 R711.23 G644.54 A062.58 B434.89
Z3830 Z3830 Z3830 Z3830 Z3830
//N00744 T23.17.59 R711.23 G644.54 A062.60 B434.92
Z3830 Z3830 Z3830 Z3830 Z3830
//N00745 T23.18.00 R711.23 G644.54 A062.63 B434.94
Z3830 Z3830 Z3830 Z3820 Z3835
//N00746 T23.18.01 R711.23 G644.54 A062.66 B434.97
Z3820 Z3825 Z3825 Z3825 Z3825
//N00747 T23.18.02 R711.22 G644.54 A062.68 B435.01
Z3815 Z3815 Z3810 Z3810 Z3810
//N00748 T23.18.03 R711.22 G644.54 A062.71 B435.04
Z3810 Z3810 Z3815 Z3815 Z3815
//N00749 T23.18.04 R711.22 G644.54 A062.74 B435.07
Z3815 Z3815 Z3810 Z3810 Z3810
//N00750 T23.18.05 R711.22 G644.54 A062.76 B435.10
Z3810 Z3815 Z3810 Z3810 Z3810
//92
```

(Courtesy Decca Survey Ltd.)

Figure 374. A listing of tape-recorded ‘Maglog’ data.

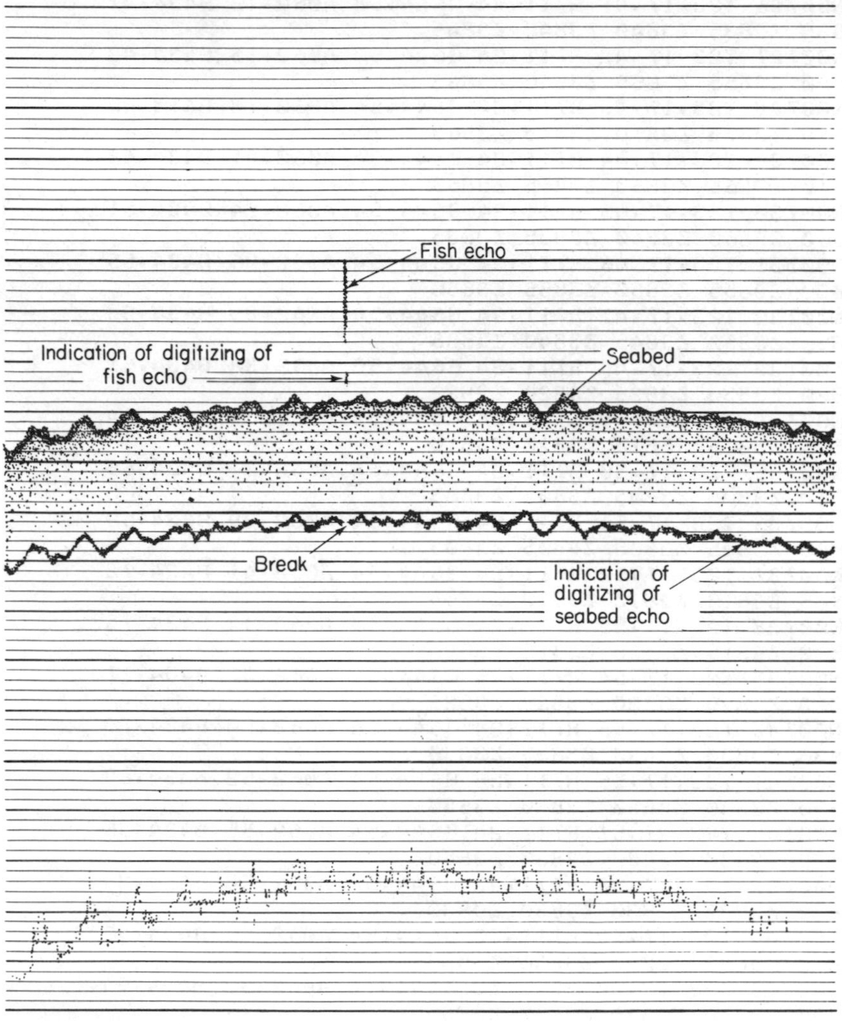

(Courtesy Fried. Krupp GmbH)

Figure 375. The Atlas Deso 10 echosounder record showing the method of indicating that soundings are being digitized.

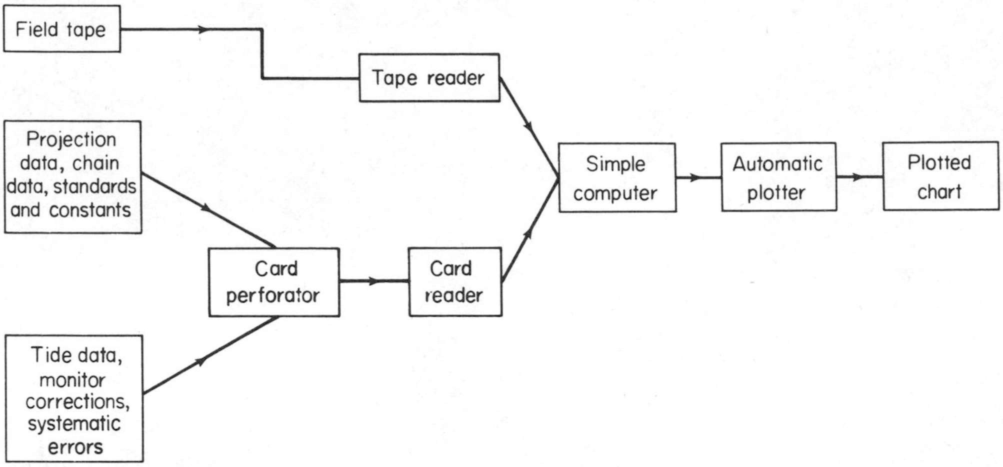

Figure 376. Automated chart production.

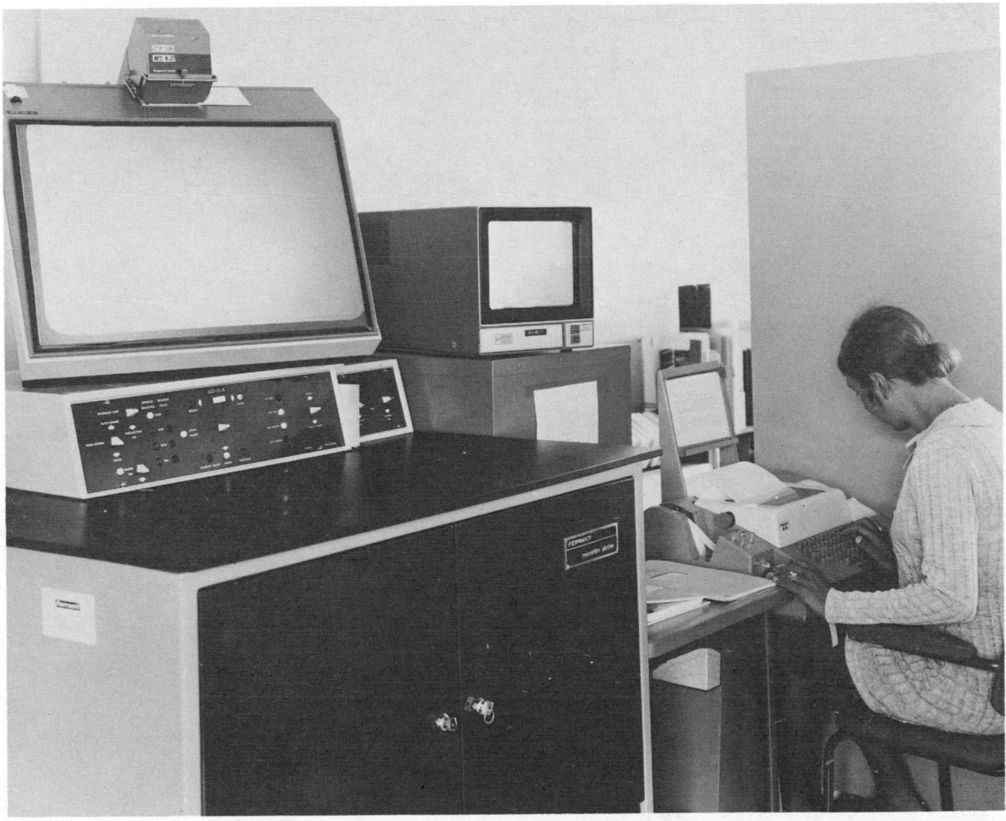

(Courtesy The Director, Institute of Oceanographic Sciences)

Figure 377. A microfilm plotter.